Springer Series on Biofilms
Volume 5

Series Editor: J. William Costerton
Pittsburgh, USA

For further volumes:
http://www.springer.com/series/7142

Hans-Curt Flemming · Jost Wingender · Ulrich Szewzyk
Editors

Biofilm Highlights

 Springer

Editors
Prof. Dr. Hans-Curt Flemming
Biofilm Centre
University of Duisburg-Essen
Faculty of Chemistry
Universitätsstr. 5
45141 Essen
Germany

Jost Wingender
Biofilm Centre
University of Duisburg-Essen
Faculty of Chemistry
Universitätsstr. 5
45141 Essen
Germany

Ulrich Szewzyk
Department of Environmental Microbiology
Berlin Institute of Technology
Franklinstrasse 29
10587 Berlin
Germany

Series Editor
Dr. J. William Costerton
Director of Microbiology Research
Department of Orthopedic Surgery
Allegheny General Hospital and
Director of Biofilm Research
Allegheny-Singer Research Institute
320 East North Ave.
Pittsburgh PA, 15212
USA

ISBN 978-3-642-19939-4   e-ISBN 978-3-642-19940-0
DOI 10.1007/978-3-642-19940-0
Springer Heidelberg Dordrecht London New York

Library of Congress Control Number: 2011930791

© Springer-Verlag Berlin Heidelberg 2011

This work is subject to copyright. All rights are reserved, whether the whole or part of the material is concerned, specifically the rights of translation, reprinting, reuse of illustrations, recitation, broadcasting, reproduction on microfilm or in any other way, and storage in data banks. Duplication of this publication or parts thereof is permitted only under the provisions of the German Copyright Law of September 9, 1965, in its current version, and permission for use must always be obtained from Springer. Violations are liable to prosecution under the German Copyright Law.

The use of general descriptive names, registered names, trademarks, etc. in this publication does not imply, even in the absence of a specific statement, that such names are exempt from the relevant protective laws and regulations and therefore free for general use.

Cover design: deblik, Berlin

Printed on acid-free paper

Springer is part of Springer Science+Business Media (www.springer.com)

Preface

Living in biofilms is the common way of life of microorganisms, transiently immobilized in their matrix of extracellular polymeric substances (EPS), interacting in many ways and using the matrix as an external digestion and protection system. This made them the oldest, most successful and ubiquitous form of life. And this is how they have organized their life in the environment, in medical context, and in technical systems, which is acknowledged by a yearly increasing number of scientific publications, which grows exponentially. It is next to impossible to keep an overview in face of all this frequently very specialized, dispersed, and detailed information. But during this entire race for data, it is very important to sometimes take a breath and have a critical look on what is really achieved and what can be learnt from it.

Biofilms are "discovered" and re-discovered in so many different fields, that it is good to refer to their common properties, principles, and processes in order not to get lost in the huge variability of their manifestations.

In this book, aspects of current biofilm research are presented in critical and sometimes provocative chapters. The purpose is double: to give an overview and to inspire further discussions. A particular intention of this book (which is supposed to be the first of a series within the Biofilm Series) is: to stimulate lateral thinking. We hope that "Biofilm Highlights" will serve these purposes.

January 2011

Hans-Curt Flemming
Jost Wingender
Ulrich Szewzyk

Acknowledgment

All graphic artworks in this book (except those in the chapter "Functional Amyloids in Biofilms") have been performed by Holtermann Grafik-Design, 29451 Danneberg, Germany, which is gratefully acknowledged.

Contents

Biofilm Dispersion 1
David G. Davies

Competition, Communication, Cooperation: Molecular Crosstalk in Multi-species Biofilms 29
Carsten Matz

Functional Bacterial Amyloids in Biofilms 41
Per Halkjær Nielsen, Morten Simonsen Dueholm,
Trine Rolighed Thomsen, Jeppe Lund Nielsen, and Daniel Otzen

Neutrophilic Iron-Depositing Microorganisms 63
Ulrich Szewzyk, Regine Szewzyk, Bertram Schmidt, and Burga Braun

Microbial Biofouling: Unsolved Problems, Insufficient Approaches, and Possible Solutions 81
Hans-Curt Flemming

Advances in Biofilm Mechanics 111
Thomas Guélon, Jean-Denis Mathias, and Paul Stoodley

Wound Healing by an Anti-Staphylococcal Biofilm Approach 141
Randall D. Wolcott, Florencia Lopez-Leban, Madanahally Divakar Kiran, and Naomi Balaban

Interfering with "Bacterial Gossip" 163
Thomas Bjarnsholt, Tim Tolker-Nielsen, and Michael Givskov

Hygienically Relevant Microorganisms in Biofilms of Man-Made Water Systems 189
Jost Wingender

Index 239

List of Contributors

Naomi Balaban Yale University, Department of Chemical Engineering, 9 Hillhouse Avenue New Haven, CT 06520-8286, USA

Thomas Bjarnsholt University of Copenhagen Faculty Of Health Sciences Department of International Health, Immunology and Microbiology Blegdamsvej 3B, DK-2200 Copenhagen, Denmark

Burga Braun Department of Environmental Microbiology, Berlin Institute of Technology, Franklinstrasse 29 10587 Berlin, Germany

David G. Davies Department of Biological Sciences, State University of New York at Binghamton, 2402 ITC Bioengineering Bldg. 85 Murray Hill Rd, Vestal NY 13850, USA

Madanahally Divakar Kiran Tufts University, Cummings School of Veterinary Medicine, Department of Biomedical Sciences, North Grafton, MA, USA

Morten Simonsen Dueholm Department of Biotechnology, Chemistry and Environmental Engineering, Aalborg University, Sohngaardsholmsvej 49, 9000 Aalborg, Denmark

Hans-Curt Flemming Biofilm Centre, University of Duisburg-Essen, Universitätsstr. 5, 45141 Essen, Germany

Michael Givskov Department of International Health, Immunology and Microbiology, University of Copenhagen, Blegdamsvej 3, DK 2200 Copenhagen, Denmark

Thomas Guélon Cemagref – LISC (Laboratory of engineering for complex systems), 24, avenue des Landais, 50 085 63 172, Aubière Cedex 1, France

Florencia Lopez-Leban Tufts University, Cummings School of Veterinary Medicine, Department of Biomedical Sciences, North Grafton, MA, USA

Jean-Denis Mathias Cemagref – LISC (Laboratory of engineering for complex systems), 24, avenue des Landais, 50 085 63 172, Aubière Cedex 1, France

Carsten Matz Helmholtz Centre for Infection Research, Inhoffenstr. 7, 38124 Braunschweig, Germany

Lund Jeppe Nielsen Department of Biotechnology, Chemistry and Environmental Engineering, Aalborg University, Sohngaardsholmsvej 49, 9000 Aalborg, Denmark

Per Halkjær Nielsen Department of Biotechnology, Chemistry and Environmental Engineering, Aalborg University, Sohngaardsholmsvej 49, 9000 Aalborg, Denmark

Daniel Otzen Department of Molecular Biology, Interdisciplinary Nanoscience Centre, Aarhus University, Gustav WiedsVej 10C, 8000 Aarhus C, Denmark

Bertram Schmidt Department of Environmental Microbiology, Berlin Institute of Technology, Franklinstrasse 29, 10587 Berlin, Germany

Paul Stoodley National Centre for Advanced Tribology, University of Southampton, University RoadSouthampton, SO17 1BJ, UK

Regine Szewzyk Department of Environmental Microbiology, Berlin Institute of Technology, Franklinstrasse 29, 10587 Berlin, Germany

Ulrich Szewzyk Department of Environmental Microbiology, Berlin Institute of Technology, Franklinstrasse 29, 10587 Berlin, Germany

Trine Rolighed Thomsen Department of Biotechnology, Chemistry and Environmental Engineering, Aalborg University, Sohngaardsholmsvej 49, 9000 Aalborg, Denmark

Tim Tolker-Nielsen Department of International Health, Immunology and Microbiology, University of Copenhagen, Blegdamsvej 3, DK 2200 Copenhagen, Denmark

Jost Wingender Biofilm Centre, University of Duisburg-Essen, Universitätsstr. 5, 45141 Essen, Germany

Randall D. Wolcott Southwest Regional Wound Care Center, Medical Biofilm Research Institute, 2002 Oxford Ave, Lubbock, TX 79410, USA

Biofilm Dispersion

David G. Davies

Abstract Biofilm dispersion has become widely recognized as a natural phenomenon associated with the terminal stage of biofilm development. As a behavioral characteristic of bacteria, this process is of major significance due to its importance in regulating biofilm structure. Biofilm dispersion is also important as a potential control point for the manipulation of biofilm development and persistence. Over the past decade, an increasing number of laboratories have focused their efforts on the study of biofilm dispersion, resulting in findings on the intracellular and extracellular mechanisms involved in this process in selected species of bacteria. The following is an overview of our current understanding of the biofilm dispersion process and the mechanisms involved in its regulation.

1 Background

It is generally accepted that bacteria have dwelled within the confines of structured biofilms since at least the time of the oldest known stromatolites, found within the 3.5-billion-year-old Dresser Formation of the Warrawoona Group, Pilbara Craton, Australia (Hofmann et al. 1999; Allwood et al. 2007). When viewed against a backdrop of more than three billion years of natural selection, it is difficult to believe that bacteria would not have developed genetic traits specific to survival

---

D.G. Davies (✉)
Department of Biological Sciences State, University of New York at Binghamton, 2402 ITC Bioengineering Bldg. 85 Murray Hill Rd, Vestal, NY 13850, USA
e-mail: dgdavies@binghamton.edu

within the biofilm niche. It is understood that most, if not all groups of bacteria live and presumably have always lived at least a part of their lifecycle within biofilms. Extensive modification and radiation of genes that would give biofilm residents even modest benefits should, therefore, be widespread throughout bacterial taxa. Thus, it is expected that residents of contemporary biofilms would possess a wide range of strategies to cope with the particular challenges posed by life within a fixed microbial community.

Living within a biofilm has its benefits, but there are also dangers for the resident microorganisms. As a biofilm grows in size, some cells will find themselves further and further from a bulk liquid interface, away from essential sources of energy or nutrients. As an additional challenge, waste products and toxins can accumulate to dangerous levels within such zones. Being trapped deep within a biofilm can, therefore, result in conditions that threaten cell survival. Furthermore, changing conditions can result in a hostile environment from which life in the biofilm provides no protection. For example, changes in nutrient loading, temperature, salinity, pH, oxygen concentration, and moisture content can threaten an entire biofilm community.

Observations of biofilms have brought to light some of the more obvious strategies that microorganisms have developed to cope with the particular challenges of being trapped within a sessile community. These include reducing metabolic activity by becoming dormant or by producing spores or endospores, the development of cross-feeding strategies to eliminate waste products and to recycle or produce new sources of energy and nutrients, and the creation of structures such as water channels to improve transport from the bulk liquid into the biofilm interior. Such strategies have successfully overcome many of the problems associated with life within a fixed microbial community, but these do not solve the problems completely. Biofilms still can become overcrowded and in many cases altered environmental conditions do not reverse themselves. Microorganisms within a biofilm need one more strategy for survival to ensure their continued existence. This strategy is one of escape. Migration out of a biofilm experiencing deteriorating environmental conditions and movement away from zones within a biofilm that are deprived of adequate mass transport are the only options that will allow survival for some microorganisms. The strong selective pressure to leave the biofilm or perish is likely to have resulted in numerous and redundant mechanisms for escape.

Two types of escape are possible from biofilms. The first, most commonly observed and best studied, is release of cells individually or in groups from a bulk liquid interface or margin of a biofilm. The second involves the escape of cells from the interior of a biofilm where they are not in direct contact with a bulk liquid interface. In the former case, it is possible for cells or clusters of cells to transition directly from the biofilm into the bulk liquid with which they are already in contact. This type of escape is referred to as detachment. In the latter case, bacteria must work their way through zones of overlying or adjacent cells to exit the biofilm. This type of escape is referred to as dispersion. The transfer of bacteria directly from a substratum to the bulk liquid is a third process known as desorption, which will not be discussed here because it is not exclusively a biofilm process.

2 Biofilm Detachment

It has long been known that within any given environment, biofilms tend to accumulate to a maximum thickness, beyond which no further increase is observed over time. Accumulation in a biofilm is a function of cell growth coupled with entrainment or attachment of cells from the surrounding medium. In flowing systems, biofilms are impacted by forces from the overlying liquid, both in normal and in tangential direction to their surface. These forces cause tensile and shear loads that the gel structure of the biofilm matrix must support in order not to break (Rittman et al. 1982). Detachment from a biofilm stems from the relationship of the external forces acting upon the biofilm and forces binding individual bacteria to the biofilm. When the external forces become sufficiently high, they will lead to detachment of biomass, transferring cells and polymer to the bulk liquid. A decrease in the forces maintaining biofilm structure will also result in increased detachment. Thus, detachment can be influenced by the role bacteria play in strengthening or weakening the cohesive strength of a biofilm. Detachment has long been considered the primary process that limits biofilm accumulation and determines the maximum thickness of biofilms in flowing systems (van Loosdrecht et al. 1997).

The rate and degree of detachment from a biofilm will be impacted by any modification of the biofilm structure. Trulear and Characklis (1982) noted that detachment from a biofilm was more frequently witnessed in thicker less dense biofilms that develop under low shear conditions. When grown under higher shear, biofilms were observed to become more compact and experience less detachment. In a follow-up study, Peyton and Characklis (1993) observed with *Pseudomonas aeruginosa* that substrate limitation resulted in a decrease in the detachment rate, presumably a result of reducing the growth rate and by some undescribed mechanism, strengthening the attachment forces within the biofilm. Alternatively, Allison and colleagues (1998) showed that following extended incubation, *Pseudomonas fluorescens* biofilms experienced increased detachment, coincident with a reduction in extracellular polymeric substances (EPS). Several other labs have reported that old cultures and cultures entering stationary phase showed increased detachment rates. For example, early work by Lamed and Bayer (1986) demonstrated that in biofilms formed by *Clostridium thermocellum*, the onset of stationary phase was correlated with increased detachment in biofilms. Ohashi and Harada (1994a, b) and Ohashi et al. (1999) have reported that the decay of cells in a biofilm due to aging or starvation leads to a decrease of the adhesive strength between cells in a biofilm. It has been postulated by O'Toole et al. (2000) that starvation of cells within a biofilm leads to detachment in order to allow bacteria to search for habitats richer in nutrients away from the parent colony. Although these observations are intriguing, they do not provide a mechanism for the observed increase or decrease in strength of biofilm matrix material.

2.1 Mechanisms of Biofilm Detachment

Breyers (1988) categorized four distinct mechanisms by which bacteria may detach from a biofilm. These are: abrasion, grazing, erosion, and sloughing (Fig. 1). These mechanisms have been described principally from the point of view of the chemical and physical environment acting upon a biofilm. Abrasion is the release of cells from a biofilm as a result of collisions with particles from the bulk liquid. Examples of systems in which abrasion is a major factor in biofilm detachment include fluidized beds and sand filters in which collisions with particles occur with a high frequency and are likely the principle factor limiting biofilm accumulation. However, all systems subjected to flowing conditions will experience some degree of abrasion depending on the size and energy of suspended particles. Collisions can also result from shed biomass from an upstream location within a system. Grazing is the act of removal of biofilm by the feeding activity of eukaryotic organisms. Depending on the number of grazing organisms and their efficiency, this mechanism of detachment can significantly reduce the overall biomass of a biofilm community. Erosion is the continuous loss of small portions of biofilm due to fluid shear in a flowing system. Organisms that are closest to the bulk water interface of a biofilm are the only cells susceptible to this form of detachment. Cells not enmeshed within the biofilm matrix and daughter cells that are produced at the interface are particularly prone to loss by erosion. Erosion can also result from the shearing effects of gas or air bubbles transported by the bulk liquid or generated from within the biofilm. Finally, the removal by fluid frictional forces of intact pieces of biofilm, containing large groupings of bacteria enmeshed within extracellular matrix material is referred to as sloughing. Typically, sloughing does not involve removal of cells attached directly to the substratum, although this can occur when attachment forces between bacteria are greater than attachment to substratum material. Sloughing can result in the removal of large sections of biofilm, and in some instances, the entire biofilm can detach and enter the bulk

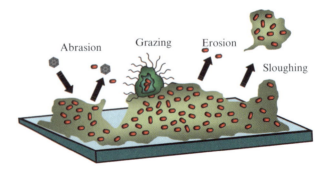

Fig. 1 Four mechanisms of biofilm detachment: abrasion, the removal of biomass due to collisions with particles; grazing, the removal of biomass due to the activity of eukaryotic organsisms; erosion, the continuous removal of biomass by fluid shear; and sloughing, the removal of intact pieces of biofilm by fluid shear

liquid. Stewart (1993) has pointed out that a distinction between sloughing and erosion may be arbitrary because in many biofilm systems there exists a continuum of detached particle sizes, from very small to very large.

3 Biofilm Dispersion

An alternative method of escape from a biofilm involves the direct and active participation of bacteria. Dispersion, as it has come to be called, is the release of live bacteria from a biofilm as a physiologically regulated response to internal or external stimulus. Dispersion, therefore, differs from detachment in that the regulating factor in the process is a change in the behavior of the participating bacteria. Dispersion can occur in the complete absence of flowing conditions and does not depend upon fluid shear to overcome binding forces within the biofilm. Fluid frictional force may enhance biofilm dispersion, but it does not initiate the process. Biofilm dispersion is, therefore, a response mechanism by bacteria to escape a biofilm, presumably due to unfavorable local conditions. This response has also been referred to as seeding dispersal as it is assumed to lead to the translocation of bacteria to new sites for colonization.

Dispersion is generally characterized as the terminal stage in biofilm development, where bacteria evacuate a mature biofilm and transition to a planktonic state. Dispersion occurs naturally with the release of cells from the interior of a biofilm through a breach in the wall of the biofilm or a microcolony, leaving behind void spaces or collapsed regions within the biofilm. In a flowing system, unless there is a change in environmental conditions, dispersion rarely involves the entire biofilm. Typically, only selected microcolonies or areas within a biofilm will undergo a dispersion event at any particular time. The biofilm as a whole is observed to undergo growth and dispersion simultaneously at different locations, which change with time. Thus, once a biofilm has begun to reach a mature state, dispersion will occur as a continuing process.

Microscopic observation of growing biofilms has revealed that before bacteria begin their escape from a biofilm, they demonstrate pre-dispersion behavior. This is characterized by the onset of swimming, twitching or floating of bacteria within a confined space within the biofilm. Bacteria closest to the bulk liquid interface do not participate in this pre-dispersion behavior and remain fixed throughout the dispersion process. With time, the region of the biofilm or microcolony within which bacteria are able to move grows in volume, typically from a central location. Eventually, a breach is made in a wall adjacent to a bulk liquid interface. The bacteria are able to swim or otherwise migrate through this breach and enter the bulk liquid, leaving behind a void within the biofilm or microcolony.

The evacuation of bacteria from within a microcolony is illustrated in Fig. 2 and was first described in 1999 (Davies et al. 1999) and expanded on by Tolker-Nielsen (2000), Sauer et al. (2002), Purevdorj-Gage and Stoodley (2005), and others (Boles et al. 2005; Davies and Marques 2009). Tolker-Nielsen described this process with

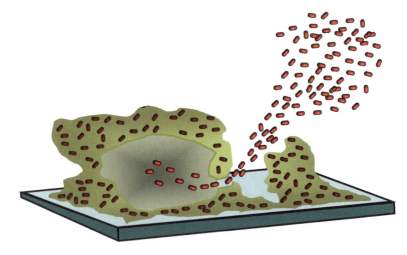

Fig. 2 Biofilm dispersion. The inside of a biofilm microcolony becomes fluid, cells within this zone begin to show signs of agitation and movement, a breach is made in the microcolony wall through which cells evacuate, entering the bulk liquid as planktonic bacteria

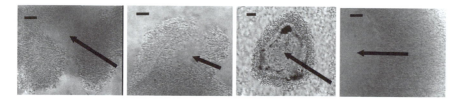

Fig. 3 Images of biofilm microcolonies following a dispersion event. In each image, the center of the microcolony has been evacuated by the resident cells, leaving behind a void space. Live bacteria are visible within these voids and their ability to swim without obstruction indicates a lack of intact matrix within this area. Size *bar* indicates 10 μm (images by David Davies)

respect to *Pseudomonas putida* OUS82 grown in continuous culture, where cells within microcolonies of the biofilm were observed to be sessile and nonmotile during the early phase of biofilm development. Following 3 days of growth, the bacteria within the microcolonies were observed to begin swimming rapidly in circles after which, the compact microcolonies were dissolved (Tolker-Nielsen et al. 2000). This behavior was also described by Boles et al. (2005), as occurring in biofilms of *P. aeruginosa* grown for 10–12 days in continuous culture. These biofilms were found to form internal cavities that eventually fractured to release motile bacteria. Observations of microcolonies of *P. aeruginosa* following a dispersion event are shown in Fig. 3. It is clear from this figure that a central void has been created within these microcolonies; the result of evacuation of cells from their interior. Prior to dispersion, the central region of each microcolony was observed to contain fixed (nonmotile) bacteria at a density similar to that which is seen in the microcolony walls in the images. During the dispersion process, bacteria from these clusters were observed to transition from an immobile state, to an agitated state, to a motile or free-swimming state, followed by evacuation of the microcolony.

Continued study of the dispersion response has revealed that biofilm microcolonies transition through repeated episodes of growth and dispersion; the same microcolony often enduring many such cycles. Multiple dispersion and regrowth events generally lead to the development of microcolonies with patterns analogous to growth rings, which can indicate the number of times that dispersion has occurred. Often cell clusters will detach from the substratum completely during a dispersion event [also observed by Stoodley et al. (2001)], presumably due to weakening of attachment structures at the base of the microcolony, allowing fluid sheer to detach the cluster. Such observations hint at detachment and dispersion as being interrelated processes. Typically, during such events, the substratum layer of cells directly attached to the surface is not involved and remains behind following the dispersion event. It is, as yet unclear whether this basement layer of cells is composed of living or dead microorganisms.

Interestingly, the propensity to undergo a dispersion event appears to be related to the size of the microcolony in question. Tolker-Nielsen et al. (2000) postulated that the size of a microcolony is the determining factor for initiation of a dispersion response. In our laboratories, we have found that dispersion is related to the size but not to the age of a cell cluster. Therefore, small microcolonies do not have a tendency to disperse, while large microcolonies do, regardless of their age. Observations of hundreds of cell clusters have demonstrated that under fixed flow conditions microcolonies must attain a specific size in order for dispersion to take place (Davies and Marques 2009). The size dependence of the dispersion response is indicated in Fig. 4. During these experiments, only microcolonies with an average diameter greater than 40 μm and a thickness of 10 μm were observed to

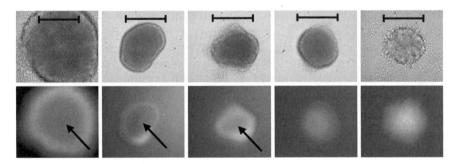

Fig. 4 Microcolonies of *P. aeruginosa* biofilms grown in continuous culture demonstrating native dispersion response. Transmitted light image (*upper panel*) and fluorescent image (*lower panel*) showing the size dependence of the dispersion response. Biofilm microcolonies growing in continuous culture with dimensions of greater than 40 μm in diameter by 10 μm thickness show dispersion (left 3). Microcolonies below this minimum dimension remain "solid" (right two photomicrographs). Fluorescence indicates presence of cells (lacZ reporter on chromosome). All images are the same relative size at X500 magnification. *Bars*, 40 μm. *Arrows* indicate void areas within a microcolony (images by David Davies)

undergo dispersion events. Microcolonies having dimensions below these minimum values contained a "solid" core of bacteria, regardless of their age. Interestingly, the microcolony size within which dispersion occurred was found to be dependent on the flow rate. When flow in the biofilm reactor was increased, the diameter and thickness of microcolonies in which dispersion occurred also increased, indicating a relationship between dispersion induction and transport. Thus, it appears that mass transport rather than size is the determining factor of whether a biofilm microcolony will initiate a dispersion response.

3.1 Mechanism of Biofilm Dispersion

In an effort to understand the underlying mechanism leading to biofilm dispersion, one commonly observed behavior has proved to be particularly informative. This is the general trend for biofilms, particularly axenic biofilms, to undergo a dispersion event following the transition from a flowing system to a batch system. In our lab, we have observed reproducible dispersion of *P. aeruginosa* biofilm cells when flow is arrested in continuous culture (Davies et al. 1999). We have also observed this behavior in additional members of the Proteobacteria as well as in members of the Low G+C Gram-Positive bacteria, under aerobic and anaerobic conditions (unpublished data). Interestingly, when biofilm dispersion occurs under these conditions, it is often not simply the center of microcolonies that disperse, but rather the entire microcolony or biofilm. Furthermore, this type of inducible dispersion will sometimes proceed from the outside of a microcolony toward the interior; a type of dispersion rarely observed naturally under continuously flowing conditions. It is interesting that motility was not a requirement for dispersion in these experiments (unpublished data).

This type of cell release was reported by Kaplan and Fine who observed that when *Neisseria subflava* was grown in batch culture on the bottom of plastic petri dishes without agitation, small visible colonies formed after 12 h of incubation. When these colonies were allowed to continue to grow in batch for an additional 12 h, they began to release single cells or small clusters of cells into the surrounding medium. They also reported that the release of cells from these colonies occurred over a discrete period of time rather than as a continuous process (Kaplan and Fine 2002). This type of behavior was also observed by Hunt et al., whose experimental work was performed using flow-cells containing pure cultures of *P. aeruginosa*. In this work, biofilms grown under continuous-flow conditions experienced a rapid release of cells after flow was stopped, and hollow voids were sometimes detectable within the interior of the biofilm microcolonies (Hunt et al. 2004). In separate work performed by Thormann et al., *Shewenella oneidensis* MR-1 biofilms were grown in a flow chamber system under oxygenated conditions for 12 h. When flow was stopped for as little as 15 min, up to 80% of the total biomass was observed to disperse from the biofilms. Similar results were obtained when *S. oneidensis* MR-1 biofilms were grown using fumerate as an alternative electron acceptor under anaerobic conditions (Thormann et al. 2005).

These and other similar observations of biofilm dispersion in a variety of laboratory reactor systems has led to the search for a common trigger that may account for dispersion induction under batch culture conditions or when flow is arrested in a continuous culture system. In the absence of flow, one or more essential nutrients within a biofilm will eventually become exhausted. This should lead to starvation conditions within the biofilm and may potentially act as an inducer for cells to disperse. One would predict that only mature biofilms should show a rapid response to stopping the supply of nutrients to a system. This is because it is only when large numbers of bacteria are actively consuming nutrients that nutrient depletion can occur over a short time interval. As expected, young biofilms with few cells tend not to undergo rapid dispersion in a continuous culture system when flow is arrested. Even under batch conditions, young biofilms commonly continue to accumulate biomass, but eventually they too will disperse. When looked at from a different point of view, biofilms of the same size in different sized batch culture systems should also show a difference in the timing of the onset of dispersion; and this is what is observed. This type of behavior can be easily witnessed in natural settings, where large-volume batch systems will support the growth of thick, often permanent biofilms on surfaces (such as in a swimming pool that is not treated with disinfectant), but small volume systems tend not to support thick biofilms or even prevent biofilms from forming (such as in a glass, small flask, or test tube).

It is easy to conclude, when switching a biofilm culture from a continuous to a batch system, that dispersion is strictly a function of nutrient availability, reflecting the relationship between nutrient supply and consumption. But additional possibilities exist which might trigger a dispersion response under such circumstances. These include the accumulation of toxins or waste-products that could act as a trigger of dispersion, the accumulation of cell-to-cell signals within a system, or the recognition by bacteria of a change in gradients within a system etc. The one common parameter that is affected by the cessation of flow is the loss of advective transport and a switch (which is typically rapid) to a system dominated by diffusion. In a large system, diffusion limitation will exert its effects following a longer time interval relative to a small system. Using these guiding principals, investigators have begun to unravel the mechanistic underpinnings of the biofilm dispersion response. Although no general consensus has emerged from this work, it is clear that multiple factors play a role in biofilm dispersion and that differences exist in the manner in which dispersion is carried out by different species of bacteria. Below is a description of our current understanding of the mechanisms, regulation, and signaling involved in biofilm dispersion.

3.2 *Role of Starvation in Biofilm Dispersion*

A number of research groups have invested considerable effort over the past decade in the examination of starvation as a likely trigger responsible for inducing a dispersion response in biofilm bacteria. It is readily apparent that in a nutrient-limited system, bacteria deep within a biofilm would be more susceptible to starvation conditions compared to bacteria residing at a location in direct contact with the bulk liquid. As would be expected, real-time microscopic observation of growing biofilms has revealed that bacteria deep within the biofilm are the ones most frequently seen to disperse. From the perspective of natural selection, it makes sense that bacteria deep within a biofilm would develop a means of escape. Individuals that acquire or develop the ability to escape a biofilm will survive at a higher rate than their conspecifics under nutrient-limited conditions, resulting in an increase in their fitness. The choice appears to be between starvation and death, or escape and survival. The argument is compelling. However, the role of starvation in biofilm dispersion, it turns out, is not so simple.

For instance, it has been observed over many years that bacteria under low nutrient or starvation conditions tend to form surface-associated biofilms rather than preferring life as planktonic cells (Bowden and Li 1997; Costerton et al. 1995; Marshall et al. 1988; Pratt and Kolter 1999). Alternatively, bacteria growing under high nutrient conditions either fail to form biofilms or form loose floc-like structures at a substratum that are easily disrupted with fluid shear (Costerton et al. 1995; Marshall et al. 1988; Danhorn et al. 2004; Jackson et al. 2002). This behavior was elegantly demonstrated in early work by Peyton and Characklis who grew *P. aeruginosa* and undefined mixed population biofilms on glucose medium under constant flow in annular biofilm reactors in which shear stress and substrate loading rate could be independently controlled. Using this system, it was demonstrated that the rate at which biofilm cells were released into the surrounding bulk liquid was directly proportional to the growth rate and substrate loading of the system. In other words, at high nutrient loads, large numbers of cells were released, and at low nutrient loads, low numbers of cells were released. Furthermore, fluid shear was shown to have a greater impact on cell release under high nutrient loads compared to low nutrient loads (Peyton and Characklis 1993). These observations appear to support the counter-hypothesis that bacteria under nutrient limiting conditions prefer to remain in a biofilm rather than return to the bulk liquid environment. Much of the early literature on biofilms supports these observations, as reviewed by Marshall (1988) and Costerton (1995). Typical of such observations is work by Dewanti and Wong (1995), who examined biofilm development by *Escherichia coli* O157:H7 on stainless steel chips in high- and low-nutrient media. In this work, it was found that biofilms developed faster with a higher number of adherent cells when grown in low-nutrient media compared to growth in tryptic soy broth. Additionally, these investigations showed that regardless of the carbon source, biofilms that were developed in minimal salts medium produced a thicker extracellular matrix, compared to biofilms grown in rich medium, with glucose as the best substrate for biofilm formation. Rochex and Lebeault (2007) investigated the release of cells from a *Pseudomonas putida* biofilm in a laminar flow cell reactor in defined mineral medium under various nutrient loading conditions. Their investigations showed that increasing the glucose concentration in the reactor from 0.1 to 1.0 gram per liter resulted in an increase in cells released from the biofilm. A similar effect was observed when the phosphorous concentration in the medium was doubled. The literature, therefore, seems to support the idea that when grown under low-nutrient conditions, biofilms experience reduced rates of cell-release compared to growth under high nutrient loads. This observation is unexpected and furthermore is contrary to the much of the literature published over the past decade.

Numerous examples have been reported supporting the alternative hypothesis that starvation leads to biofilm dispersion, and this has become the prevailing view among biofilm researchers. An early report relating to starvation as a trigger for dispersion was given by Delaquis et al. (1989) examining the effects of glucose and nitrogen depletion on the colonization of glass petri plates by *P. fluorescens*. These experiments were performed by allowing *P. fluorescens* to attach to the petri dish in rich medium and then the medium was removed and replaced with fresh medium at a lower glucose or nitrogen concentration. When this was done, the depletion of either glucose or nitrogen was correlated with the active detachment of cells from the biofilm (Delaquis et al. 1989). In another investigation, Sawyer and Hermanowicz examined the growth and detachment rates of an environmental isolate of *Aeromonas hydrophila* attached to a surface under varying nutrient supply conditions in a complex medium. Growth and detachment of cells were observed in real time under constant shear, using phase contrast microscopy in glass parallel-plate flow chambers. This work demonstrated that specific detachment rates increased as the nutrient supply decreased, indicating that nutrient limitation causes this bacterium to be released from a biofilm (Sawyer and Hermanowicz 1998). In another series of investigations by Hunt et al., *P. aeruginosa* biofilms were nutrient starved under continuous-flow conditions in a drip-flow reactor by switching from medium containing glucose to medium without glucose. A significant amount of cell release was observed following the switch, suggesting that starvation was responsible for triggering detachment in this particular system. Because no change in flow rate or shear was involved in these experiments, it was concluded that waste product or metabolite accumulation did not play a role in the outcome of the study (Hunt et al. 2004). Using a once-through continuous culture system, Thormann et al. examined the effect of nutrient depletion on the dispersion of *Shewenella oneidensis* MR-1 biofilms. In these experiments, biofilms were grown for 12 h in flow-cells and the influent lines were switched from lactate medium to buffered water or lactate medium depleted in oxygen. These experiments resulted in the release of up to 80% of the cell mass within 15 min, following the change in nutrient conditions. From this work, it was concluded that a rapid change in oxygen concentration was the determining factor resulting in dispersion (Thormann et al. 2005).

It is difficult to draw general conclusions from the existing literature relating to the role of starvation or nutrient limitation on biofilm dispersion. Experiments performed using different systems and different organisms under different nutrient conditions have yielded conflicting results, with some sets of experiments leading to dispersion under nutrient limiting conditions and others leading to dispersion under nutrient-rich conditions. These observations indicate that different organisms likely use different mechanisms to regulate whether to remain in a biofilm or leave. It is interesting to note, however, that there is one apparent consistency with the experimental procedures used in these studies. This common theme arises from the problem associated with switching from high-nutrient conditions to low-nutrient conditions in contrast to the other way around. It is not possible to remove nutrients from medium, it is only possible to add nutrients to medium. For technical reasons, therefore, the transition from high-nutrient to low-nutrient medium is typically carried out by the replacement of one medium with another, resulting in a dramatically shift in environmental conditions over a short time interval. It is possible to transition gradually from high- to low-nutrient conditions by diluting the high-nutrient medium, but this practice requires additional technical and material costs and is rarely practiced. Thus, most studies on the role of starvation in biofilm dispersion have been done by running a system under high nutrient loads to allow biofilm development and then abruptly changing conditions to a low-nutrient medium to induce starvation. It may be possible that the rapid change in nutrient loading, as indicated by James et al. (1995), Sauer et al. (2004) and Thormann et al. (2005) is acting as a trigger for dispersion rather than starvation itself. For instance, in the work performed by Peyton and Characklis (1993), nutrient conditions were changed from high load to low load through a gradient by diluting low-nutrient medium into a large reactor vessel (an annular reactor). These experiments actually resulted in decreased cell release from the biofilm, even under increased shear force. A rapid transition or step change in environmental conditions is typically associated with major changes in cell physiology, such as a transition from aerobic to anaerobic growth. Perhaps, it is the change that is important rather than the final environmental conditions.

Unfortunately, there is insufficient data at this time to draw a firm conclusion regarding the role of starvation in the dispersion response. However, work with changing nutrient conditions has resulted in the development of techniques that have been used to reproducibly induce dispersion. This has permitted the study of the mechanisms, regulation, and signaling involved in the biofilm dispersion response. In these areas, there has been considerable progress over the past decade. As a rule, it has been observed that any dramatic shift in environmental conditions will result in a dispersion response, up or down. Interestingly, the one known exception to this rule is the switch to low-nutrient conditions. It is widely known but not generally reported that certain bacteria will not demonstrate a dispersion response even to abrupt depletion of nutrients in a system. However, it is almost universally true that at least for heterotrophic bacteria, a dramatic shift to high nutrients results in dispersion. Other environmental parameters that have shown a positive effect on biofilm dispersion include a shift up in temperature and any dramatic change in pH or ionic strength in a medium.

3.3 *Release of Extracellular Enzymes*

Bacteria within biofilms are enmeshed within a gel matrix composed of EPS. In order to affect their release from a biofilm, bacteria must, therefore, break their attachments to the EPS and at least partially degrade this material in order to move from the biofilm into the surrounding bulk liquid. The regulation of the production and release of extracellular enzymes and the activity of these enzymes on EPS form important areas of inquiry in research on biofilm dispersion.

In an early attempt to understand the mechanism by which bacteria affect their release from biofilms, Boyd and Chakrabarty (1994) focused attention in the mid-1990s on the processing of alginate by various mucoid (alginate overproducing) and nonmucoid strains of *P. aeruginosa*. Alginate is an acidic polysaccharide that is believed to comprise the majority of the EPS matrix of *P. aeruginosa* growing in restricted environments, such as in lung infections of certain patients afflicted with cystic fibrosis. Using a simplified approach for their studies, Boyd and Chakrabarty grew *P. aeruginosa* on solid medium on petri dishes, which were then rinsed to collect released bacteria in their various experiments. This technique was used to examine the effects of alginate lyase on the alginate EPS, forming the matrix of their surface colonies. Alginate lyase is an endolytic enzyme synthesized by *P. aeruginosa* which cleaves alginate polymer strands by the β-elimination of the 4-*O*-glycosidic bond to yield two shorter saccharide polymers (Gacesa 1987). These investigations showed that the alginate overproducing strain of *P. aeruginosa* released relatively few bacteria when colonies were rinsed, compared to large numbers of bacteria released from a strain defective in alginate production, indicating that alginate serves as the major polymer anchoring *P. aeruginosa* within the matrix. When a plasmid-encoded gene for alginate lyase was activated by IPTG induction in colonies of mucoid *P. aeruginosa*, this led to alginate degradation, allowing the release of large numbers of cells during the wash step. This work was important in that it demonstrated the potential for bacteria to mediate their escape from a biofilm by the release of EPS degrading enzymes into the surrounding matrix. Previously, work with the methanogenic archaebacterium *Methanosarcina mazei* showed that this organism produces a disaggregatase enzyme with activity similar to *P. aeruginosa* alginate lyase (Xun et al. 1990). The disaggregatase caused cellular aggregates to break apart into separate cells by hydrolytic cleavage of the heteropolysaccharide capsule. Conditions that are generally unfavorable for growth were associated with disaggregatase activity.

More recently, the ability of exoenzymes to release matrix-bound bacteria was examined using cultures of *Streptococcus mutans*. In a study by Vats and Lee, a monolayer of *S. mutans* was allowed to form over a period of 1 h on the surface of epon-hydroxyapatite rods. These bacteria were then treated with an extracellular protease called surface protein-releasing enzyme (SPRE), obtained from *S. mutans* cultures. Results showed that adherent cells detached in response to the addition of this preparation, and that the response was blocked when pronase was used to pre-treat the SPRE prior to addition to the adhered cells. In addition, cells treated with SPRE prepared from *S. mutans* defective in SPRE production failed to be released from the surface (Vats and Lee 2000). These experiments were among the first to evaluate the role of extracellular enzymes on desorption of bacteria from a surface.

While disaggregation of surface colonies and flocs, and desorption of cells from attachment to a substratum are not directly biofilm processes, work with these systems, nonetheless, demonstrates principals that are likely applicable to mature biofilms and biofilm dispersion. One well-studied system that has direct applicability to the release of bacteria from mature biofilms has been performed on the periodontal pathogen, *Actinobacillus actinomycetemcomitans*. Cells of this organism form tenaciously adherent biofilm colonies on surfaces such as plastic and glass and disperse naturally after prolonged incubation. By performing transposon mutagenesis, Kaplan and co-workers were able to isolate a strain of *A. actinomycetemcomitans* defective in dispersion when grown on 96-well microtiter plates. The transposon insertion in the mutant strain mapped to a gene, designated *dspB*, that encoded a hydrolytic protein that was named dispersin-B (DspB). When this enzyme was purified and added to biofilms of *A. actinomycetemcomitans* defective in *dspB*, it was shown to disperse these biofilms, while heat-inactivated DspB failed to cause dispersion (Kaplan et al. 2003). Itoh and co-workers showed that the activity of DspB is to cleave polymeric ß-1,6-*N*-acetyl-D-glucosamine (poly-ß-1,6-GlcNAc), which has been implicated as an *E. coli* and *Staphylococcus epidermidis* biofilm adhesin. Enzymatic hydrolysis of poly-ß-1,6-GlcNAc was also shown to result in the dispersion of biofilms formed by *Yersinia pestis* and *P. fluorescens* (Itoh et al. 2005).

It appears clear that in at least some instances (if not all), extracellular enzymes are involved in orchestrating the release of bacteria from biofilms. While the studies presented here indicate that a single enzyme may be involved in the dispersion process for specific bacteria, it is not clear what occurs in biofilms formed by multiple species. Furthermore, it is known that many (if not most) species of bacteria elaborate many types of matrix polymer during biofilm formation. For instance, *P. aeruginosa* is known to produce at least three types of polysaccharides (alginate, Psl, and Pel), polynucleotides and several types of protein into its matrix (May et al. 1991; Ma et al. 2006, 2007; Friedman and Kolter 2004a, b; Whitchurch et al. 2002; Matsukawa and Greenberg 2004). Work in our own lab has shown that when *P. aeruginosa* is induced to disperse it releases extracellular enzymes that have been shown to include lyases, nucleases, proteases, and lipases. Information regarding the role and regulation of extracellular enzymes in the biofilm dispersion response is at present very limited. For instance, it is unknown at this time whether degradative enzymes are produced exclusively to cleave a cell's own polymers, polymers of its own and other conspecifics, or whether these enzymes additionally or exclusively cleave the polymers produced by members of other species within a biofilm. Additionally, it is unknown whether the release of polymer degrading enzymes is coordinated within a biofilm. If one takes the position that starvation is the principal cue for dispersion in at least some bacteria, how do these cells manage to escape from a biofilm if they are not surrounded by cells of other species also experiencing starvation at the same time? The answers to these and many other questions will have to await further research on this interesting and important topic.

3.4 Role of Biosurfactants in Biofilm Dispersion

The finding that *P. aeruginosa* produces and secretes a rhamnose-containing glyco-lipid biosurfactant called rhamnolipid has led many to speculate on the function of this compound in biofilms (Hisatsuka et al. 1971). It has been hypothesized that biosurfactants may play a role in biofilm dispersion by interacting with the matrix, either directly or in combination with released enzymes. In work performed by Boles and co-workers, it was observed that biofilms formed by strains of *P. aeruginosa* having increased rhamnolipid production, dispersed after two days of growth in culture, compared to ten days for wild-type biofilms under the same conditions. These dispersion events demonstrated similar behavior to the wild type, with the development of microcolonies followed by dispersion from the interior of these structures, leaving behind central voids. A similar result was obtained when purified rhamnolipid or sodium dodecyl sulfate was added exogenously to growing biofilms (Boles et al. 2005). A role for rhamnolipid has also been suggested by Davey and co-workers who reported that biofilms produced by mutants deficient in rhamnolipid synthesis do not maintain the spaces or channels between microcolonies in a biofilm. When rhamnolipid production was prevented, biofilms that formed were found to be lacking in these structures (Davey et al. 2003). This behavior was also noted by Purevdorj-Gage et al. (2005).

Thus, rhamnolipid and perhaps other biosurfactant molecules appear to play a role in the maintenance of the structure of the EPS matrix of biofilms. Whether this role is direct, or is the result of an enhancing effect on the degradation of matrix material by extracellular enzymes, is unclear at this time. Interestingly, Purevdorj-Gage noted that in a rhamnolipid defective mutant, biofilms continued to grow to a significantly greater thickness compared to wild-type biofilms. However, normal biofilm dispersion with the resultant hollow microcolonies was also observed when rhamnolipids were not produced. These findings indicate that while rhamnolipid may enhance biofilm dispersion (in *P. aeruginosa*), it is not required.

3.5 *Intracelluar Regulation of Biofilm Dispersion*

Little is known about the intracellular regulation of the biofilm dispersion response. From the work presented above, it is clear that the response is under physiological regulation; however, no complete story has emerged to describe how biofilm dispersion is triggered by the cell. Despite this, important regulatory points have been identified which when disrupted will prevent or reduce either biofilm development or dispersion in selected species.

Central carbon flux and catabolism appear to play a major role in biofilm regulation. Csr (carbon storage regulator) is global regulatory system that controls bacterial gene expression at the posttranscriptional level. Its effector is a small RNA-binding protein referred to as CsrA in *E. coli*, or RsmA (repressor of stationary phase metabolites) in phytopathogenic *Erwinia* species (Romeo 1998). CsrA is responsible for repression of several key metabolic pathways induced in stationary phase growth. For instance, CsrA has been shown to dramatically affect the biosynthesis of intracellular glycogen in *E. coli* through its negative control of the expression of two glycogen biosynthetic operons and the gluconeogenic gene *pckA*

(Romeo et al. 1993). Examination of the effects of *csrA* on several enzymes, genes, and metabolites of central carbohydrate metabolism has established a more extensive role for *csrA* in directing intracellular carbon flux (Sabnis et al. 1995). This activity is carried out by the binding of CsrA to mRNA transcripts, either increasing or decreasing their decay rates (Jackson et al. 2002, 2004). In experiments carried out in the laboratory of Tony Romeo, it was found that when *csrA* was placed under the control of a *tac* promoter, biofilms of *E. coli* could be induced to disperse upon activation of the *csrA* gene by the addition of IPTG. Control experiments showed that IPTG did not affect biofilm formation or dispersal in strains lacking the inducible *csrA* gene. These experiments indicated that *csrA* expression can serve as a general signal for biofilm dispersal in *E. coli*. Interestingly, when medium was amended with 0.2% glucose, biofilm thickness increased in both the induced and uninduced cultures, suggesting that glucose can override the regulatory effect of CsrA on biofilm dispersal (Jackson et al. 2002).

Another regulator of carbon metabolism, the Crc (catabolite repression control) protein has been shown to prevent the utilization of key sugars, such as glucose when tricarboxylic acid (TCA) cycle intermediates are present in the cell. Disruption of *crc* was shown by O'Toole et al. (2000) to dramatically decrease biofilm formation by *P. aeruginosa*, indicating that nutritional cues are integrated by Crc as part of a signal transduction pathway that regulates biofilm development. Sauer et al. have shown that *P. aeruginosa* biofilm dispersion can be induced by key sugars and TCA cycle intermediates. By repeating these experiments with a *crc* mutant of *P. aeruginosa*, strain PAO 8023, it was shown that there was no dispersion response following changes in nutrient concentration. Neither the addition of alternate carbon substrates (glutamate, succinate, citrate, or glucose) nor the addition of ammonium chloride induced dispersion. It was noted that the lack of a dispersion response may have been related to a biofilm formation defect within the strain resulting in a thin unstructured biofilm (Sauer et al. 2004). These findings hint at a potential role of Crc in regulating biofilm dispersion in *P. aeruginosa*.

A third regulator of carbon metabolism, CcpA, (catabolite control protein) has been shown to have an effect on biofilm formation by *Bacillus subtilis*. In a study by Stanley et al., it was shown that biofilm formation was inhibited when *B. subtilis* was grown in the presence of a rapidly metabolized carbon source and stimulated upon depletion of glucose. It was shown that CcpA mediated biofilm inhibition in *B. subtilis*, and *ccpA* mutants formed thicker biofilms than wild-type strains in the presence of glucose. These results suggested that in the presence of a preferred carbon source, *B. subtilis* cells demonstrate a preference for planktonic growth. Stanley et al. (2003) proposed one possible model to explain the regulation of the depth of the biofilm is that, under conditions of catabolite repression, CcpA represses a gene that either decreases the rate of attachment of cells to a biofilm or increases the rate of dispersion of cells from the biofilm.

In recent years, the involvement of a second messenger molecule bis-(3'-5')-cyclic dimeric guanosine monophosphate (c-di-GMP) has been implicated as playing an important role in the adaptation of many different bacterial species to their environment. In particular, c-di-GMP has been cited as a central regulator of biofilm formation and the main switch between biofilm and planktonic modes of existence in gram-negative bacteria (Karatan and Watnick 2009). Cyclic nucleotides represent second messenger molecules in all kingdoms of life (Roger et al. 2004). The second messenger c-di-GMP was first identified in *Gluconacetobacter xylinus*, where it regulates production of cellulose through modulation of cellulose synthase activity (Ross et al. 1990). It has been demonstrated that proteins having GGDEF and EAL domains are involved in the turnover of c-di-GMP in vivo (Simm et al. 2004). The GGDEF domain acts as a cyclase, leading to the production of c-di-GMP, and the EAL domain acts as a phosphodiesterase leading to c-di-GMP degradation (Christen et al. 2005; Ferreira et al. 2008; Kim and McCarter 2007; Tal et al. 1998). Recent work has indicated that increased levels of c-di-GMP in a variety of gram-negative bacteria are correlated with increased cell aggregation and surface attachment, leading to enhanced biofilm formation. It has been shown that in *Salmonella enterica* serovar Typhimurium, *P. aeruginosa*, *P. putida* and *E. coli*, GGDEF and EAL domains mediate similar phenotypic changes related to the transition between attached and planktonic modes of existence (Hickman et al. 2005; Güvener and Harwood 2007). Furthermore, it has also been reported that by modulating the intracellular c-di-GMP pool via overexpression of a diguanylate cyclase or phosphodiesterase it is possible to induce biofilm dispersion in at least one organism, *Shewanella oneidensis* (Thormann et al. 2006). In *P. aeruginosa*, it has been noted that nutrient-induced biofilm dispersion is dependent on c-di-GMP signaling through a chemotaxis transducer protein, BdlA. It has been shown that in a *bdlA* mutant, the ability to disperse in response to step changes in nutrient concentrations was lost. It is hypothesized that the PAS domains of BdlA initiate a chemosensory signaling cascade, which activates a phosphodiesterase that may result in lowering intracellular levels of c-di-GMP, leading to dispersion (Morgan et al. 2006).

Currently, the factors that regulate c-di-GMP levels and the modes of action of c-di-GMP are unknown (Merritt et al. 2007). However, the findings given above clearly point to a role for c-di-GMP in the formation and maintenance of biofilms, with high levels of c-di-GMP being correlated with biofilm formation and low levels of c-di-GMP correlated with planktonic growth. Further work is required before we will understand the full story behind the relationship between EPS production, global carbon regulation, and biofilm dispersion.

3.6 *Extracellular Inducers of Biofilm Dispersion*

The search for an extracellular signal responsible for biofilm dispersion has uncovered a range of factors that have been shown to stimulate biofilm disruption. The role of changing environmental conditions as, such as the depletion of oxygen or a sudden increase in carbon load has been discussed above. Additional factors have been shown to act as inducers of biofilm dispersion. For instance, Chen and Stewart demonstrated that addition of chemicals such as antimicrobial agents to a mixed biofilm of *P. aeruginosa* and *Klebsiella pneumoniae* resulted in significant removal of cells from these biofilms. Additional treatments leading to a loss of more than 25% of the biomass from these biofilms included NaCl and CaCl2; chelating agents; surfactants such as sodium dodecyl sulfate (SDS), Tween 20, and Triton X-100; a pH increase; and lysozyme, hypochlorite, monochloramine, and concentrated urea. Some treatments caused significant killing but little cell release, while other treatments resulted in release with little killing (Chen and Stewart 2000). Since this study focused on the remaining biofilm, the detachment mechanisms involved in these processes were not determined.

The identification of a cell-to-cell communication molecule responsible for biofilm dispersion has been the focus of a number of researchers over the past decade. Recently, indole has been shown to act as an intercellular messenger, inhibiting biofilm formation in *E. coli*, but it was also shown to enhance biofilm formation in *P. aeruginosa* (Lee et al. 2007a, b). While effective in preventing biofilm formation, Indole has not yet been shown to activate a dispersion response in existing biofilms. Another molecular inducer that has been identified to play a role in the induction of biofilm dispersion is nitric oxide (NO). Barraud and co-workers have demonstrated that dispersion of *P. aeruginosa* biofilms is inducible with low, sublethal concentrations (25–500 nM) of the NO donor sodium nitroprusside (SNP). Moreover, a *P. aeruginosa* mutant lacking the only enzyme capable of generating metabolic NO through anaerobic respiration did not disperse, whereas an NO reductase mutant exhibited greatly enhanced dispersal (Barraud et al. 2006). It is not clear whether NO is generated by *P. aeruginosa* as a specific signal to induce biofilm formation or whether it acts as a metabolite indicator of unfavorable environmental conditions.

The quorum sensing inducer molecules known as acyl-homoserine lactones (AHLs) have also been implicated as extracellular signals having the potential to stimulate a dispersion response in biofilm bacteria (David et al. 1998). Allison and co-workers have reported that *P. fluorescens* B52 biofilms grown on glass coverslips in spent medium were reduced in total biomass compared to biofilms grown in fresh medium. It was suspected that AHLs present in the spent medium may be responsible for inhibiting biofilm development in this strain. Although bioassays with an Agrobacterium AHL-reporter failed to detect the presence of AHL in the spent medium preparation, the exogenous addition of N-acyl-hexanoyl homoserine lactone to fresh growth medium was shown to reduce biofilm formation and had a similar effect as using spent culture medium (Allison et al. 1998). Quorum sensing using N-butanoyl-L-homoserine lactone was also found to be involved in the sloughing of the filamentous biofilm formed by *Serratia marcescens*. Biofilms formed by this bacterium were observed to consistently slough from the substratum after approximately 75–80 h of development. While a mutant defective in quorum-sensing, when supplemented with exogenous signal, formed a wild-type filamentous biofilm and sloughed at the same time as the wild type. When the AHL signal from the quorum-sensing mutant was removed prior to the time of sloughing, the biofilm did not undergo significant detachment. Together, these data suggested that biofilm formation and sloughing by *S. marcescens* are regulated by AHL-mediated quorum-sensing (Rice et al. 2005). Other investigators have also demonstrated that acyl-homoserine lactones may play a role in the production and regulation of matrix material produced by bacteria in biofilms and of cell aggregation in culture. Lynch and co-workers reported a decrease in biofilm formation in quorum sensing-defective *Aeromonas hydrophila* cultured on a steel surface in continuous culture, indicating a potential inverse relationship between quorum sensing and biofilm dispersion. In a separate study, Puskas and colleagues reported that inactivation of quorum sensing in *Rhodobacter sphaeroides* results in the formation of cell aggregates in liquid culture. Addition of *R. spaeroides* acyl-homoserine lactone to cultures of cells defective in quorum sensing resulted in disaggregation of cell clumps, indicating an inverse relationship between aggregation and quorum sensing in this organism (Puskas et al. 1997; Lynch et al. 2002).

Autoinducer molecules have also been implicated in the induction of dispersion of biofilms formed by gram-positive bacteria. The autoinducing peptide AIP-I is produced by *S. aureus*. Examination with confocal laser scanning microscopy (CLSM) showed that when biofilms of *S. aureus* were treated with AIP-1, they sloughed from the glass surface of a continuous culture flow cell over a period of 1–2 days, suggesting that AIP-I activated a detachment or desorption mechanism. The concentration of released bacteria in the effluent increased markedly 24–36 h after AIP-I addition. In contrast, the number of bacteria in the biofilm effluent of untreated cultures remained relatively constant. Computational analysis of the detachment phenotype indicated a $91.3 \pm 4.3\%$ reduction in biomass within 48 h of AIP-I addition. Consistent with a protease-mediated mechanism, increased levels of serine proteases were detected in the biofilm reactor effluents. Furthermore, it was shown that quorum-sensing activation increased the production of extracellular proteases. The authors concluded that quorum-sensing is required to form a biofilm, and quorum-sensing reactivation in established biofilms is required to detach or slough these biofilms (Boles and Horswill 2008).

Recently, it has been shown that a small messenger fatty acid molecule, *cis*-2-decenoic acid, produced by *P. aeruginosa* in batch and biofilm cultures induces a true dispersion response in biofilms formed by *P. aeruginosa*, as well as a range of gram-negative and gram-positive bacteria and yeast (Davies and Marques 2009). Figure 5 shows time course photomicrographs of biofilm microcolonies of *P. aeruginosa* grown in continuous culture and treated with purified spent medium from *P. aeruginosa* cultures containing 10 μM *cis*-2-decenoic acid (A–C), or 10 μM synthetic *cis*-decenoic acid (D–F). Both the purified natural inducer from spent medium and the synthetic compound were shown to induce a dispersion response in *P. aeruginosa*, which began after approximately 5 min of exposure (observed as rapid twitching of cells enmeshed in the microcolony matrix), and resulted in complete dispersion of the microcolonies within 30 min.

Dispersion that is induced naturally is observed to originate at the interior of biofilm microcolonies and results in the formation of void spaces where bacteria have evacuated the biofilm. Conversely, exogenous dispersion induction results in the complete disaggregation of microcolonies. This observation indicates that biofilm dispersion is initiated where the accumulation of inducer is highest. In nature, presumably all cells in a culture produce the inducer at a constant rate.

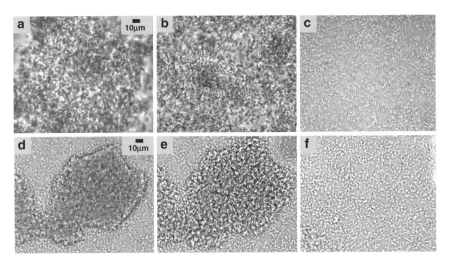

Fig. 5 Effect of *cis*-2-decenoic acid on developed biofilm microcolonies of *P. aeruginosa* grown in continuous culture in flow cells. Following 4 days of culture, biofilms were treated with 10 μM *cis*-decenoic acid under flowing conditions for 30 min. (**a–c**) shows addition of medium amended with purified *P. aeruginosa* spent culture medium containing *cis*-decenoic acid: (**a**), microcolony before treatment; (**b**), microcolony 7 min after treatment; (**c**), microcolony 30 min after treatment. (**d–f**) shows addition of medium containing synthetic *cis*-2-decenoic acid: (**d**), microcolony before treatment; (**e**), microcolony 11 min after treatment; (**f**) microcolony 30 min after treatment (from David Davies)

Under these conditions, the inducer will become most concentrated at those locations having the greatest diffusion limitation; the center of a microcolony, near the substratum. It is at this location that endogenous dispersion is always detected in flowing systems. When the inducer is delivered to the cells along with the medium, it has been shown to act first upon cells, which are in close proximity to the bulk liquid. It is relevant to note that bacteria attached directly to the substratum do not respond to exogenous dispersion induction with *cis*-2-decenoic acid.

It is interesting that the activity of *cis*-2-decenoic acid is neither restricted to *P. aeruginosa* nor is it restricted to gram-negative bacteria. Figure 6 is a series of photomicrographs showing the response of a gram-positive *Streptococcus* to treatment with *cis*-2-decenoic acid. Following approximately 40 min of exposure to inducer-containing medium, the microcolony was completely disaggregated. Upon restarting flow to the system, the cells from the disaggregated microcolony were observed to completely disperse. Bacteria directly adhered to the substratum were not observed to participate in the dispersion response. It is notable that this organism was able to undergo a dispersion response despite being nonmotile and under quiescent conditions.

Cell-to-cell signaling mediated by fatty acid derivatives has previously been described for a number of bacterial species, including *Xanthomonas* sp., *P. aeruginosa*, *Mycobacterium* sp., *Stenotrophomonas maltophilia*, *Xylella fastidiosa*, and *Burkholderia cenocepacia* (Boon et al. 2007; Chatterjee et al.

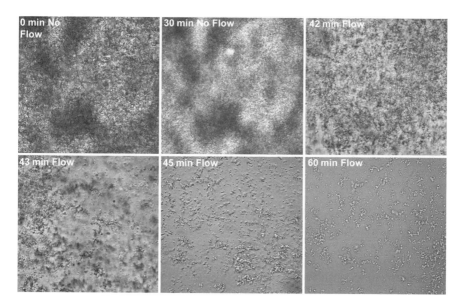

Fig. 6 Effect of *cis*-2-decenoic acid on developed biofilm microcolonies of a *Streptococcus* sp. grown in continuous culture in flow cells. Following 6 days of culture, the biofilm was treated with medium amended with purified *P. aeruginosa* spent culture fluid containing 10 µM *cis*-decenoic acid under quiescent conditions for 60 min. At 30 min of treatment, the cells of the microcolony in the image are seen to begin to separate from the matrix, and at 40 min the microcolony has disaggregated. When flow was reinstated, the microcolony dispersed leaving behind only bacteria attached directly to the substratum (from David Davies)

2008; Huang and Wong 2007; Wang et al. 2004). The best characterized of these low-molecular-weight fatty acids is DSF, which is responsible for the regulation of virulence in *Xanthomonas campestris* (Barber et al. 1997). DSF signaling has also been shown to regulate the production of extracellular proteases and exopolysaccharide production, aggregative behavior, biofilm formation, flagellum synthesis, resistance to toxins, activation of oxidative stress, and activation of aerobic respiration (Chatterjee et al. 2008; Huang and Wong 2007; Fouhy et al. 2007). DSF is structurally related to *cis*-2-decenoic acid, having a double bond at position 2, but it contains a 12-carbon chain with a branched methyl group at the number 11 position. The synthesis and detection of DSF have been shown to require products of the *rpf* (for regulation of pathogenicity factors) gene cluster in *X. campestris* (Barber et al. 1997). Synthesis of DSF requires RpfF, a putative enoyl coenzyme A (CoA) hydratase, and RpfB, a putative long-chain fatty acyl CoA ligase. DSF perception is dependent on the two-component system comprising the sensor kinase, RpfC, and the response regulator, RpfG (Barber et al. 1997; Dow et al. 2003; Ryan et al. 2006; Slater et al. 2000). While a BLAST search of the *P. aeruginosa* genome does not reveal the presence of an *rpf* gene cluster or protein closely related to RpfF, it does show 12 enoyl CoA hydratases with some sequence homology to RpfF, indicating the potential for synthesis of a DSF-related fatty acid signal. More closely related to *cis*-2-decenoic acid is *cis*-2-dodecenoic acid, a functional analog of DSF produced by *B. cenocepacia* and named BDSF (Boon et al. 2007). Sensing of DSF has been shown by Slater et al. to be mediated in *X. campestris* by the sensor kinase RpfC. A structural analog to this sensor in *P. aeruginosa* is PA1396, which was shown by Ryan et al. (2006) to respond to DSF produced by *S. maltophilia* (Slater et al. 2000).

The similarity between *cis*-2-decenoic acid, DSF and BDSF suggests that as a class of molecules, small messenger *cis*-monounsaturated fatty acids have activity across a wide range of bacteria as extracellular signals. As an inducer of biofilm dispersion, there are clear advantages to having an extracellular signal that is recognized by diverse species. In order to release cells from the biofilm matrix during a dispersion response, microorganisms must rely on the degradation of extracellular polymers produced by neighboring microorganisms of other species, as well as their own species. It is unlikely that a bacterium will produce enzymes to degrade all of the matrix polymers in which it may be enmeshed within a multispecies biofilm. Consequently, in order to disperse from a multispecies biofilm, it must be necessary for the resident organisms to release enzymes in a coordinated fashion. This can be effectively achieved when a cell-to-cell communication molecule is used to orchestrate a coordinated response by unrelated organisms. In addition to prokaryotes, many biofilms also contain eukaryotes, indicating an advantage to cross-kingdom activity for an extracellular inducer of biofilm dispersion. Cross-kingdom activity has been proposed previously for fatty acid messengers from evidence that DSF is recognized by *C. albicans* binding to the receptor of farnesoic acid, leading to an arrest in filamentation (Wang et al. 2004). This is further supported by the observation that *cis*-2-decenoic acid is able to induce the dispersion of *C. albicans* (Davies and Marques 2009).

The broad-spectrum activity of *cis*-2-decenoic acid suggests that this and other short-chain *cis*-2-monounsaturated fatty acids likely have deep evolutionary roots. Therefore, the discovery of additional small fatty acid messengers is anticipated in other organisms. It is interesting that fatty acid communication has been found to be present in many plant and animal species, and the connection to cell dispersion in these systems may be an interesting area for future investigation.

4 Conclusions

It is likely that the biofilm dispersion response is a mechanism to escape starvation conditions or overcrowding within a sessile population, allowing fixed cells the opportunity to migrate to a more favorable environment and thin out the population that remains. Figure 7 shows a schematic representation of the biofilm lifecycle with photomicrographs from each stage in the process given below. Following initial interaction with a surface by reversible (1) and irreversible attachment (2), bacteria are simulated to commence production and excretion of matrix polymers (2) that allow cells to become cemented to the surface. Prolonged growth results in

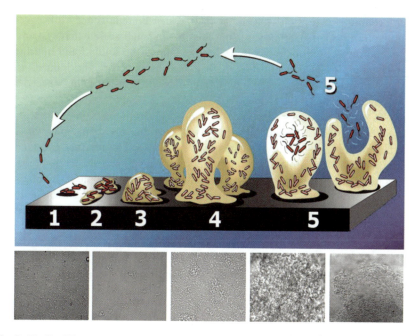

Fig. 7 Biofilm lifecycle. Stages in the development and dispersion of biofilm are shown proceeding from right to left. *Lower panel* shows photomicrographs of bacteria at each of the five stages shown in the schematic above (from left to right) (from David Davies)

the development of microcolonies (3), composed of cells enmeshed in a gel-like EPS matrix. As the microcolonies continue to increase in size (4), cells in the interior of the microcolonies will experience overcrowding, decreased availability of nutrients, increased concentrations of waste-products, toxins and excreted metabolites including cell-to-cell signaling molecules, along with changes in their physicochemical environment (pH, ionic strength, buffering capacity, etc.). Cells recognize these changes (5), through the transduction of external or internal signals, for instance, by recognizing autoinducers and signaling molecules (such as oxygen or NO) through interaction with sensors anchored to the membrane or located inside the cell. These sensing systems interact, by undefined mechanisms, with second messenger molecules such as c-di-GMP, resulting in global changes in cell physiology. In at least some instances, this involves interaction with global carbon regulators. An important component of the response is the release of enzymes into the extracellular environment where they interact with and degrade polymers forming the matrix. Whether this is a gated release of existing enzymes, or results from the synthesis of new proteins, is unclear at this time. As the matrix within a microcolony is digested, cells become free to move by active motility or browning motion. Eventually, a breach is made in the matrix at a margin of the microcolony through which the bacteria are able to escape into the surrounding bulk liquid.

The sensing systems involved in regulating the dispersion response are not clearly defined at present. When biofilm microcolonies are small, an inducer, which accumulates in the extracellular matrix, can be removed by diffusive and advective transport. This removal is not possible in small-scale batch systems and may explain why biofilms typically fail to develop at the solid–liquid interface in these systems. When cell clusters attain a dimension where the inducer is not adequately washed out from the interior (the rate of diffusion being exceeded by the rate of production), the inducer is able to attain a concentration necessary for activation of the dispersion response, releasing cells from the biofilm.

The identification of signaling systems responsible for dispersion will have important implications in the management of biofilms. The unusual resistance of biofilm bacteria to treatment with antimicrobial agents, and the persistence and chronic nature of biofilm infections could potentially be overcome if, in treatment, biofilm bacteria could be forced to transition to a planktonic state. The induction of biofilm dispersion prior to, or in combination with therapy by antimicrobial agents should provide a practical mechanism for improving the outcome of antimicrobial treatments. In situations where microbicides are unwanted or unnecessary, dispersion induction could be used as an alternative to toxic compounds or reactive chemicals. Biofilm dispersion thus holds significant promise a mechanism for the management and control of resistant or recalcitrant bacterial populations.

References

Allison DG, Ruiz B, SanJose C, Jaspe A, Gilbert P (1998) Extracellular products as mediators of the formation and detachment of *Pseudomonas fluorescens* biofilms. FEMS Microbiol Lett 167:179–184

Allwood A, Walter MR, Burch IW, Kamber BS (2007) 3.43 billion-year-old stromatolite reef from the Pilbara Craton of Western Australia: ecosystem-scale insights to early life on Earth. Precambrian Res 158:198–227

Barber CE et al (1997) A novel regulatory system required for pathogenicity of *Xanthomonas campestris* is mediated by a small diffusible signal molecule. Mol Microbiol 24:555–566

Barraud N et al (2006) Involvement of nitric oxide in biofilm dispersal of *Pseudomonas aeruginosa*. J Bacteriol 188:7344–7353

Boles BR, Horswill AR (2008) *Agr*-mediated dispersal of *Staphylococcus aureus* biofilms. PLoS Pathog 4:e1000052

Boles BR, Thoendel M, Singh PK (2005) Rhamnolipids mediate detachment of *Pseudomonas aeruginosa* from biofilms. Mol Microbiol 57:1210–1223

Boon C et al (2007) A novel DSF-like signal from *Burkholderia cenocepacia* interferes with Candida albicans morphological transition. ISME J 2:27–36

Bowden GH, Li YH (1997) Nutritional influences on biofilm development. Adv Dent Res 11:81–99

Boyd A, Chakrabarty AM (1994) Role of alginate lyase in cell detachment of *Pseudomonas aeruginosa*. Appl Environ Microbiol 60:2355–2359

Breyers JD (1988) Modeling biofilm accumulation. In: Bazin MJ, Prosser JI (eds) Physiology models in microbiology, vol 2. Boca Raton, FL, pp 109–144

Chatterjee S, Newman KL, Lindow SE (2008) Cell-to-cell signaling in *Xylella fastidiosa* suppresses movement and xylem vessel colonization in grape. Mol Plant-Microbe Interact 21:1309–1315

Chen X, Stewart PS (2000) Biofilm removal caused by chemical treatments. Water Res 34:4229–4233
Christen M, Christen B, Folcher M, Schauerte A, Jenal U (2005) Identification and characterization of a cyclic di-GMP-specific phosphodiesterase and its allosteric control by GTP. J Biol Chem 280:30829–30837
Costerton JW, Lewandowski Z, Caldwell DE, Korber DR, Lappin-Scott HM (1995) Microbial biofilms. Annu Rev Microbiol 49:711–745
Danhorn T, Hentzer M, Givskov M, Parsek MR, Fuqua C (2004) Phosphorus limitation enhances biofilm formation of the plant pathogen *Agrobacterium tumefaciens* through the PhoR-PhoB regulatory system. J Bacteriol 186:4492–4501
Davey ME, Caiazza NC, O'Toole GA (2003) Rhamnolipid surfactant production affects biofilm architecture in *Pseudomonas aeruginosa* PAO1. J Bacteriol 185:1027–1036
David GA, Begoña R, Carmen S, Almudena J, Peter G (1998) Extracellular products as mediators of the formation and detachment of *Pseudomonas fluorescens* biofilms. FEMS Microbiol Lett 167:179–184
Davies DG, Marques CNH (2009) A fatty acid messenger is responsible for inducing dispersion in microbial biofilms. J Bacteriol 191:1393–1403
Davies DG (1999) Regulation of matrix polymer in biofilm formation and dispersion. In: Wingender J, Neu TR, Flemming H-C (eds) Microbial extrapolymeric substances, characterization, structure and function. Springer, Berlin, pp 93–112
Delaquis PJ, Caldwell DE, Lawrence JR, McCurdy AR (1989) Detachment of *Pseudomonas fluorescens* from biofilms on glass surfaces in response to nutrient stress. Microb Ecol 18:199–210
Dewanti R, Wong ACL (1995) Influence of culture conditions on biofilm formation by *Escherichia coli* O157:H7. Int J Food Microbiol 26:147–164
Dow JM et al (2003) Biofilm dispersal in *Xanthomonas campestris* is controlled by cell-cell signaling and is required for full virulence to plants. Proc Natl Acad Sci 100:10995–11000
Ferreira RBR, Antunes LCM, Greenberg EP, McCarter LL (2008) *Vibrio parahaemolyticus* ScrC modulates cyclic dimeric GMP regulation of gene expression relevant to growth on surfaces. J Bacteriol 190:851–860
Fouhy Y et al (2007) Diffusible signal factor-dependent cell-cell signaling and virulence in the nosocomial pathogen *Stenotrophomonas maltophilia*. J Bacteriol 189:4964–4968
Friedman L, Kolter R (2004a) Two genetic loci produce distinct carbohydrate-rich structural components of the *Pseudomonas aeruginosa* biofilm matrix. J Bacteriol 186:4457–4465
Friedman L, Kolter R (2004b) Genes involved in matrix formation in *Pseudomonas aeruginosa* PA14 biofilms. Mol Microbiol 51:675–690
Gacesa P (1987) Alginate-modifying-enzymes: a proposed unified mechanism of action for the lyases and epimerases. FEBS Lett 212:199–202
Güvener ZT, Harwood CS (2007) Subcellular location characteristics of the *Pseudomonas aeruginosa* GGDEF protein, WspR, indicate that it produces cyclic-di-GMP in response to growth on surfaces. Mol Microbiol 66:1459–1473
Hickman JW, Tifrea DF, Harwood CS (2005) A chemosensory system that regulates biofilm formation through modulation of cyclic diguanylate levels. Proc Natl Acad Sci USA 102:14422–14427
Hisatsuka K, Nakahara T, Sano N, Yamada K (1971) Formation of rhamnolipid by *Pseudomonas aeruginosa* and its function in hydrocarbon fermentation. Agric Biol Chem 35:686–692
Hofmann HJ, Grey K, Hickman AH, Thorpe RI (1999) Origin of 3.45 Ga coniform stromatolites in Warrawoona group, Western Australia. Geol Soc Am Bull 111:1256–1262
Huang T-P, Wong ACL (2007) A cyclic AMP receptor protein-regulated cell-cell communication system mediates expression of a FecA homologue in *Stenotrophomonas maltophilia*. Appl Environ Microbiol 73:5034–5040
Hunt SM, Werner EM, Huang B, Hamilton MA, Stewart PS (2004) Hypothesis for the role of nutrient starvation in biofilm detachment. Appl Environ Microbiol 70:7418–7425

Itoh Y, Wang X, Hinnebusch BJ, Preston JF III, Romeo T (2005) Depolymerization of {beta}-1,6-N-acetyl-D-glucosamine disrupts the integrity of diverse bacterial biofilms. J Bacteriol 187:382–387

Jackson DW, Simecka JW, Romeo T (2002) Catabolite repression of *Escherichia coli* biofilm formation. J Bacteriol 184:3406–3410

Jackson KD, Starkey M, Kremer S, Parsek MR, Wozniak DJ (2004) Identification of psl, a locus encoding a potential exopolysaccharide that is essential for *Pseudomonas aeruginosa* PAO1 biofilm formation. J Bacteriol 186:4466–4475

James GA, Korber DR, Caldwell DE, Costerton JW (1995) Digital image analysis of growth and starvation responses of a surface-colonizing Acinetobacter sp. J Bacteriol 177:907–915

Kaplan JB, Fine DH (2002) Biofilm dispersal of neisseria subflava and other phylogenetically diverse oral bacteria. Appl Environ Microbiol 68:4943–4950

Kaplan JB, Ragunath C, Ramasubbu N, Fine DH (2003) Detachment of *Actinobacillus actinomycetemcomitans* biofilm cells by an endogenous {beta}-hexosaminidase activity. J Bacteriol 185:4693–4698

Karatan E, Watnick P (2009) Signals, regulatory networks, and materials that build and break bacterial biofilms. Microbiol Mol Biol Rev 73:310–347

Kim Y-K, McCarter LL (2007) ScrG, a GGDEF-EAL protein, participates in regulating swarming and sticking in *Vibrio parahaemolyticus*. J Bacteriol 189:4094–4107

Lamed R, Bayer EA (1986) Contact and cellulolysis in *Clostridium thermocellum* via extensive surface organelles. Experientia 42:72–73

Lee J, Bansal T, Jayaraman A, Bentley WE, Wood TK (2007a) Enterohemorrhagic *Escherichia coli* biofilms are inhibited by 7-hydrozyindole and stimulated by isatin. Appl Environ Microbiol 73:4100–4109

Lee J, Jayaraman A, Wood TK (2007b) Indole is an inter-species biofilm signal mediated by SdiA. BMC Microbiol 7:1–15

Lynch MJ et al (2002) The regulation of biofilm development by quorum sensing in *Aeromonas hydrophila*. Environ Microbiol 4:18–28

Ma L, Jackson KD, Landry RM, Parsek MR, Wozniak DJ (2006) Analysis of *Pseudomonas aeruginosa* conditional Psl variants reveals roles for the Psl polysaccharide in adhesion and maintaining biofilm structure postattachment. J Bacteriol 188:8213–8221

Ma L, Lu H, Sprinkle A, Parsek MR, Wozniak D (2007) *Pseudomonas aeruginosa* Psl is a galactose- and mannose-rich exopolysaccharide. J Bacteriol. doi:10.1128/JB.00620-07

Marshall JC (1988) Adhesion and growth of bacteria at surfaces in oligotrophic habitats. Can J Microbiol 34:503–506

Matsukawa M, Greenberg EP (2004) Putative exopolysaccharide synthesis genes influence *Pseudomonas aeruginosa* biofilm development. J Bacteriol 186:4449–4456

May TB et al (1991) Alginate synthesis by *Pseudomonas aeruginosa*: a key pathogenic factor in chronic pulmonary infections of cystic fibrosis patients. Clin Microbiol Rev 4:191–206

Merritt JH, Brothers KM, Kuchma SL, O'Toole GA (2007) SadC reciprocally influences biofilm formation and swarming motility via modulation of exopolysaccharide production and flagellar function. J Bacteriol 189(22):8154–8164

Morgan R, Kohn S, Hwang S-H, Hassett DJ, Sauer K (2006) BdlA, a chemotaxis regulator essential for biofilm dispersion in *Pseudomonas aeruginosa*. J Bacteriol 188:7335–7343

Ohashi A, Harada H (1994a) Adhesion strength of biofilm developed in an attached-growth reactor. Water Sci Technol 20:10–11

Ohashi A, Harada H (1994b) Characterization of detachment mode of biofilm developed in an attached-growth reactor. Water Sci Technol 30:35–45

Ohashi A, Koyama T, Syutsubo K, Harada H (1999) A novel method for evaluation of biofilm tensile strength resisting erosion. Water Sci Technol 39:261–268

O'Toole G, Kaplan HB, Kolter R (2000) Biofilm formation as microbial development. Annu Rev Microbiol 54:49–79

Peyton BM, Characklis WG (1993) A statistical analysis of the effect of substrate utilization and shear stress on the kinetics of biofilm detachment. Biotechnol Bioeng 41:728–735

Pratt LA, Kolter R (1999) Genetic analyses of bacterial biofilm formation. Curr Opin Microbiol 2:598–603

Purevdorj-Gage B, Costerton WJ, Stoodley P (2005) Phenotypic differentiation and seeding dispersal in non-mucoid and mucoid *Pseudomonas aeruginosa* biofilms. Microbiology 151:1569–1576

Puskas A, Greenberg EP, Kaplan S, Schaefer AL (1997) A quorum-sensing system in the free-living photosynthetic bacterium *Rhodobacter sphaeroides*. J Bacteriol 179:7530–7537

Rice SA et al (2005) Biofilm formation and sloughing in *Serratia marcescens* are controlled by quorum sensing and nutrient cues. J Bacteriol 187:3477–3485

Rittman BR (1982) The effect of shear stress on biofilm loss rate. Biotechnol Bioeng 24:501–506

Rochex A, Lebeault JM (2007) Effects of nutrients on biofilm formation and detachment of a *Pseudomonas putida* strain isolated from a paper machine. Water Res 41:2885–2992

Roger S, Michael M, Abdul K, Manfred N, Ute R (2004) GGDEF and EAL domains inversely regulate cyclic di-GMP levels and transition from sessility to motility. Mol Microbiol 53:1123–1134

Romeo T (1998) Global regulation by the small RNA-binding protein CsrA and the non-coding RNA molecule CsrB. Mol Microbiol 29:1321–1330

Romeo T, Gong M, Liu MY, Brun-Zinkernagel AM (1993) Identification and molecular characterization of *csrA*, a pleiotropic gene from *Escherichia coli* that affects glycogen biosynthesis, gluconeogenesis, cell size, and surface properties. J Bacteriol 175:4744–4755

Ross P et al (1990) The cyclic diguanylic acid regulatory system of cellulose synthesis in *Acetobacter xylinum*. Chemical synthesis and biological activity of cyclic nucleotide dimer, trimer, and phosphothioate derivatives. J Biol Chem 265:18933–18943

Ryan RP et al (2006) Cell-cell signaling in *Xanthomonas campestris* involves an HD-GYP domain protein that functions in cyclic di-GMP turnover. Proc Natl Acad Sci USA 103:6712–6717

Sabnis NA, Yang H, Romeo T (1995) Pleiotropic regulation of central carbohydrate metabolism in *Escherichia coli* via the gene *csrA*. J Biol Chem 270:29096–29104

Sauer K, Camper AK, Ehrlich GD, Costerton JW, Davies DG (2002) *Pseudomonas aeruginosa* displays multiple phenotypes during development as a biofilm. J Bacteriol 184:1140–1154

Sauer K et al (2004) Characterization of nutrient-induced dispersion in *Pseudomonas aeruginosa* PAO1 biofilm. J Bacteriol 186:7312–7326

Sawyer LK, Hermanowicz SW (1998) Detachment of biofilm bacteria due to variations in nutrient supply. Water Sci Technol 37:211–214

Simm R, Morr M, Kader A, Nimtz M, Romling U (2004) GGDEF and EAL domains inversely regulate cyclic di-GMP levels and transition from sessility to motility. Mol Microbiol 53:1123–1134

Slater H, Alvarez-Morales A, Barber CE, Daniels MJ, Dow JM (2000) A two-component system involving an HD-GYP domain protein links cell-cell signalling to pathogenicity gene expression in *Xanthomonas campestris*. Mol Microbiol 38:986–1003

Stanley NR, Britton RA, Grossman AD, Lazazzera BA (2003) Identification of catabolite repression as a physiological regulator of biofilm formation by *Bacillus subtilis* by use of DNA microarrays. J Bacteriol 185:1951–1957

Stewart PS (1993) A model of biofilm detachment. Biotechnol Bioeng 41:111–117

Stoodley P et al (2001) Growth and detachment of cell clusters from mature mixed-species biofilms. Appl Environ Microbiol 67:5608–5613

Tal R et al (1998) Three cdg operons control cellular turnover of cyclic di-GMP in *Acetobacter xylinum*: genetic organization and occurrence of conserved domains in isoenzymes. J Bacteriol 180:4416–4425

Thormann KM, Saville RM, Shukla S, Spormann AM (2005) Induction of rapid detachment in *Shewanella oneidensis* MR-1 biofilms. J Bacteriol 187:1014–1021

Thormann KM et al (2006) Control of formation and cellular detachment from *Shewanella oneidensis* MR-1 biofilms by cyclic di-GMP. J Bacteriol 188:2681–2691

Tolker-Nielsen et al T (2000) Development and dynamics of *Pseudomonas* sp. biofilms. J Bacteriol 182:6482–6489

Trulear MG, Characklis WG (1982) Dynamics of biofilm processes. J Water Pollut Control Fed 9:1288–1301

van Loosdrecht MCM, Picioreanu C, Heijnen JJ (1997) A more unifying hypothesis for the structure of microbial biofilms. FEMS Microbiol Ecol 24:181–183

Vats N, Lee SF (2000) Active detachment of *Streptococcus mutans* cells adhered to epon-hydroxylapatite surfaces coated with salivary proteins in vitro. Arch Oral Biol 45:305–314

Wang LH et al (2004) A bacterial cell-cell communication signal with cross-kingdom structural analogues. Mol Microbiol 51:903–912

Whitchurch CB, Tolker-Nielsen T, Ragas PC, Mattick JS (2002) Extracellular DNA required for bacterial biofilm formation. Science 295:1487

Xun L, Mah RA, Boone DR (1990) Isolation and characterization of disaggregatase from *Methanosarcina mazei* LYC. Appl Environ Microbiol 56:3693–3698

Competition, Communication, Cooperation: Molecular Crosstalk in Multi-species Biofilms

Carsten Matz

Abstract Many microorganisms exist in the environment as multicellular communities, so-called biofilms. Chemical communication is an essential part of the way in which biofilm populations coordinate their behavior and respond to environmental challenges. Recent research has been unravelling a complex web of chemical crosstalk mediating microbial symbiosis, competition and defense against predators and pathogens. Understanding the molecular basis of biofilm interactions in their ecological context bears the potential of refining natural product discovery and the development of biofilm-derived biotechnologies.

Microbial biofilms constitute the major proportion of bacterial biomass and activity in many natural and man-made systems. At the same time, biofilms serve as important environmental reservoirs for pathogenic microbes (Flanders and Yildiz 2004) and are the causative agents for many persistent bacterial infections (Costerton et al. 1999). Rules governing biofilm assembly, function and evolution have been largely unexplored but as with communities of macro-organisms, local interactions between component organisms in spatially structured environments are likely to be of central importance (Hibbing et al. 2010). Structure, composition and function of biofilm communities are predicted to be determined by synergistic and antagonistic interactions among component species (Hassell et al. 1994; Kerr et al. 2002; Battin et al. 2003). Despite the consensus that natural biofilms represent multi-species communities of diverse micro-organisms, studies to date have addressed the physiology and regulation of biofilm functions almost to the exclusion of species–species interactions.

Biofilms in the natural environment are very complex entities that potentially consist of many hundreds of different species. There are real challenges in understanding how different bacteria interact with their own and other species. To study

C. Matz (✉)
Helmholtz Centre for Infection Research, Inhoffenstr. 7, 38124 Braunschweig, Germany
e-mail: carsten.matz@web.de

organismic interactions in multi-species biofilms, marine biofilms have become a leading model system (Egan et al. 2008). Biofilms in the sea are most evident as slimy surface growth or "Aufwuchs" on docks, boat hulls or intertidal rocks. A glimpse through the microscope reveals, however, that even marine animals and plants or inconspicuous sediments may be covered with biofilms. Marine and freshwater biofilms are complex structures that comprise both bacteria and eukaryotic microbes (Fig. 1). Following the initial attachment of bacteria, surfaces get further colonized by fungi, cyanobacteria, unicellular algae and bacteria-eating protozoa, thus creating a dynamic and diverse community of micro-organisms. In the absence of large shear forces, such biofilm communities can grow extensive mats or slimes. In the process of the successive surface colonization, aquatic macro-organisms such as macroalgae and invertebrates (barnacles, worms, mussels, snails) arrive, by which aquatic surface communities often become visible to the naked eye. These biofilm communities may not quite reach the grandeur of coral reefs, but they may be comparable in their complexity of the interactions between component

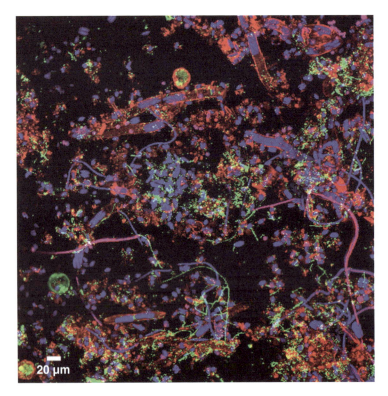

Fig. 1 Laser scanning microscopy image of a marine biofilm community. Bacteria (*green*) were stained with SYTO9 and extracellular polymeric substances (*red*) were stained with Alexa568-conjugated lectin AAL (*Aleuria aurantia*). Pink and blue fluorescence is due to the autofluorescence of cyanobacteria and microalgae (diatoms, chlorophytes), respectively. Circular cells (*green*, *red*) represent heterotrophic protists. Image courtesy of Dr. B. Zippel

species. Owing to their densely packed nature and limited capacity for diffusion, chemical communication might be an ideal way for species to interact within a biofilm as distances are small enough for the diffusion of signal and effector molecules (Decho et al. 2010). This chapter summarizes first insights into the biomolecular complexity underlying interspecies communication in marine biofilm communities.

1 Competition and Defence by Chemical Weapons

Life in a biofilm is similar to that in a city (Watnick and Kolter 2000): the dense settlement not only allows symbiotic relationships between component species, but also tightens the competition for limited resources (space, nutrients, light) or may attract predators. Relatively high population density and limited diffusibility of the exopolymer matrix support the notion that biofilm residents interact and communicate by small chemical compounds. In the last few years, it has become evident that marine bacteria harbour a broad arsenal of biologically active metabolites, which are predicted to function as chemical weapons and signals (Jensen and Fenical 1994).

The marine gamma-proteobacterium *Pseudoalteromonas tunicata* illustrates the range of target-specific molecules that biofilm bacteria may use to grow and survive in multi-species biofilm communities (Thomas et al. 2008). In the interaction with competing fungi, *P. tunicata* makes use of a secreted inhibitory tambjamine alkaloid (Fig. 2a) (Franks et al. 2005, 2006). The biosynthetic pathway is coded by a cluster of 19 genes (*tamA* to *tamS*) encoding proteins with homology to prodigiosin biosynthetic genes in various other bacteria (Burke et al. 2007).

Microcolonies of *P. tunicata* also produce the antibacterial AlpP lysine oxidase (James et al. 1996; Mai-Prochnow et al. 2008), which has been proposed to keep competing bacterial species in check by the production of H_2O_2. In photic habitats, unicellular algae such as diatoms are among the strongest competitors of bacteria

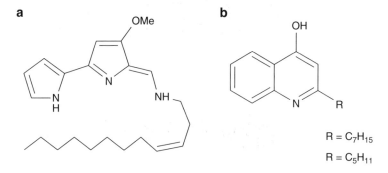

Fig. 2 Metabolites with antifungal and antialgal activities: Tambjamine (**a**) and alkyl-chinolinols (**b**) isolated from marine *Pseudoalteromonas* spp.

for micronutrients. In *P. tunicata*, a not-yet-characterized 3–10 kDa peptide is suspected to inhibit the growth of diatoms and the germination of macroalgal spores. Other bacteria of the genus *Pseudoalteromonas* are known for the production of alkyl-chinolinols (Fig. 2b) (Long et al. 2003), which were found to inhibit the growth but not to kill various diatom. Generally, the increasing number of reports on allelopathic effects elicited by biofilm-derived bioactives suggests that much of the spatial and temporal dynamics found in natural biofilm communities is controlled by chemical compounds.

A life-threatening ecological factor for attached bacteria is the grazing pressure elicited by bacteria-consuming protists, the protozoa (amoebae, ciliates, flagellates). While bacterial biofilms without grazing defence can be rapidly eliminated by protozoa, biofilm bacteria capable of chemical defence are resistant against grazing and may grow to high cell densities (Fig. 3).

Again, *P. tunicata* provides a good example for the production of an effective antipredator molecule, the purple indole alkaloid violacein (Fig. 4) (Matz et al. 2008). Violacein is synthesized by a number of marine bacteria and stored in the periplasm between the inner and outer membrane of the bacterial cell (Matz et al. 2008; Hakvåg et al. 2009). Once a bacterium gets phagocytized by a protozoan predator, violacein is thought to be released into the phagolysosome upon the digestion of the outer membrane.

Although the molecular mechanism has not been elucidated in detail, there is evidence that violacein as a redox-active molecule disrupts the membrane potential of mitochondria. This triggers the eukaryotic suicide program (apoptosis) of the protozoan cell, thus leading to the complete lysis of the protozoan predator within a

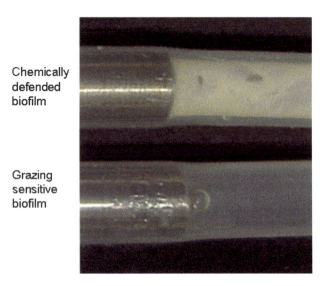

Fig. 3 Biofilms exposed to grazing by *Acanthamoeba polyphaga*. Chemically defended biofilms (*top*) maintain high biomass, while undefended biofilms are cleared from the tube (*bottom*). Image courtesy of Dr. M. Weitere

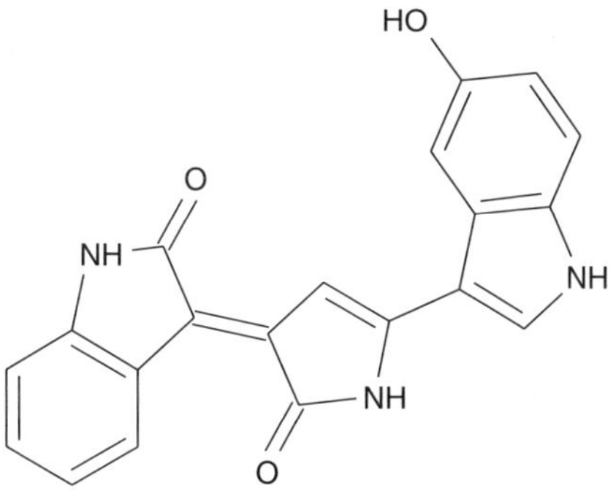

Fig. 4 Predator-active compound: violacein isolated from a range of marine bacteria, including *Pseudoalteromonas* spp. and *Microbulbifer* sp.

few hours (Matz et al. 2008). The effectiveness of this defence mechanism is illustrated by the fact that a single bacterium, containing about one femtogram of violacein, is sufficient to stop the feeding activity of the protozoan cell within minutes. Other recent studies suggest that such chemically mediated grazing resistance is not uncommon in biofilms (Matz et al. 2004a, 2005, 2008), and could be an explanation for the accumulation and persistence of biofilms, despite the presence and feeding activity of their natural consumers.

2 Cooperation by Chemical Communication

It is striking that the biosynthesis of defence molecules such as violacein is significantly increased in biofilm bacteria in comparison with free-swimming bacterial cells (Matz et al. 2008). One reason for this is that within biofilms bacteria reach relatively high local cell densities even at low nutrient levels. High cell densities in turn facilitate chemical communication between bacterial cells using pheromone-like signalling molecules (Williams et al. 2007). In many Gram-negative bacteria, cell–cell communication occurs via *N*-acyl homoserine lactones (AHLs, Fig. 5), while autoinducing peptides represent common signal molecules in Gram-positive bacteria. Moreover, furanosylesters that are derived from dihydroxypentanedione are described as the AI-2 signalling system for both Gram-negative and Gram-positive bacteria.

By using quorum sensing molecules, bacteria obtain information on their population size and may coordinate a group behaviour similar to a multi-cellular organism. Temporal patterns of gene expression during bacterial growth indicate that most genes are activated by quorum sensing during the transition from logarithmic phase to stationary phase (Schuster et al. 2003; Wagner et al. 2003); that is, when cell densities markedly increase and nutrients become increasingly depleted. By enabling cooperation and synchronization, quorum sensing is thought to adjust the population response to changing environmental conditions and increase the fitness of dense bacterial assemblages in late logarithmic and early stationary phase (Keller and Surette 2006).

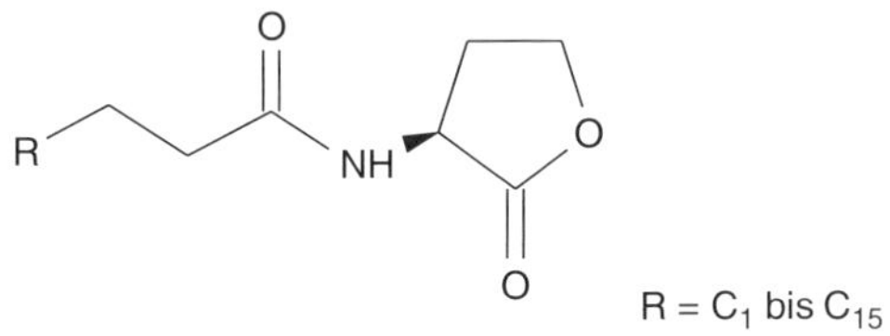

Fig. 5 Quorum sensing molecules: *N*-acyl homoserine lactones produced by Gram-negative bacteria

R = C$_1$ bis C$_{15}$

Predation and the necessity of defence exemplify one facet of the selective advantages in the evolution of bacterial cell-to-cell communication systems: in exponentially growing bacterial populations, predation is not an immediate threat (because bacterial growth rates are usually high enough to compensate for grazing losses) but grazing becomes a serious problem with the onset of nutrient depletion (stationary phase). In fact, evidence has been gathered from studies of four bacterial species, *Pseudomonas aeruginosa*, *Chromobacterium violaceum*, *Vibrio cholerae* and *Serratia marcescens* that quorum sensing mutants have a significantly reduced antipredator fitness compared with their isogenic wild-type strains (Matz et al. 2004a, b, 2005; Queck et al. 2006). In addition to their role in monitoring population density, quorum sensing signals have also been proposed to function as diffusion sensors (Redfield 2002). In the context of grazing resistance, synchronization by a quorum might be required to reach extracellular toxin levels high enough to ward off predators, whereas the secretion of inhibitors by a single cell is likely to be ineffective.

Upon reaching the critical quorum size, quorum-sensing signals induce the expression of grazing defence genes, as in the case of the violacein gene cluster *vioABCDE*. It is well known for plant-associated bacteria that already microcolonies of less than 40 bacteria are enough to turn on the AHL-mediated communication (Dulla and Lindow 2008). Indirectly, the feeding activity of protozoa appears to promote the achievement of the minimum quorum size. In particular, the size selective feeding of protozoa promotes the formation of microcolonies (Matz et al. 2004a), which may lead to the accelerated induction of quorum sensing, and more indirectly to the induction of chemical defence. To what extent the formation of microcolonies, or the synthesis and release of defence molecules also is directly inducible by predator-derived signals (kairomones), remains a question for future investigations.

3 Eukaryotic Response to Biofilm Signals

The colonization of marine surfaces by microbial biofilms entails the settlement of macro-organisms in the form of algal spores or invertebrate larvae. When choosing the appropriate attachment site, the temporarily free-swimming spores and larvae of macroalgae and invertebrates may respond to chemical cues or signals released by biofilms. Deterrents often appear to be nonpolar secondary metabolites (see, e.g.,

P. tunicata), while there seems to be mainly primary metabolites such as carbohydrates and peptides among the water-soluble attractants (Qian et al. 2007). For example, larvae of the tubeworm *Hydroides elegans* searching for suitable attachment habitat rely on the specific composition of the biofilm EPS, and as-yet-unidentified metabolites produced by associated bacteria or diatoms (Harder et al. 2002; Lam et al. 2005).

In addition, macro-organisms are capable of eavesdropping on the chemical communication of biofilm bacteria. Spores of the green alga *Ulva intestinalis* exploit different long-chained AHL signals of communicating biofilm bacteria for their own orientation (Joint et al. 2002). When *Ulva* zoospores reach the surface, they slow down their swimming movement at micromolar hotspots of AHLs and attach directly to the AHL-producing biofilm bacteria (Joint et al. 2007). It has been demonstrated that the detection of AHLs results in calcium influx into the zoospore. Currently, we can only speculate about the ecological and evolutionary benefits of biofilm-mediated habitat choice for spores and larvae: The colonization of abiotic surfaces by "biotic" biofilms could serve as a general indicator/proxy for habitable conditions, or facilitate specific associations with bacterial symbionts.

4 Biofilm Inhibition by Molecular Mimicry

In addition to sediments and rocks, biofilms colonize the "living" surfaces of many marine animals and plants, a phenomenon termed epibiosis. Marine animals and plants are exposed to the constant risk of being literally overgrown by epibionts. One strategy for the host organism to block epibiont settlement is the production of antifoulants. The Australian red alga *Delisea pulchra*, for instance, employs molecular mimicry by releasing halogenated furanones (Fig. 6a), which is remarkably similar in its structure to the AHLs that many biofilm bacteria use for intraspecific

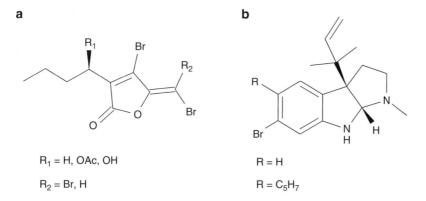

Fig. 6 Host-derived quorum sensing antagonists: brominated furanones and alkaloids isolated from the red alga *Delisea pulchra* (**a**) and the bryozoan *Flustra foliacea* (**b**)

communication (Givskov et al. 1996; Manefield et al. 1999). Furanones were found to modulate the activity of the AHL-dependent transcriptional activator LuxR in the bacterial quorum-sensing systems by reducing the half-life of the LuxR receptor proteins (Manefield et al. 2002). Blocking communication between biofilm bacteria leads to the failure of bacterial survival functions. Among other functions, biofilm bacteria may be compromised in their defence against predators in the presence of QS antagonists and thus rapidly grazed by protozoa.

Likewise, marine animals have been reported to produce AHL-antagonists to reduce the formation of epibiotic biofilm. One example may be the brominated alkaloids, which were isolated from the North Sea bryozoan *Flustra foliacea* (Fig. 6b) (Peters et al. 2003). However, the QS-specific antagonism of *Delisea* furanones and *Flustra* alkaloids is limited to a specific concentration range. At higher concentrations, these molecules possess a general biocidal effect, so that – in addition to compromising biofilm homeostasis – the settlement of invertebrate larvae and algal spores is directly inhibited.

5 Host Defence Mediated by Epibiotic Biofilms

Epibiotic biofilms may not only be detrimental to the animal or plant host. In recent years, it has become increasingly clear that many of the natural products isolated from marine plants and animals are produced by epibiotic or symbiotic bacteria (König et al. 2006; Piel 2009). These molecules may assist host organisms, for example, in the defence against parasitic bacteria and fungi. Biofilm bacteria which inhibit the growth of parasitic fungi through the production of isatin (Fig. 7a) and the fungus-specific QS signal tyrosol have been identified on crustacean embryos (Fig. 7b) (Gil-Turnes et al. 1989; Gil-Turnes and Fenical 1992).

Substances produced by epibiotic bacteria include complex polyketides from the class of bryostatins (Fig. 8). Bryostatin was originally isolated from the bryozoan *Bugula neritina* and are in clinical testing phase due to their promising therapeutic properties. It is now known that bryostatin is actually produced by the bacterium *Endobugula sertula* that forms biofilm-like cell clusters on the surface of bryozoan larvae (Sudek et al. 2007; Sharp et al. 2007). One reported ecological benefit of hosting these epibiotic bacteria is that bryozoan larvae deter predatory fish by the bacterial production of bryostatin.

Moreover, selective tolerance, attraction or transmission of primary bacterial epibionts may assist the host to control the composition of the secondary community of epibionts and the intensity of colonization. A good example may be a bacterium such as *P. tunicata* (see above), which has originally been isolated from the surface of the green alga *Ulva lactuca* and the sea squirt *Ciona intestinalis* and which can influence directly and indirectly the colonization of bacteria, fungi, protozoa, algae and invertebrates by a variety of low-molecular-weight inhibitors. Interestingly, violacein which protects *P. tunicata* and other biofilm bacteria from predatory protozoa shows also high activity against herbivorous invertebrates.

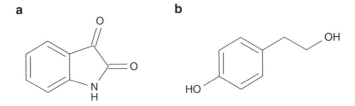

Fig. 7 Fungus-inhibiting compounds derived from epibiotic bacteria: Isatin (**a**) and tyrosol (**b**)

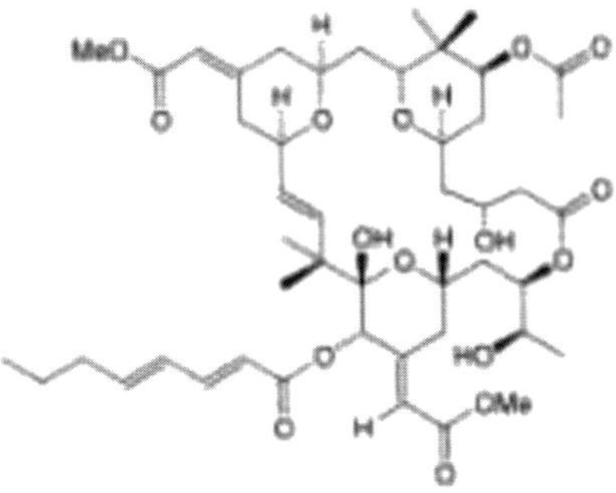

Fig. 8 Antifeedant of bryozoan host is produced by epibiotic bacteria: bryostatin

Recent studies suggest that violacein-producing biofilms could protect not only themselves but also deter grazers of the host algae *U. lactuca*.

6 Conclusions

Marine micro-organisms are versatile producers of secondary metabolites. Although the sea has yielded thousands of bioactive metabolites over the past two decades, we are only beginning to explore the natural functions of these molecules. Many micro-organisms exist in the environment as surface-associated biofilm communities. The close spatial proximity of micro-organisms at surfaces drives specific interspecies interactions and generates complex and highly differentiated microbial communities. Chemical communication is recognized to be an essential part of the way in which biofilm organisms coordinate their behaviour and respond to environmental challenges. Recent studies have been unravelling first aspects of the complex chemical crosstalk mediating microbial symbiosis, competition and defence against predators and pathogens (Fig. 9). Future research is anticipated to provide insights into how interspecies communication may shape the structural and functional dynamics of biofilm communities.

Marine biofilms play a central role as hotspots of biological diversity and molecular complexity. From the progressive description of new marine natural products arises the question of their function in the natural context. The study of

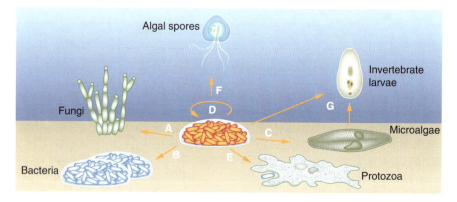

Fig. 9 Chemical warfare and communication in a natural multi-species biofilm. Biofilm communities are shaped by antagonistic and synergistic interactions between microbial species. Bacterial populations may establish in a competitive environment by secreting specific inhibitors that target competing fungi (A), bacteria (B), or microalgae (C). Clonal growth in microcolonies allows bacteria to reach cell densities high enough for quorum sensing (D), which may induce the production of antifeedants against protozoa (E). Secondary colonizers, such as macroalgal spores (F) and invertebrate larvae (G) may be attracted by biofilm-derived cues. Biofilm communities associated with living surfaces may further by modulated by host-derived compounds

chemical interactions in natural biofilms, however, faces the dilemma of dealing with the almost unlimited complexity of communities and the limited culturability of their members. This requires a close integration of analytical chemistry and microbial genetics with innovative cultivation approaches and realistic bioassays. Promising novel approaches also include the development and application of global analytical tools such as metabolomics and metagenomics. Understanding the molecular basis of biofilm interactions in their ecological context bears the potential of refining natural product discovery and the development of biofilm-derived biotechnologies.

References

Battin TJ, Kaplan LA, Denis NJ, Hansen CM (2003) Contributions of microbial biofilms to ecosystem processes in stream mesocosms. Nature 426:439–442

Burke C, Thomas T, Egan S, Kjelleberg S (2007) The use of functional genomics for the identification of a gene cluster encoding for the biosynthesis of an antifungal tambjamine in the marine bacterium *Pseudoalteromonas tunicata*. Environ Microbiol 9:814–818

Costerton JW, Stewart PS, Greenberg EP (1999) Bacterial biofilms: a common cause of persistent infections. Science 284:1318–1322

Decho AW, Norman RS, Visscher PT (2010) Quorum sensing in natural environments: emerging views from microbial mats. Trends Microbiol 18:73–80

Dulla G, Lindow SE (2008) Quorum size of *Pseudomonas syringae* is small and dictated by water availability on the leaf surface. Proc Natl Acad Sci USA 105:3082–3087

Egan S, Thomas T, Kjelleberg S (2009) Unlocking the diversity and biotechnological potential of marine surface associated microbial communities. Curr Opin Microbiol 11:219–225

Flanders JR, Yildiz FH (2004) Biofilms as reservoirs for disease. In: Ghannoum M, O'Toole GA (eds) Microbial biofilms. ASM, Washington, DC, pp 314–331

Franks A, Haywood P, Holmström C, Egan S, Kjelleberg S, Kumar N (2005) Isolation and structure elucidation of a novel yellow pigment from the marine bacterium *Pseudoalteromonas tunicata*. Molecules 10:1286–1291

Franks A, Egan S, Holmström C, James S, Lappin-Scott H, Kjelleberg S (2006) Inhibition of fungal colonization by *Pseudoalteromonas tunicata* provides a competitive advantage during surface colonization. Appl Environ Microbiol 72:6079–6087

Gil-Turnes MS, Fenical W (1992) Embryos of *Homarus americanus* are protected by epibiotic bacteria. Biol Bull 182:105–108

Gil-Turnes MS, Hay ME, Fenical W (1989) Symbiotic marine bacteria chemically defend crustacean embryos from a pathogenic fungus. Science 246:116–118

Givskov MG, de Nys R, Manefield M, Gram L, Maximilien R, Eberl L, Molin S, Steinberg PD, Kjelleberg S (1996) Eukaryotic interference with homoserine lactone-mediated prokaryotic signalling. J Bacteriol 178:6618

Hakvåg S, Fjærvik E, Klinkenberg G, Borgos SE, Josefsen KD, Ellingsen TE, Zotchev SB (2009) Violacein-producing *Collimonas* sp. from the sea surface microlayer of costal waters in Trøndelag, Norway. Mar Drugs 12:576–588

Harder T, Lau SCK, Dahms H-U, Qian P-Y (2002) Isolation of bacterial metabolites as natural inducers for larval settlement in the marine polychaete *Hydroides elegans*. J Chem Ecol 28:2029–2043

Hassell MP, Comins HN, May RM (1994) Species coexistence and self-organizing spatial dynamics. Nature 370:290–292

Hibbing ME, Fuqua C, Parsek MR, Peterson SB (2010) Bacterial competition: surviving and thriving in the microbial jungle. Nat Rev Microbiol 8:15–25

James SG, Holmström C, Kjelleberg S (1996) Purification and characterization of a novel antibacterial protein from the marine bacterium D2. Appl Environ Microbiol 62:2783–2788

Jensen PR, Fenical W (1994) Strategies for the discovery of secondary metabolites from marinebacteria: ecological perspective. Annu Rev Microbiol 48:559–584

Joint I, Tait K, Callow ME, Callow JA, Milton D, Williams P, Camara M (2002) Cell-to-cell communication across the prokaryote-eukaryote boundary. Science 298:1207

Joint I, Tait K, Wheeler G (2007) Cross-kingdom signalling: exploitation of bacterial quorum sensing molecules by the green seaweed *Ulva*. Philos Trans R Soc Lond B Biol Sci 362:1223–1233

Keller L, Surette MG (2006) Communication in bacteria: an ecological and evolutionary perspective. Nat Rev Microbiol 4:249–258

Kerr B, Riley MA, Feldman MW, Bohannan BJM (2002) Local dispersal promotes biodiversity in a real-life game of rock-paper-scissors. Nature 418:171–174

König GM, Kehraus S, Seibert S, Abdel-Lateff A, Müller D (2006) Natural products from marine organisms and their associated microbes. Chembiochem 7:229–238

Lam C, Harder T, Qian P-Y (2005) Induction of larval settlement in the polychaete *Hydroides elegans* by extracellular polymers of benthic diatoms. Mar Ecol Prog Ser 286:145–154

Long RA, Qureshi A, Faulkner DJ, Azam F (2003) 2-n-pentyl-4-quinolinol produced by a marine *Alteromonas* sp. and its potential ecological and biogeochemical roles. Appl Environ Microbiol 69:568–576

Mai-Prochnow A, Lucas-Elio P, Egan S, Thomas T, Webb JS, Sanchez-Amat A, Kjelleberg S (2008) Hydrogen peroxide linked to lysine oxidase activity facilitates biofilm differentiation and dispersal in several gram-negative bacteria. J Bacteriol 190:5493–5501

Manefield M, de Nys R, Kumar N, Read R, Givskov M, Steinberg P, Kjelleberg S (1999) Evidence that halogenated furanones from Delisea pulchra inhibit acylated homoserine lactone (AHL)-mediated gene expression by displacing the AHL signal from its receptor protein. Microbiology 145:283–291

Manefield M, Rasmussen TB, Henzter M, Andersen JB, Steinberg PD, Kjelleberg S, Givskov MG (2002) Halogenated furanones inhibit quorum sensing through accelerated LuxR turnover. Microbiology 148:1119–1127

Matz C, Bergfeld T, Rice SA, Kjelleberg S (2004a) Microcolonies, quorum sensing and cytotoxicity determine the survival of *Pseudomonas aeruginosa* biofilms exposed to protozoan grazing. Environ Microbiol 6:218–226

Matz C, Deines P, Boenigk J, Arndt H, Eberl L, Kjelleberg S, Jürgens K (2004b) Impact of violacein-producing bacteria on survival and feeding of bacterivorous nanoflagellates. Appl Environ Microbiol 70:1593–1599

Matz C, McDougald D, Moreno AM, Yung PY, Yildiz FH, Kjelleberg S (2005) Biofilm formation and phenotypic variation enhance predation-driven persistence of *Vibrio cholerae*. Proc Natl Acad Sci USA 102:16819–16824

Matz C, Webb JS, Schupp PJ, Phang SY, Penesyan A, Egan S, Steinberg PD, Kjelleberg S (2008) Marine biofilms evade eukaryotic predation by targeted chemical defense. PLoS ONE 3:e2744

Peters L, König GM, Wright AD, Pukall R, Stackebrandt E, Eberl L, Riedel K (2003) Secondary metabolites of *Flustra foliacea* and their influence on bacteria. Appl Environ Microbiol 69:3469

Piel J (2009) Metabolites from symbiotic bacteria. Nat Prod Rep 26:338–362

Queck SY, Weitere M, Moreno AM, Rice SA, Kjelleberg S (2006) The role of quorum sensing mediated developmental traits in the resistance of Serratia marcescens biofilms against protozoan grazing. Environ Microbiol 8:1017–25

Qian PY, Lau SCK, Dahms H-U, Dobretsov S, Harder T (2007) Marine biofilms as mediators of colonization by marine macroorganisms: Implications for antifouling and aquaculture. Mar Biotechnol 9:399–410

Redfield RJ (2002) Is quorum sensing a side effect of diffusion sensing? Trends Microbiol 10:365–370

Schuster M et al (2003) Identification, timing, and signal specificity of *Pseudomonas aeruginosa* quorum-controlled genes: a transcriptome analysis. J Bacteriol 185:2066–2079

Sharp KH, Davidson SK, Haygood MG (2007) Localization of 'Candidatus Endobugula sertula' and the bryostatins throughout the life cycle of the bryozoan *Bugula neritina*. ISME J 1:693–702

Sudek S, Lopanik NB, Waggoner LE, Hildebrand M, Anderson C, Liu H, Patel A, Sherman DH, Haygood MG (2007) Identification of the putative bryostatin polyketide synthase gene cluster from "Candidatus Endobugula sertula", the uncultivated microbial symbiont of the marine bryozoan *Bugula neritina*. J Nat Prod 70:67–74

Thomas T, Evans FF, Schleheck D, Mai-Prochnow A, Burke C, Penesyan A, Dalisay DS, Stelzer-Braid S, Saunders N, Johnson J, Ferriera S, Kjelleberg S, Egan S (2008) Analysis of the *Pseudoalteromonas tunicata* genome reveals properties of a surface-associated life style in the marine environment. PLoS ONE 3:e3252

Wagner VE et al (2003) Microarray analysis of *Pseudomonas aeruginosa* quorum-sensing regulons: effects of growth phase and environment. J Bacteriol 185:2080–2095

Watnick P, Kolter R (2000) Biofilm, city of microbes. J Bacteriol 182:2675–2679

Williams P, Winzer K, Chan WC, Cámara M (2007) Look who's talking: communication and quorum sensing in the bacterial world. Philos Trans R Soc Lond B Biol Sci 362:1119–1134

Functional Bacterial Amyloids in Biofilms

Per Halkjær Nielsen, Morten Simonsen Dueholm, Trine Rolighed Thomsen, Jeppe Lund Nielsen, and Daniel Otzen

Abstract Functional bacterial amyloids constitute a group of important proteinaceous surface structures. Most amyloids are highly insoluble in water and resistant to most enzymes and thermal and chemical denaturants. Their functions in bacteria are still not well described but seem to include fimbriae and other cell appendages for adhesion and biofilm formation, cell envelope components, spore coating, formation of large extracellular structures, amyloids acting as cytotoxins and probably several others, as yet unknown. Very few bacterial amyloids have been purified and investigated in depth. Details about the biophysical properties and ecological significance are restricted to *E. coli*, some pseudomonads and a few other bacteria. Recently, we have found that functional amyloids are widespread among microorganisms in biofilms in nature and in engineered systems, indicating that these surface structures are substantially more diverse in structure and function than hitherto anticipated. In this chapter, we highlight some of these recent results and discuss the ecological importance of amyloid surface structures in bacteria.

1 Introduction

Many different bacterial surface structures are important for bacterial adhesion to surfaces and further development into biofilms. These surface components primarily include not only various proteinaceous adhesins, both different fimbriae and nonfimbriated structures (Klemm and Schembri 2000; Barnhart and Chapman

---

P.H. Nielsen (✉) • M.S. Dueholm • T.R. Thomsen • J.L. Nielsen
Department of Biotechnology, Chemistry and Environmental Engineering, Aalborg University, Sohngaardsholmsvej 49, 9000 Aalborg, Denmark
e-mail: phn@bio.aau.dk

D. Otzen
Department of Molecular Biology, Interdisciplinary Nanoscience Centre, Aarhus University, Gustav Wieds Vej 10C, 8000 Aarhus, Denmark

2006; Latasa et al. 2006), but also other structures such as extracellular DNA and polysaccharides (Whitchurch et al. 2002; Ryder et al. 2007).

Functional bacterial amyloids constitute one such group of important proteinaceous surface structures. These structures were first described in *Escherichia coli* (Chapman et al. 2002), where curli fimbriae, described by Olsen et al. (1989), were shown to have amyloid properties. The length of curli can vary from about 0.1 to 10 μm with a width of 4–12 nm. Curli consist mainly of the CsgA protein, which is assembled in a fibrillar tertiary structure on the outer bacterial membrane (Hammar et al. 1995). Most amyloids are highly insoluble in water, resistant to most enzymes and thermal and chemical denaturants. This extreme resistance is a result of their quaternary structure; the proteins are folded as β-sheets stacked perpendicular to the fibril axis (Chapman et al. 2002). This is a general property of amyloid proteins regardless of the structure or composition of the original protein. Amyloid deposits occur in neurodegenerative diseases such as Alzheimer's and Parkinson's diseases in humans and are also formed by prion proteins in different mammals (Dobson 2005; Fowler et al. 2007; Hammer et al. 2008; Goedert and Spillantini 2006). However, the disease-associated amyloids are a result of protein misfolding, whereas the bacterial amyloids are made with a purpose, and are therefore called functional amyloids. This designation is based on the implicit assumption that proteins produced in copious quantities by bacteria must have a biological function; otherwise, the bacteria would not have a competitive advantage in their production. Assembly of functional amyloids is regulated by the organisms to prevent toxicity and to ensure the right polymerization into fibrils. Examples of other functional amyloids are hydrophobins covering fungal spores and hyphae, rendering them hydrophobic (Wosten et al. 1999; Wosten 2001) and amyloids fibrils coating fish eggs to protect them from dehydration (Podrabsky et al. 2001).

Curli fibrils enhance adhesion to a variety of surfaces and are involved in biofilm formation (reviewed by Barnhart and Chapman 2006; Otzen and Nielsen 2008) and may also increase resistance to chlorine (Ryu and Beuchat 2005) and resistance to chemical and enzymatic digestion (White et al. 2006). They are also described as having different roles in pathogenesis, owing mainly to their specific binding to fibronectin, laminin, plasminogen, and human contact phase proteins. This may allow these pathogens to colonize and invade host tissues. Curli-like fibrils are also described for other members of the family *Enterobacteriaceae* in the *Gammaproteobacteria* such as *Salmonella typhimurium*, *Citrobacter* spp., and *Enterobacter sakazaki* (Collinson et al. 1993; Zogaj et al. 2003) (Fig. 1). Other types of amyloid-like adhesions have been described on the surface of bacteria (Gebbink et al. 2005). Gram-positive *Streptomyces* secrete hydrophobic proteins forming amyloid-like fibrils called chaplins covering the cell surface, which are believed to be important for the formation of aerial hyphae at the water–air interface (Claessen et al. 2003).

Recently, we have found functional amyloids to be widespread among microorganisms in biofilms in nature and in engineered systems such as wastewater treatment plants (Larsen et al. 2007; 2008). Using various staining techniques for culture-independent detection of amyloids by fluorescence microscopy, a large fraction of bacteria from many phyla is observed to express amyloids in mixed

Fig. 1 Atomic Force Microscopy of isolated gammaproteobacterial *Aeromonas salmonicida*. *Lower part* represents a close-up of the interface between cell surface and the substratum (aluminum). Substances in this interface are amyloid-positive with ThT-staining and conformational antibodies. *Bar* equals 0.2 μm

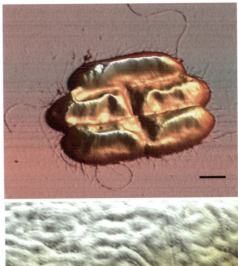

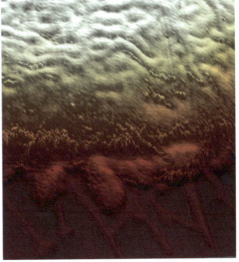

communities. A more detailed study of a number of Gram-positive isolates (Jordal et al. 2009) strongly indicates that these surface structures are widespread among bacteria and have substantially more diverse structure and function than hitherto anticipated. In this chapter, we will highlight some of these recent results, show some novel results, and discuss the ecological importance of amyloid surface structures in bacteria.

2 Amyloids: Structure

Amyloids were originally defined, based on their histopathological traits, as extracellular proteinaceous aggregates that have a β-sheet structure and bind the aniline dye Congo red in such a fashion, that it produces a characteristic apple-green birefringence when viewed under cross polarized light. To qualify as amyloids, the aggregates furthermore needed to have fibrillar morphology, upon analysis with electron microscopy (Sipe and Cohen 2000).

The classical amyloid definition was challenged due to a growing interest in amyloids from a structural and mechanistic viewpoint and furthermore from the fact that some classical amyloids had been observed intracellularly (Lin et al. 2007; Fandrich 2007). The new biophysical definition of amyloids is broader and defines amyloids as any fibrillar polypeptide aggregate with a cross-β quaternary structure (Fandrich 2007). In the cross-β structure, the polypeptide folds in a regular manner back on itself so that adjacent chain segments are laterally arranged perpendicular to the fibril axis (Fig. 2a) (Sunde et al. 1997; Jimenez et al. 1999; Nelson et al. 2005). Sometimes multiple amyloid filaments (protofilaments) are twined as ropes to make up the mature amyloid fibril (Fandrich 2007). This is, however, not yet known to be the case for functional amyloids among bacteria.

The gold standard used to determine whether a fibril consists of amyloid structure is x-ray fiber diffraction. The amyloid fold is here characterized by an intense meridional reflection at 4.7–4.8Å (Fig. 2a), which results from the mean separation of the hydrogen-bonded β-strands that are arranged perpendicular to the fiber axis, and a weaker equatorial reflection at approximately 10Å, which originates from the spacing between the β-sheets (Sunde et al. 1997) orthogonal to the β-strands.

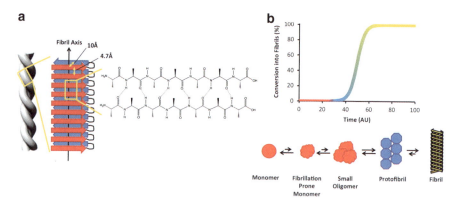

Fig. 2 Structure model and formation of amyloid fibrils. (**a**) Schematic representation of an amyloid fibril, composed of two intertwined protofilaments. In this structure, each protein subunit form a strand-turn-strand motif, which are stacked to make parallel β-sheets. Other arrangements are also possible. The protein subunits could form a single β-strand motif or could be arranged into an anti-parallel β-sheet. The characteristic lengths giving rise to reflections in X-ray fiber diffraction are shown by *arrows*. (**b**) Conversion from monomeric precursor to amyloid fibrils. A mechanistic description of the conversion from monomeric precursor into amyloid fibrils though a lag, growth, and stationary phase is shown in the *top*. The species involved in the fibrillation process are shown at the *bottom*. The native monomers are assumed to be in equilibrium with a fibrillation prone monomer structure. The latter may associate to form small oligomers, which in turn aggregate to produce structured protofibrils. The protofibrils act as nuclei for further growth by incorporation of additional precursors and associate to form the mature amyloid fibrils. Colors of the different species are used to indicate where in the fibrillation process they are present

Fourier transform infrared (FTIR) spectroscopy has also proved to be useful as a biophysical tool for the identification of amyloid structures. A comparison of the FTIR spectra of various globular β-sheet proteins and amyloids has shown that for globular proteins the amide I maximum for β-sheets lies within the range 1,630–1,643 cm^{-1}, whereas it shifts to 1,611–1,630 cm^{-1} for amyloid fibrils (Zandomeneghi et al. 2004). The shift of the β-strand absorbance maximum to lower wave numbers for amyloids results from the very strong hydrogen bonds formed in the cross-β protein fold (Gasset et al. 1992).

High-resolution structures of amyloids have only been solved for a few amyloids. These amyloids were all formed *in vitro* and in most cases from small synthetic peptides. Very little is therefore known of the structure of functional amyloids. A preliminary study of the structure of the curli fimbriae has been made (Shewmaker et al. 2009). In this study, solid-state NMR was used to show that the curli amyloid did not contain an in-register parallel β-sheet architecture, which is common to many human disease-associated amyloids and the yeast prion amyloids. Instead, the curli amyloid fibrils seem to be built by protein subunits, which are folded in a β-helical structure (Fig. 3) (Shewmaker et al. 2009). A similar structure has also been suggested for the Tafi fimbria found in *Salmonella*. In this model, the repeated motifs within the protein subunits are suggested to form a strand-loop-strand motif in the β-helix (Collinson et al. 1999; White et al. 2001).

3 Fibrillation and Aggregation

In vitro experiments have shown that the conversion of soluble proteins into amyloid fibrils typically follows a nucleation-dependent mechanism containing a lag phase, a growth phase, and a stationary phase (Fig. 2b). The lag phase is usually taken to represent the formation of fibrillation prone nuclei (although it may also represent

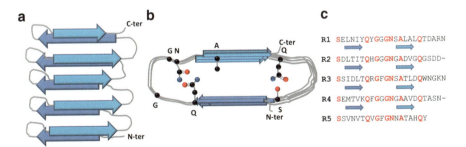

Fig. 3 Structural model of CsgA in the curli amyloid. CsgA folds in a parallel β-helix conformation where each repeated unit makes up a strand-loop-strand motif. (**a**) Ribbon diagram of the structure viewed perpendicular to the fiber axis. (**b**) Same model viewed parallel with the fiber axis. Conserved residues in the repeats are represented as *ball-and-sticks*. Individual atoms are colored as *red* (oxygen), *blue* (nitrogen), and *black* (carbon). (**c**) Primary structure of the repeats in CsgA. Conserved residues are colored *red* and predicted β-strands are shown below the sequence (after Collinson et al. 1999; White et al. 2001; Shewmaker et al. 2009)

the early stage of a cascade of fibrillation growth, known as secondary nucleation) (Ferrone 1999). Formation of nuclei is followed by an elongation reaction (the growth phase), where fibrils are formed by incorporation of additional precursors at the growing ends of the nuclei. The stationary phase is reached when the precursor monomers are depleted from solution (Nielsen et al. 2001; Jarrett and Lansbury 1993; Ban et al. 2003; Lomakin et al. 1996). The nature of the fibrillation nuclei is still debated, but they are generally considered to be structured oligomers, termed protofibrils (Chiti and Dobson 2006; Jarrett and Lansbury 1993). These small oligomers or protofibrils, rather than amyloid fibrils, are nowadays considered as the real cytotoxic species in relation to many amyloid diseases in humans (McLean et al. 1999; Nilsberth et al. 2001; Volles and Lansbury 2003; Sousa et al. 2001; Bucciantini et al. 2002) and also when used as cytotoxins by bacteria (see below). Elegant work predominantly from the Chapman lab (Chapman et al. 2002; Hammer et al. 2007; 2008; Wang et al. 2007; Wang and Chapman 2008) and other groups (Hammar et al. 1996) has demonstrated a deceptively simple way in which *E. coli* avoids the accumulation of these potentially deleterious toxic oligomers in the formation of its amyloid curli fibrils, which are predominantly composed of the CsgA protein in combination with smaller amounts of CsgB. CsgB is exported to the bacterial cell surface where an amphipathic helix at the C-terminus of the protein anchors it at the membrane–water interface. A fibrillogenic region consisting of four imperfect repeats in the N-terminal part of the protein serves as a nucleation point for CsgA, converting it to a β-sheet rich detergent-insoluble state that can incorporate additional soluble CsgA monomers into the growing ends of the fibril. Although CsgA can fibrillate spontaneously *in vitro* (Chapman et al. 2002), this is prevented in the cytosol, most likely because the export machinery keeps it in an unfolded state until it reaches the outer surface of the cell. There is some evidence that the imperfect repeats present in CsgA allow a single monomer to act as a structural nucleus for fibrillation due to their ability to fold back on themselves within one molecule (Wang et al. 2007).

The amyloid forming propensity of a polypeptide is not linked with a single amino acid sequence or defining motif. Rather, two main interactions drive the fibrillation and stabilize the amyloid structure besides the ubiquitous main chain hydrogen bonds. The first is the polar interaction formed between glutamine and asparagine residues and the polypeptide backbone (Chan et al. 2005). Curli fimbriae and the prion proteins both rely strongly on this interaction (Fig. 3) (Wang and Chapman 2008; Toyama et al. 2007). The second is the hydrophobic interaction, which favors the formation of a hydrophobic fibril core (Madine et al. 2008). Sophisticated computer algorithms have been developed to predict the fibrillation propensity of a given sequence based on these interactions as well as a number of other competing or supporting interactions, such as α-helical and β-sheet propensity, the existence of alternating hydrophilic/hydrophobic residue patterns (which will favor β-sheet formation) and electrostatic charge (DuBay et al. 2004; Pawar et al. 2005; Fernandez-Escamilla et al. 2004). Factors such as ionic strength, peptide concentration as well as intrinsic thermodynamic (Chiti et al. 2000) and kinetic (Pedersen et al. 2004) stability etc. can also be expected to contribute.

CsgA and its homologue CsgB both contain several repeats rich in glutamine and asparagine residues (Fig. 3) (Collinson et al. 1999; White et al. 2001; Hammar et al. 1996), which are critical for aggregation (Wang and Chapman 2008), in contrast to the aromatic residues which otherwise are thought to stabilize amyloid fibrils by stacking interactions (Gazit 2002). Different Asn/Gln-to-Ala mutations in CsgA interfere to different extents with CsgA self-polymerization and nucleation by CsgB, demonstrating that nucleation and fibrillation are distinct mechanisms (Gazit 2002). Furthermore, the exquisite sensitivity of specific sites to mutagenesis (even to conservative Asn/Gln switches such as Gln49Asn/Asn144Gln) indicates that the curli are arranged in a very specific and nondegenerative structural pattern, which does not accept alternative arrangements (Wang and Chapman 2008). This is consistent with our own observations that CsgA fibrils form one well-defined fold to the exclusion of other structures, making it remarkably invariant to changes in solvent conditions, temperature, ionic strength, etc. (Dueholm et al. 2011). This contrasts strongly with the known polymorphism of nonfunctional amyloid fibrils, leading to remarkably different fibrillar structures by the same peptide as a result of relatively modest changes in solvent conditions. Polymorphism is exhibited by peptides as diverse as glucagon (Pedersen et al. 2006; Pedersen and Otzen 2008), calcitonin (Bauer et al. 1995), and the Alzheimer Aβ peptide (Petkova et al. 2005).

4 Detection of Amyloids

Detection of bacterial amyloid adhesins in pure culture studies is usually based on the binding of the dye Congo red to the β-sheets of the proteins. Bacteria producing amyloid-like components can be identified by red colonies when grown on agar plates containing Congo red (Collinson et al. 1993). However, neither this method nor the use of the fluorescent dye Thioflavin T, which undergoes a shift in its fluorescence emission spectrum upon binding fibrils (Krebs et al. 2005), is particularly reliable since other biopolymers such as DNA and cellulose may elicit similar spectroscopic responses (Groenning et al. 2007). Besides dyes, antibodies can be used to detect amyloids (Glabe 2004). Particularly, the use of conformationally specific antibodies is promising as it specifically targets the fibril structure described above, without binding to monomers of the same proteins (O'Nuallain and Wetzel 2002). By combining antibodies or ThT with fluorescence *in situ* hybridization (FISH) using rRNA-targeting oligonucleotides for identification of specific bacteria, it is also possible to investigate directly in biofilms whether specific uncultured bacteria produce amyloids (Figs. 4, 5, and 6) (Larsen et al. 2007; Larsen et al. 2008). However, all these staining techniques may suffer from insufficient specificity. If possible, transmission electron microscopy (TEM) studies in combination with purification in hot/boiling SDS of the amyloids and a biophysical characterization of their properties should be combined for a complete proof of their amyloidic nature (Jordal et al. 2009). Such biophysical characterization

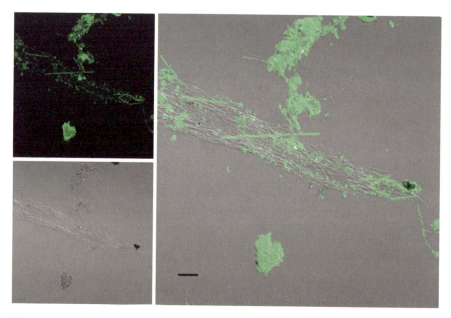

Fig. 4 Staining of biofilm with conformationally specific amyloid antibody (WO1, *green*, *upper left*), phase contrast (*lower left*), and overlay. *Scale bar* 20 μm

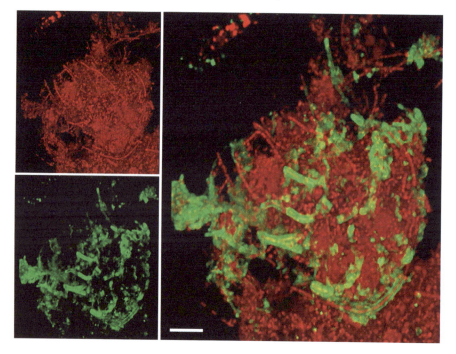

Fig. 5 Simultaneous staining of biofilm with conformationally specific amyloid antibody (WO1, *green*, *lower left*) and staining of all cells with propidium iodide (*red*, *upper left*). The right image shows the overlay. *Scale bar* 20 μm

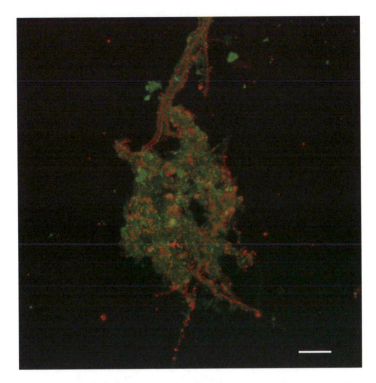

Fig. 6 Simultaneous staining of biofilm with conformationally specific amyloid antibody (WO1, *red*) and staining of all cells with FISH with the EUBmix oligonucleotide probes (*green*). *Scale bar* 20 μm

includes a verification of the structure by X-ray fiber diffraction, FTIR spectroscopy, circular dichroism, ThT fluorescence and – provided isotope-labeled amyloid can be prepared in sufficient quantity and quality – solid-state NMR (Shewmaker et al. 2009). Furthermore, treatment with strong acid such as 90–100% formic acid should release monomers that can be further purified and characterized by SDS-PAGE and various spectroscopic and microscopic techniques and sequenced by MS/MS. Ultimately, the amino acid sequence can be determined and, if the genome is available, used for the identification of the gene(s). These procedures require pure cultures, vigorous expression of amyloids, which typically only takes place under certain growth conditions, and an efficient purification protocol, which in our experience needs to be optimized for each species.

5 Bacterial Amyloids: Diversity in Structure and Function

Almost all studies of functional amyloids in bacteria have been carried out on curli-like fibrils in a few bacterial species of importance for adhesion and biofilm formation in relation to human infections. *E. coli* curli have been and are still a very important genetic and molecular tool set to study mechanisms of the controlled formation of functional amyloid. However, today we know that many other – and perhaps the majority of all bacteria – are able to produce a range of multifunctional amyloids not only in relation to human infections but also in all aspects of bacterial life in nature. The list of bacteria shown capable to produce amyloids is growing rapidly and encompasses representatives from several phyla, including *Proteobacteria, Actinobacteria, Firmicutes*, and *Bacteriodetes*. Our recent study of biofilm samples from a variety of different habitats, including fresh water lakes, brackish water, drinking reservoirs, and wastewater treatment plants shows that amyloids are present in all habitats and are often associated with a large fraction of the bacterial species (Larsen et al. 2007; Larsen et al. 2008). The amyloids have revealed a large diversity in structure and apparent function such as fimbriae and other cell appendages for adhesion and biofilm formation, cell envelope components, for example, for ensuring hydrophobic surfaces, coating of spores, formation of large extracellular structures, and amyloids acting as cytotoxins. However, at present very few bacterial amyloids have been purified and investigated in depth, so many of the proposed functions presented below are still speculations, and the final proofs, which includes experiments by appropriate mutants, need to be carried out in future studies. The diversity of possible functions are highlighted and discussed below.

5.1 *Fimbriae and Pili for Adhesion and Biofilm Formation*

E. coli, Salmonella typhimurium and other bacteria in *Enterobacteriaceae* produce thin fimbriae called curli or tafi (thin aggregative fimbriae). They are 3–12 nm wide and may be up to a few micrometers long. Together with cellulose fibers, which are also produced by many of these bacteria (Barnhart and Chapman 2006; Saldana et al. 2009), curli (or tafi) are of key importance for adhesion to surfaces and biofilm formation and thus in pathogenesis by attachment, invasion of host cells, interaction with host proteins, and activation of the immune system. Curli promote binding to a variety of different surfaces, including plant cells, stainless steel, glass and plastics, and can substantially enhance the resistance to chlorine. The importance of these structures is well described in numerous papers and several reviews, for example, Barnhart and Chapman (2006) and Otzen and Nielsen (2008). The curli operons in *E. coli* (*csgBA* and *csgDEFG*) have homologous operons in *Salmonella* (*agfBA* and *agfDEFG*) and the fimbriae are biochemical and functional analogs. The major protein components in the final curli fibril are CsgA and CsgB, and at least six proteins are involved in the formation of curli. Details about the different proteins and their function and regulation are described in a number of recent publications (Hammer et al. 2008; Epstein et al. 2009; Nenninger et al. 2009; Wang et al. 2010). In short, a nucleation protein CsgB is attached to the cell surface where it initiates CsgA polymerization and both subunits are incorporated into the final fiber. To avoid precocious fibril assembly in the periplasm, the lipoprotein CsgG acts as a platform for exporting both CsgB and CsgA to the surface and curli production takes place only from distinct areas (foci) on the cell surface. A chaperone-like assembly protein CsgF mediates that CsgB is an efficient nucleator.

The regulation of the expression of curli is carried out through a complex regulatory network with c-di-GMP and CsgD as important components (e.g., Gerstel and Romling 2003; Jubelin et al. 2005; Prigent-Combaret et al. 2001; Brombacher et al. 2003, see overview in Barnhart and Chapman 2006; Verstraeten et al. 2008). A number of environmental factors are known to induce curli formation, such as environmental stresses, but it seems to vary with different strains.

Interestingly, the biogenesis of curli can be inhibited by certain ring-fused pyridones called curlicides (Cegelski et al. 2009). These compounds are peptidomimetics that can target protein–protein interaction in macromolecular assembly. The curlicides inhibit curli formation *in vivo* and *in vitro*, attenuate biofilm formation by *E. coli*, and infections in mouse models showed that the virulence is significantly reduced (Cegelski et al. 2009). These compounds may be extremely valuable for control of infection by various enterobacteria.

Pseudomonads also produce amyloids. Amyloid fimbriae were purified from an environmental isolate (*Pseudomonas* strain UK4) closely related to *P. fluorescens* (Larsen et al. 2007). The amyloid-like structure of the native fimbriae was confirmed by FTIR spectroscopy, circular dichroism, and ThT fluorescence. Formic acid depolymerized the fimbriae and itself assembled into a native-like structure *in vitro*. Sequencing of the 25 kDa amyloid monomer revealed that it contains features also seen for CsgA of the curli system, including repeat motifs with conserved glutamine and asparagine residues. However, there does not seem to be any evolutionary relationship between the two systems based on sequence homology (Dueholm et al. 2010). Full genome sequencing of the UK4 strain indicated that the amyloid is produced from a conserved operon that includes six components (*fapA-F*), among which we find a close homologue to the amyloid protein in analogy to *E. coli*'s CsgA–CsgB pair (Dueholm et al. 2010). Heterologues expression of the whole operon in *E. coli* BL21(DE3) from a single inducible promoter resulted in a highly aggregative phenotype, which was able to form thick biofilms on polypropylene surfaces. This property was confirmed to be a result of natively folded amyloid fimbriae (FapC) expressed by the heterologues host (Dueholm et al. 2010). Blast searches showed that the operon was conserved in many, but not all, sequenced pseudomonads. Conformationally specific antibodies as well as promoter activity studies using *fap*-promoter-*lacZ* fusions further revealed that other pseudomonads including *P. aerugenosa* also produce amyloid fimbriae. Presence of these amyloids may also be involved in the pathogenesis of this opportunistic pathogen. Other gammaproteobacterial environmental isolates also produce amyloids including strains closely related to *Aeromonas hydrophila* and *Chryseobacterium* sp. (Larsen et al. 2007). Finally, bacteria belonging to the phylum *Bacteroidetes* have been found to produce amyloids (Larsen et al. 2007).

The Gram-positive *Mycobacterium tuberculosis* causes tuberculosis in humans. The bacterium produces thin pili, 2–3 nm in width called MTP (*Mycobacterium tuberculosis* pili) (Alteri et al. 2007), and they are assumed to be critical for adherence, colonization, and infection of the host. The pili bind to the human extracellular matrix protein laminin and share a number of properties with curli, although their amyloid nature is not yet conclusively proven. Mutants without the gene *mtp* have reduced laminin-binding capabilities, consistent with the fact that pili mediate specific recognition of host–cell receptors facilitating close contact and tissue colonization. *M. tuberculosis* preferentially attaches to, and invades, damaged areas of the human respiratory mucosa in an organ culture system. The *mtp* gene is present in only few mycobacteria and seems to be limited to the pathogenic species, underlining its role in cell invasion. Very similar fibrils have also been seen in *Mycobacterium avium* subsp. *Paratuberculosis* (Rulong et al. 1991; Takeo et al. 1984) and might well be amyloid.

Among the Gram-positive bacteria several other Mycolata species and phylogenetically distant non-Mycolata actinobacteria (from genera *Actinospica*, *Geodermatophilus*, and *Streptosporangium*) most likely also produce fimbria-like surface components, but this has not yet been verified by TEM or biophysical studies (Jordal et al. 2009). Among the *Firmicutes*, an isolate belonging to the *Paenibacillus* produces thin fimbria-like amyloid (Larsen et al. 2007). *Bacillus subtilus* produces biofilms, which are stabilized by protein fibrils composed of the protein TasA. These fibrils are likely amyloids as they bind Congo red and Thioflavin T and have a high content of β-strand structure. Purified TasA is furthermore able to form amyloid fibrils *in vitro* (Romero et al. 2010).

5.2 Coating Cell Surfaces

The filamentous bacterium *Streptomyces coelicolor* and other streptomycetes undergo complex morphological differentiation, forming a submerged mycelium that may grow into air and septate into spores. *S. coelicolor* produce so-called chaplins, which is a class of eight hydrophobic proteins (ChpA–H) that spontaneously self-assemble extracellularly into amyloid fibrils forming a amphipathic membrane (Claessen et al. 2004; Claessen et al. 2006). This hydrophobic membrane mediates attachment to other hydrophobic surfaces, facilitates penetration of the liquid–air interface by lowering surface tension and thus allows formation of aerial hyphae. If the chaplins is not produced, the development of aerial hyphae is impaired. The chaplins are also responsible for making the spores hydrophobic, which probably facilitates wind dispersal. Atomic force microscopy (AFM) studies on the surface of living *S. coelicolor* indicate that the amyloid fibers can be present also on the cell surface during submerged vegetative growth (Del Sol et al. 2007). Presence of such structures is also supported by visualization via conformationally specific antibodies after saponification of an outer layer of exopolymers (Jordal et al. 2009).

Other spore-forming bacteria also seem to produce amyloids as an integrated part of the spore's surface, for example, *Nocardia*, *Enterococcus* and *Bacillus* (Jordal et al. 2009). Spores covered by amyloids are well described in various fungi, where they are known as hydrophobins (Gebbink et al. 2005). This coating facilitates their dispersal by wind, enhances attachment to surfaces, and possibly also their pathogenic properties. Presence of amyloid-like structures on spores of

Bacillus atrophaeus has recently been observed by detailed AFM studies (Plomp et al. 2007). Spores from *B. mycoides* are very hydrophobic as verified by AFM force measurements, possibly due to the presence of amyloids (Bowen et al. 2002). Biofilm formation by *B. cereus* also primarily takes place at the air–liquid surface, where they sporulate and may be dispersed (Wijman et al. 2007). Presently, it seems that spores produced by many spore-forming Gram-positive bacteria are covered by amyloids, which, besides promoting wind dispersal, surface attachment and pathogenicity, possibly also explain their extreme resistance to environmental stresses.

5.3 Amyloids Integrated into the Cell Envelope

A number of different types of bacteria may contain amyloids as an integrated part of their cell envelope. Filamentous bacteria belonging to phylum *Chloroflexi* in biofilm systems from wastewater treatment plants appear to have amyloids as a part of their cell wall separating two adjacent cells (Fig. 7) (Larsen et al. 2008). The specific function is unknown but also other bacteria have amyloids integrated into their cell envelope. Bacteria belonging to the genus *Gordonia* in Mycolata have amyloids embedded in the cell envelope (Larsen et al. 2008; Jordal et al. 2009). The cell wall in many Mycolata consists of an outer layer consisting of mycolic acids, lipids, proteins, and polysaccharides, and an inner electron-dense cell wall core consisting of peptidoglycan and arabinogalactan. Fibril-like structures were visible

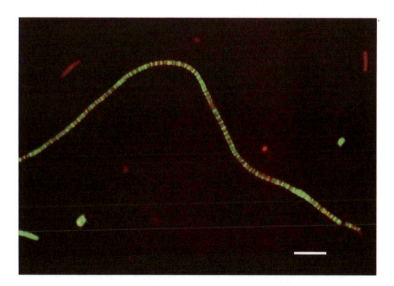

Fig. 7 Simultaneous labeling of uncultured filamentous *Chloroflexi* with antibody (*green*) and oligonucleotide probes (*red*)

in *G. amarae* after removal of the lipid layer by saponification. The fibrils were purified and partly characterized as amyloids. Such a layer of lamellar amyloid in the *G. amarae* envelope could explain the bacterium's remarkable resistance to permeabilization and disruption. Also, some actinobacterial non-Mycolata and bacteria from the phylum *Firmicutes* have cell envelope amyloids (Jordal et al. 2009). Thus, it seems to be a more universal property in many Gram-positive bacteria, and more detailed studies are needed to reveal the exact nature and function of these amyloids.

5.4 Cytotoxins

Cytotoxic proteins with amyloid properties are purposely produced by some bacteria and thus an example of an amyloid designed to be lethal. In these cases, however, prefibrillar aggregates rather than the final amyloid are most likely the cytotoxic species. *Klebsiella pneumoniae* produces Microcin E492 (also called Mcc), which is a small peptide bacteriocin that kills *Enterobacteriaceae* bacteria by forming pores in the cytoplasmic membrane (de Lorenzo 1984; Bieler et al. 2005). Microcin activity is highest when the protein is in prefibrillar aggregated state before it assembles into amyloid-like fibrils without toxicity. The finding that Microcin E492 naturally exists both as functional toxic pores and as harmless fibrils suggests that protein aggregation into amyloid fibrils can be used to deposit or sequester potentially lysing proteins when their aggressive properties are no longer needed.

Harpins represent another example of how bacteria can benefit from the cytotoxic nature of the amyloids or protofibrils. Many *Xanthomonas* species and other Gram-negative plant pathogens produce harpins, which are heat-stable glycine-rich type III-secreted proteins that elicit a hypersensitive response in plants (Oh et al. 2007). The gene *hrp* (hypersensitive response (HR) and pathogenicity) encodes the protein involved. This hypersensitive response is a plant defense mechanism that shows the presence of intracellular pathogens and elicits plant cell death. Harpins from other plant pathogens, such as *Erwinia amylovora* and *Pseudomonas syringae*, also form amyloid-like protofibrils. It is at present not clear whether the formation of amyloid takes place on the surface of the bacteria or whether the monomers only or primarily self-assemble in the plant cells. The encoded protein aggregates in a multistage process starting with bead- and ring-like structures, followed by curvilinear protofibrils and finally mature fibrils. Remarkably, all aggregated states of the *Xanthomonas* Hrp protein (but not the monomer, as indicated by the lack of response of a nonaggregating variant) appear to elicit an HR in plants (Oh et al. 2007). This contrasts with the specific cytotoxicity residing in the prefibrillar oligomer observed for proteins involved in neurodegenerative diseases. The phenomenon highlights an interesting possibility, namely that harpins have evolved another mechanism of cytotoxicity, which simply requires a threshold level of protein aggregation. It is noteworthy that a region of Hrp is homologous to the yeast prion protein (Kim et al. 2004) whose mechanism of toxicity remains unknown.

5.5 Extracellular Large Amyloid Structures

In pure cultures of some Mycolata (e.g., *M. avium*), large amounts of extracellular amyloid have been described. Extracellular fibrils with a length of more than 50 μm are observed, often integrated in an extracellular polymeric substances (EPS) matrix also containing extracellular DNA (Fig. 8) suggesting that comprehensive extracellular fibrillation has taken place (Jordal et al. 2009). However, it may be a special case for these pure culture studies where special growth conditions may provoke a very high extracellular self-assembly. *In vitro* studies with CsgA and CsgB have shown that formation of branched fibrils can take place (White et al. 2001), supporting the hypothesis that such large structures can be formed.

Interestingly, *in vivo* studies have shown that curli fibrils can be formed on *E. coli csgA* deletion mutants containing a functional *csgB* gene from CsgA monomers added exogeneous or excreted by *E. coli csgB* mutants, a process called interbacterial complementation (Hammer et al. 2008). In mixed communities, this may result in interspecies fiber formation (Wang et al. 2010). As mentioned above, CsgA and CsgB share high sequence homology among the enterobacteria, so cross-species nucleation may occur in mixed biofilms because the subunits are excreted into the extracellular environment (Wang et al. 2010). If some bacteria produce CsgA-like molecules and others CsgB-like nucleators, it may facilitate interspecies fiber formation. So far, however, it is uncertain how similar amyloid monomers actually are, and how well they can complement each other, if they do not originate from closely related species. A study of seeded lysozyme fibrillation showed that for efficient cross seeding, proteins with a sequence identity of more than 36% is needed (Krebs et al. 2004). Similar results were obtained in a study on the fibrillation of hetero-dimerized human immunoglobulin domains. When the linked domains had a more than 70% sequence identity, they were highly prone to co-aggregation, whereas no interaction was seen when the sequence identity was below 30–40% (Wright et al. 2005).

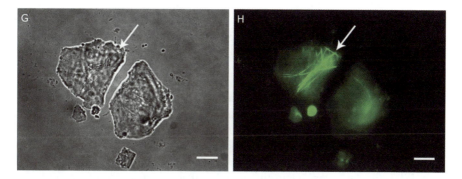

Fig. 8 Extracellular amyloids in pure cultures of the Gram-positive *Tsukamurella spumae*. Antibodies (WO2, *green*) bind to long fibrils (*right* image, *arrows*). Bar, 10 μm. The *left* image shows same field in bright field

6 Each Species, Its Own Amyloid

Curli-like fibrils are found in several members of the family *Enterobacteriaceae* in the *Gammaproteobacteria* such as *E. coli*, *S. typhimurium*, *Citrobacter* spp., and *Enterobacter sakazaki* (Collinson et al. 1993; Zogaj et al. 2003). A more detailed investigation of the *csgA* gene in various genomes, the protein structure of CsgA and its homologs and the phylogenetic relationships of various bacteria based on 16S rDNA phylogeny yield several interesting results.

The *csgA* gene sequence of *E. coli* K12 is conserved and relatively similar to other *E. coli* strains (91–100% on amino acid level, 91–98% on gene level), but is rather different from salmonella species (73–74% similarity on both gene and amino acids level). When more distantly related gammaproteobacteria such as pseudomonads are included in the analysis, it clearly appears that the *csgA* gene is not highly conserved across bacterial taxa. For pseudomonads, the gene similarity is below 20% to *E. coli* K12 (with an amino acid similarity of 30–35%). This is illustrated in Fig. 9, where a phylogenetic tree based on the *csgA* gene is depicted. Furthermore, the *16S rRNA* gene tree shows high similarity to the tree based on the *csgA* gene and forms similar clusters. This indicates that horizontal gene transfer has not taken place, and highlights an important point: the development of amyloids has happened many times during evolution so numerous different amyloids are presumably distributed across the bacterial kingdom. In a popular sense, it can be expressed as "each species, its own amyloid".

This *csgA* gene can be used as a suitable target for the identification of amyloid-containing enterobacteria and closely related bacteria in various samples. Zogaj et al. (2003) designed several primers to detect presence of *csgA* gene in enterobacterial isolates by polymerase chain reaction (PCR). In complex samples, this may be used for the identification of enterobacterial bacteria instead of (or in combination with) the routinely used *16S rRNA* gene. We have designed several sets of primers in order to target selected bacteria (all with genomes in the databases containing putative *csgA* gene sequences as shown in Fig. 9). The species-specific primers targeting putative *csgA* genes were used on different water, biofilm and wastewater treatment samples, and it was possible to detect the *csgA* gene in most samples (data not shown) showing that genes for curli-like fibrils can be detected in complex environmental samples using PCR. Furthermore, by PCR, fingerprinting techniques, cloning, sequencing and analyzing these fragments, it was possible to identify a number of the organisms carrying *csgA* homologs in these samples; these were mostly species related to *E. coli*, *Pseudomonas*, and *Shewanella*.

When the *csgA* gene from different microorganisms with relatively high similarity is investigated in more detail, it appears that some of them have deletions or insertions in the gene. For example, *Shigella flexneri* carries an insertion sequence (called IS600) in the *csgA* gene and the genome of *Pseudomonas putida* it described as having one or more premature stops or frame shifts (Sakellaris et al. 2000; Altschul et al. 1997). It is also very interesting to notice that although most of the *csg* genes have homologs in *Pseudomonas*, no homologs of the transcriptional activator, *csgD*, can be found. So although these bacteria have genes with homology to *csgA*, it remains at present unknown whether they express curli-like fibrils on their surfaces.

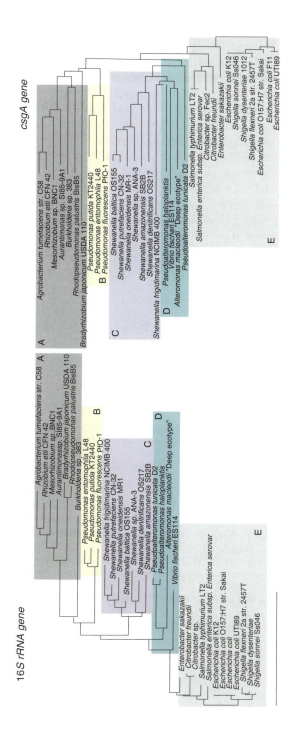

Fig. 9 Comparison of phylogenetic trees based on *16S rRNA* gene (*left*) and *csgA* gene (*right*). The trees based on aligned nucleic data were estimated by using distance matrix, parsimony, maximum likelihood and resulted in congruent tree topologies. Maximum likelihood trees are shown. The *shaded areas* indicate the different clusters identified. Group A consists of *Agrobacterium*, *Rhizobium*, *Mesorhizobium*, *Aurantimonas*, *Burkholderia*, *Rhodopseudomonas*, and *Bradyrhizobium*. Group B: *Pseudomonas*. Group C: *Shewanella*. Group D: *Pseudoalteromonas*, *Alteromonas*, and *Vibrio*. Group E is a cluster of *E. coli*, *Shigella*, *Citrobacter*, *Salmonella*, and *Enterobacter*. The *scale bars* represent 0.1 changes per nucleotide

7 Concluding Remarks

We have just seen the tip of the iceberg related to diversity and function of bacterial amyloids. The functions range from fimbriae and other cell appendages for adhesion and biofilm formation, cell envelope components, spore coating, formation of large extracellular structures, amyloids acting as cytotoxins and probably several more, so far unknown. Some of these proposed functions are still speculations, but more and more evidence is accruing to shed light on these issues. Importantly, very few bacterial amyloids have been purified and investigated in depth today, so details about the biophysical details such as diversity in amino acid sequence, potential repetitive motifs, and other details of nucleation mechanisms, branching, mechanical strength, and ultrastructure is still missing. The results obtained so far strongly indicate that only very closely related bacteria share similar types of amyloids, so we can expect to find an exciting diversity across the various phyla. Hopefully, such insights can uncover more general biological principles in the formation of amyloids and their ecological functions.

Acknowledgements We are grateful to Poul Larsen, Aleida A. V. van Steenwijk, and Peter B. Jordal for their many experimental and intellectual contributions to our research on bacterial amyloid, and L. Gurevich for AFM imaging. PHN and DO acknowledge support from the Lundbeck foundation and the Villum Kann Rasmussen Foundation (BioNET).

References

Alteri CJ, Xicohtencatl-Cortes J, Hess S, Caballero-Olin G, Giron JA, Friedman RL (2007) *Mycobacterium tuberculosis* produces pili during human infection. Proc Natl Acad Sci USA 104:5145–5150

Altschul SF, Madden TL, Schaffer AA, Zhang J, Zhang Z, Miller W, Lipman DJ (1997) Gapped BLAST and PSI-BLAST: a new generation of protein database search programs. Nucleic Acids Res 25:3389–3402

Ban T, Hamada D, Hasegawa K, Naiki H, Goto Y (2003) Direct observation of amyloid fibril growth monitored by Thioflavin T fluorescence. J Biol Chem 278:16462–16465

Barnhart MM, Chapman MR (2006) Curli biogenesis and function. Annu Rev Microbiol 60:131–147

Bauer HH, Aebi U, Haner M, Hermann R, Muller M, Merkle HP (1995) Architecture and polymorphism of fibrillar supramolecular assemblies produced by *in vitro* aggregation of human calcitonin. J Struct Biol 115:1–15

Bieler S, Estrada L, Lagos R, Baeza M, Castilla J, Soto C (2005) Amyloid formation modulates the biological activity of a bacterial protein. J Biol Chem 280:26880–26885

Bowen WR, Fenton AS, Lovitt RW, Wright CJ (2002) The measurement of *Bacillus mycoides* spore adhesion using atomic force microscopy, simple counting methods, and a spinning disk technique. Biotechnol Bioeng 79:170–179

Brombacher E, Dorel C, Zehnder AJ, Landini P (2003) The curli biosynthesis regulator CsgD co-ordinates the expression of both positive and negative determinants for biofilm formation in *Escherichia coli*. Microbiology 149:2847–2857

Bucciantini M, Giannoni E, Chiti F, Baroni F, Formigli L, Zurdo J, Taddei N, Ramponi G, Dobson CM, Stefani M (2002) Inherent toxicity of aggregates implies a common mechanism for protein misfolding diseases. Nature 416:507–511

Cegelski L, Pinkner JS, Hammer ND, Cusumano CK, Hung CS, Chorell E, Aberg V, Walker JN, Seed PC, Almqvist F, Chapman MR, Hultgren SJ (2009) Small-molecule inhibitors target *Escherichia coli* amyloid biogenesis and biofilm formation. Nat Chem Biol 5:913–919

Chan JC, Oyler NA, Yau WM, Tycko R (2005) Parallel beta-sheets and polar zippers in amyloid fibrils formed by residues 10-39 of the yeast prion protein Ure2p. Biochemistry 44: 10669–10680

Chapman MR, Robinson LS, Pinkner JS, Roth R, Heuser J, Hammar M, Normark S, Hultgren SJ (2002) Role of *Escherichia coli* curli operons in directing amyloid fiber formation. Science 295:851–855

Chiti F, Dobson CM (2006) Protein misfolding, functional amyloid, and human disease. Annu Rev Biochem 75:333–366

Chiti F, Taddei N, Bucciantini M, White P, Ramponi G, Dobson CM (2000) Mutational analysis of the propensity for amyloid formation by a globular protein. EMBO J 19:1441–1449

Claessen D, Rink R, de Jong W, Siebring J, de Vreugd P, Boersma FG, Dijkhuizen L, Wosten HA (2003) A novel class of secreted hydrophobic proteins is involved in aerial hyphae formation in *Streptomyces coelicolor* by forming amyloid-like fibrils. Genes Dev 17:1714–1726

Claessen D, Stokroos I, Deelstra HJ, Penninga NA, Bormann C, Salas JA, Dijkhuizen L, Wosten HA (2004) The formation of the rodlet layer of streptomycetes is the result of the interplay between rodlins and chaplins. Mol Microbiol 53:433–443

Claessen D, de Jong W, Dijkhuizen L, Wosten HA (2006) Regulation of *Streptomyces* development: reach for the sky! Trends Microbiol 14:313–319

Collinson SK, Doig PC, Doran JL, Clouthier S, Trust TJ, Kay WW (1993) Thin, aggregative fimbriae mediate binding of *Salmonella enteritidis* to fibronectin. J Bacteriol 175:12–18

Collinson SK, Parker JM, Hodges RS, Kay WW (1999) Structural predictions of AgfA, the insoluble fimbrial subunit of *Salmonella* thin aggregative fimbriae. J Mol Biol 290:741–756

de Lorenzo V (1984) Isolation and characterization of microcin E492 from *Klebsiella pneumoniae*. Arch Microbiol 139:72–75

Del Sol R, Armstrong I, Wright C, Dyson P (2007) Characterization of changes to the cell surface during the life cycle of *Streptomyces coelicolor*: atomic force microscopy of living cells. J Bacteriol 189:2219–2225

Dobson CM (2005) Structural biology: prying into prions. Nature 435:747–749

DuBay KF, Pawar AP, Chiti F, Zurdo J, Dobson CM, Vendruscolo M (2004) Predicting absolute aggregation rates of amyloidogenic polypeptide chains. J Mol Biol 341:1317–1326

Dueholm M, Hein KL, Christiansen G, Otzen DE (2011) Invariant fibrillation of the major curli subunit CsgA under changing conditions implies robust design of aggregation (submitted)

Dueholm MS, Petersen SV, Sønderkær M, Larsen P, Christiansen G, Enghild JJ, Nielsen JL, Nielsen KL, Nielsen PH, Otzen DE (2010) Functional amyloids in *Pseudomonas*. Mol Microbiol 77:1009–1020

Epstein EA, Reizian MA, Chapman MR (2009) Spatial clustering of the curlin secretion lipoprotein requires curli fiber assembly. J Bacteriol 191:608–615

Fandrich M (2007) On the structural definition of amyloid fibrils and other polypeptide aggregates. Cell Mol Life Sci 64:2066–2078

Fernandez-Escamilla AM, Rousseau F, Schymkowitz J, Serrano L (2004) Prediction of sequence-dependent and mutational effects on the aggregation of peptides and protein. Nat Biotechnol 22:1302–1306

Ferrone F (1999) Analysis of protein aggregation kinetics. Meth Enzymol 309:256–274

Fowler DM, Koulov AV, Balch WE, Kelly JW (2007) Functional amyloid: from bacteria to humans. Trends Biochem Sci 32:217–224

Gasset M, Baldwin MA, Lloyd DH, Gabriel JM, Holtzman DM, Cohen F, Fletterick R, Prusiner SB (1992) Predicted alpha-helical regions of the prion protein when synthesized as peptides form amyloid. Proc Natl Acad Sci USA 89:10940–10944

Gazit E (2002) A possible role for pi-stacking in the self-assembly of amyloid fibrils. FASEB J 16:77–83

Gebbink MF, Claessen D, Bouma B, Dijkhuizen L, Wosten HA (2005) Amyloids: a functional coat for microorganisms. Nat Rev 3:333–341

Gerstel U, Romling U (2003) The *csgD* promoter, a control unit for biofilm formation in *Salmonella typhimurium*. Res Microbiol 154:659–667

Glabe CG (2004) Conformation-dependent antibodies target diseases of protein misfolding. Trends Biochem Sci 29:542–547

Goedert M, Spillantini MG (2006) A century of Alzheimer's disease. Science 314:777–781

Groenning M, Olsen L, van de Weert M, Flink JM, Frokjaer S, Jorgensen FS (2007) Study on the binding of Thioflavin T to beta-sheet-rich and non-beta-sheet cavities. J Struct Biol 158:358–369

Hammar M, Arnqvist A, Bian Z, Olsen A, Normark S (1995) Expression of two *csg* operons is required for production of fibronectin- and congo red-binding curli polymers in *Escherichia coli* K-12. Mol Microbiol 18:661–670

Hammar M, Bian Z, Normark S (1996) Nucleator-dependent intercellular assembly of adhesive curli organelles in *Escherichia coli*. Proc Natl Acad Sci USA 93:6562–6566

Hammer ND, Schmidt JC, Chapman MR (2007) The curli nucleator protein, CsgB, contains an amyloidogenic domain that directs CsgA polymerization. Proc Natl Acad Sci USA 104:12494–12499

Hammer ND, Wang X, McGuffie BA, Chapman MR (2008) Amyloids: friend or foe? J Alzheimers Dis 13:407–419

Jarrett JT, Lansbury PT Jr (1993) Seeding "one-dimensional crystallization" of amyloid: a pathogenic mechanism in Alzheimer's disease and scrapie? Cell 73:1055–1058

Jimenez JL, Guijarro JI, Orlova E, Zurdo J, Dobson CM, Sunde M, Saibil HR (1999) Cryo-electron microscopy structure of an SH3 amyloid fibril and model of the molecular packing. EMBO J 18:815–821

Jordal PB, Dueholm MS, Larsen P, Petersen SV, Enghild JJ, Christiansen G, Hojrup P, Nielsen PH, Otzen DE (2009) Widespread abundance of functional bacterial amyloid in mycolata and other gram-positive bacteria. Appl Environ Microbiol 75:4101–4110

Jubelin G, Vianney A, Beloin C, Ghigo JM, Lazzaroni JC, Lejeune P, Dorel C (2005) CpxR/OmpR interplay regulates curli gene expression in response to osmolarity in *Escherichia coli*. J Bacteriol 187:2038–2049

Kim JG, Jeon E, Oh J, Moon JS, Hwang I (2004) Mutational analysis of *Xanthomonas* harpin HpaG identifies a key functional region that elicits the hypersensitive response in nonhost plants. J Bacteriol 18:6239–6247

Klemm P, Schembri MA (2000) Bacterial adhesins: function and structure. Int J Med Microbiol 290:27–35

Krebs MR, Morozova-Roche LA, Daniel K, Robinson CV, Dobson CM (2004) Observation of sequence specificity in the seeding of protein amyloid fibrils. Protein Sci 13:1933–1938

Krebs MR, Bromley EH, Donald AM (2005) The binding of Thioflavin-T to amyloid fibrils: localisation and implications. J Struct Biol 149:30–37

Larsen P, Nielsen JL, Dueholm MS, Wetzel R, Otzen D, Nielsen PH (2007) Amyloid adhesins are abundant in natural biofilms. Environ Microbiol 9:3077–3090

Larsen P, Nielsen JL, Otzen D, Nielsen PH (2008) Amyloid-like adhesins produced by floc-forming and filamentous bacteria in activated sludge. Appl Environ Microbiol 74:1517–1526

Latasa C, Solano C, Penades JR, Lasa I (2006) Biofilm-associated proteins. CR Biol 329:849–857

Lin CY, Gurlo T, Kayed R, Butler AE, Haataja L, Glabe CG, Butler PC (2007) Toxic human islet amyloid polypeptide (h-IAPP) oligomers are intracellular, and vaccination to induce anti-toxic oligomer antibodies does not prevent h-IAPP-induced beta-cell apoptosis in h-IAPP transgenic mice. Diabetes 56:1324–1332

Lomakin A, Chung DS, Benedek GB, Kirschner DA, Teplow DB (1996) On the nucleation and growth of amyloid beta-protein fibrils: detection of nuclei and quantitation of rate constants. Proc Natl Acad Sci USA 93:1125–1129

Madine J, Jack E, Stockley PG, Radford SE, Serpell LC, Middleton DA (2008) Structural insights into the polymorphism of amyloid-like fibrils formed by region 20-29 of amylin revealed by solid-state NMR and X-ray fiber diffraction. J Am Chem Soc 130:14990–15001

McLean CA, Cherny RA, Fraser FW, Fuller SJ, Smith MJ, Beyreuther K, Bush AI, Masters CL (1999) Soluble pool of Abeta amyloid as a determinant of severity of neurodegeneration in Alzheimer's disease. Ann Neurol 46:860–866

Nelson R, Sawaya MR, Balbirnie M, Madsen AO, Riekel C, Grothe R, Eisenberg D (2005) Structure of the cross-beta spine of amyloid-like fibrils. Nature 435:773–778

Nenninger AA, Robinson LS, Hultgren SJ (2009) Localized and efficient curli nucleation requires the chaperone-like amyloid assembly protein CsgF. Proc Natl Acad Sci USA 106:900–905

Nielsen L, Khurana R, Coats A, Frokjaer S, Brange J, Vyas S, Uversky VN, Fink AL (2001) Effect of environmental factors on the kinetics of insulin fibril formation: elucidation of the molecular mechanism. Biochemistry 40:6036–6046

Nilsberth C, Westlind-Danielsson A, Eckman CB, Condron MM, Axelman K, Forsell C, Stenh C, Luthman J, Teplow DB, Younkin SG, Naslund J, Lannfelt L (2001) The 'Arctic' APP mutation (E693G) causes Alzheimer's disease by enhanced Abeta protofibril formation. Nat Neurosci 4:887–893

Oh J, Kim JG, Jeon E, Yoo CH, Moon JS, Rhee S, Hwang I (2007) Amyloidogenesis of type III-dependent harpins from plant pathogenic bacteria. J Biol Chem 282:13601–13609

Olsen A, Jonsson A, Normark S (1989) Fibronectin binding mediated by a novel class of surface organelles on *Escherichia coli*. Nature 338:652–655

O'Nuallain B, Wetzel R (2002) Conformational Abs recognizing a generic amyloid fibril epitope. Proc Natl Acad Sci USA 99:1485–1490

Otzen D, Nielsen PH (2008) We find them here, we find them there: functional bacterial amyloid. Cell Mol Life Sci 65:910–927

Pawar AP, DuBay KF, Zurdo J, Chiti F, Vendruscolo M, Dobson CM (2005) Prediction of "aggregation-prone" and "aggregation-susceptible" regions in proteins associated with neurodegenerative diseases. J Mol Biol 350:379–392

Pedersen JS, Otzen DE (2008) Amyloid-a state in many guises: survival of the fittest fibril fold. Protein Sci 17:2–10

Pedersen JS, Christiansen G, Otzen DE (2004) Modulation of S6 fibrillation by unfolding rates and gatekeeper residues. J Mol Biol 341:575–588

Pedersen JS, Dikov D, Flink JL, Hjuler HA, Christiansen G, Otzen DE (2006) The changing face of glucagon fibrillation: structural polymorphism and conformational imprinting. J Mol Biol 355:501–523

Petkova AT, Leapman RD, Guo Z, Yau WM, Mattson MP, Tycko R (2005) Self-propagating, molecular-level polymorphism in Alzheimer's beta-amyloid fibrils. Science 307:262–265

Plomp M, Leighton TJ, Wheeler KE, Hill HD, Malkin AJ (2007) *In vitro* high-resolution structural dynamics of single germinating bacterial spores. Proc Natl Acad Sci USA 104:9644–9649

Podrabsky JE, Carpenter JF, Hand SC (2001) Survival of water stress in annual fish embryos: dehydration avoidance and egg envelope amyloid fibers. Am J Physiol 280:R123–R131

Prigent-Combaret C, Brombacher E, Vidal O, Ambert A, Lejeune P, Landini P, Dorel C (2001) Complex regulatory network controls initial adhesion and biofilm formation in *Escherichia coli* via regulation of the *csgD* gene. J Bacteriol 183:7213–7223

Romero D, Aguilar C, Losick R, Kolter R (2010) Amyloid fibers provide structural integrity to *Bacillus subtilis* biofilms. Proc Natl Acad Sci USA 107:2230–2234

Rulong S, Aguas AP, da Silva PP, Silva MT (1991) Intramacrophagic *Mycobacterium avium* bacilli are coated by a multiple lamellar structure: freeze fracture analysis of infected mouse liver. Infect Immun 59:3895–3902

Ryder C, Byrd M, Wozniak DJ (2007) Role of polysaccharides in *Pseudomonas aeruginosa* biofilm development. Curr Opin Microbiol 10:644–648

Ryu JH, Beuchat LR (2005) Biofilm formation by *Escherichia coli* O157:H7 on stainless steel: effect of exopolysaccharide and curli production on its resistance to chlorine. Appl Environ Microbiol 71:247–254

Sakellaris H, Hannink NK, Rajakumar K, Bulach D, Hunt M, Sasakawa C, Adler B (2000) Curli loci of *Shigella* spp. Infect Immun 68:3780–3783

Saldana Z, Xicohtencatl-Cortes J, Avelino F, Phillips AD, Kaper JB, Puente JL, Giron JA (2009) Synergistic role of curli and cellulose in cell adherence and biofilm formation of attaching and effacing *Escherichia coli* and identification of Fis as a negative regulator of curli. Environ Microbiol 11:992–1006

Shewmaker F, McGlinchey RP, Thurber KR, McPhie P, Dyda F, Tycko R, Wickner RB (2009) The functional curli amyloid is not based on in-register parallel beta-sheet structure. J Biol Chem 284:25065–25076

Sipe JD, Cohen AS (2000) Review: history of the amyloid fibril. J Struct Biol 130:88–98

Sousa MM, Cardoso I, Fernandes R, Guimaraes A, Saraiva MJ (2001) Deposition of transthyretin in early stages of familial amyloidotic polyneuropathy: evidence for toxicity of nonfibrillar aggregates. Am J Pathol 159:1993–2000

Sunde M, Serpell LC, Bartlam M, Fraser PE, Pepys MB, Blake CC (1997) Common core structure of amyloid fibrils by synchrotron X-ray diffraction. J Mol Biol 273:729–739

Takeo K, Kimura K, Kuze F, Nakai E, Nonaka T, Nishiura M (1984) Freeze-fracture observations of the cell walls and peribacillary substances of various mycobacteria. J Gen Microbiol 130:1151–1159

Toyama BH, Kelly MJ, Gross JD, Weissman JS (2007) The structural basis of yeast prion strain variants. Nature 449:233–237

Verstraeten N, Braeken K, Debkumari B, Fauvart M, Fransaer J, Vermant J, Michiels J (2008) Living on a surface: swarming and biofilm formation. Trends Microbiol 16:496–506

Volles MJ, Lansbury PT Jr (2003) Zeroing in on the pathogenic form of alpha-synuclein and its mechanism of neurotoxicity in Parkinson's disease. Biochemistry 42:7871–7878

Wang X, Chapman MR (2008) Sequence determinants of bacterial amyloid formation. J Mol Biol 380:570–580

Wang X, Smith DR, Jones JW, Chapman MR (2007) *In vitro* polymerization of a functional *Escherichia coli* amyloid protein. J Biol Chem 282:3713–3719

Wang X, Zhou Y, Ren JJ, Hammer ND, Chapman MR (2010) Gatekeeper residues in the major curlin subunit modulate bacterial amyloid fiber biogenesis. Proc Natl Acad Sci USA 107:163–168

Whitchurch CB, Tolker-Nielsen T, Ragas PC, Mattick JS (2002) Extracellular DNA required for bacterial biofilm formation. Science 295:1487

White AP, Collinson SK, Banser PA, Gibson DL, Paetzel M, Strynadka NC, Kay WW (2001) Structure and characterization of AgfB from *Salmonella enteritidis* thin aggregative fimbriae. J Mol Biol 311:735–749

White AP, Gibson DL, Kim W, Kay WW, Surette MG (2006) Thin aggregative fimbriae and cellulose enhance long-term survival and persistence of Salmonella. J Bacteriol 188:3219–3227

Wijman JG, de Leeuw PP, Moezelaar R, Zwietering MH, Abee T (2007) Air-liquid interface biofilms of *Bacillus cereus:* formation, sporulation, and dispersion. Appl Environ Microbiol 73:1481–1488

Wosten HA (2001) Hydrophobins: multipurpose proteins. Annu Rev Microbiol 55:625–646

Wosten HA, van Wetter MA, Lugones LG, van der Mei HC, Busscher HJ, Wessels JG (1999) How a fungus escapes the water to grow into the air. Curr Biol 9:85–88

Wright CF, Teichmann SA, Clarke J, Dobson CM (2005) The importance of sequence diversity in the aggregation and evolution of proteins. Nature 438:878–881

Zandomeneghi G, Krebs MR, McCammon MG, Fandrich M (2004) FTIR reveals structural differences between native beta-sheet proteins and amyloid fibrils. Protein Sci 13:3314–3321

Zogaj X, Bokranz W, Nimtz M, Romling U (2003) Production of cellulose and curli fimbriae by members of the family *Enterobacteriaceae* isolated from the human gastrointestinal tract. Infect Immun 71:4151–4158

Neutrophilic Iron-Depositing Microorganisms

Ulrich Szewzyk, Regine Szewzyk, Bertram Schmidt, and Burga Braun

Abstract Neutrophilic iron-depositing microorganisms include various groups of bacteria, algae, and protozoa. The most striking feature of these microorganisms is their ability to precipitate ferric iron around their cells and colonies in many different forms. Growth of these microorganisms has various practical implications, for example, formation of iron ore in many parts of the world, aging of water wells, and clogging of drinking water pipes. Morphological description of many genera and species of iron-depositing bacteria by microscopy dates back to the nineteenth century, but only very few pure cultures of bacteria such as *Leptothrix discophora* and *Gallionella ferruginea* have been obtained in the last decades. Therefore, little has been known on the physiology or phylogeny of these bacteria.

Using a combination of different cultivation techniques and molecular methods we were able to demonstrate a large diversity of iron-depositing bacteria in natural habitats as well as in drinking water systems. Pure cultures were obtained for many microscopically defined morphotypes belonging to well-known iron-depositing genera such as *Leptothrix*, *Pedomicrobium*, *Pseudomonas*, and *Hyphomicrobium*. In addition, many cultures isolated were not closely related to known iron bacteria according to phylogenetic analysis. Clones obtained by clone libraries from natural habitats also indicated that known genera of iron-depositing bacteria are much more diverse than assumed so far. With the pure cultures and clones in hand, we are now able to study the physiology of iron-depositing bacteria and their possible role in natural and technical habitats.

U. Szewzyk (✉) · R. Szewzyk · B. Schmidt · B. Braun
Department of Environmental Microbiology, Berlin Institute of Technology, Franklinstrasse 29, 10587 Berlin, Germany
e-mail: ulrich.szewzyk@tu-berlin.de

1 Introduction

The relevance of biochemical processes for the deposition of oxidized iron and manganese compounds has been recognized almost 200 years ago. Iron and/or manganese-depositing microorganisms include various groups of bacteria, algae, and protozoa that deposit oxidized iron minerals intracellular or extracellular. The oxidized iron may result from different physiological or chemical processes, for example, anoxygenic photosynthesis and nitrate- or oxygen-dependent redox reactions. The magnetotactic bacteria which deposit oxidized iron minerals (magnetite crystals) inside their cells will not be dealt with in this review.

The main focus in this review is on iron-depositing organisms living in habitats at neutral to slightly acidic pH conditions where oxygen is present. Such habitats can be found in wetlands, podsolic soils, stratified lakes, and rivers as well as in marine environments (Drabkova 1971; Dubinina 1978; Jones 1981; Glathe and Ottow 1972; Schmidt and Overbeck 1984; Lünsdorf et al. 1997; Emerson and Revsbech 1994a, b) given that iron is present in a reduced form. Also in technical systems where water is treated, conditions may occur supporting the development of iron-depositing bacteria (Beger 1952; Ralph and Stevenson 1995).

1.1 Iron-Depositing Microorganisms in Geology

Banded iron formation (BIF) – that is, layers of oxidized iron – is mostly encountered in precambrian sedimentary rock formations. Usually, it is assumed that the oxidized iron originates from oxidation by free oxygen released from oxygenic photosynthetic processes. With the knowledge of the physiology of certain iron-oxidizing bacteria an alternative origin of BIF appears possible: deposition of BIF mediated by anoxygenic phototrophic bacteria, using ferrous iron as electron donor for their photosynthesis and depositing ferric iron (Kappler and Newman 2004; Kappler et al. 2005a). Although it has been assumed that these bacteria live lithoautotrophically, it has been shown recently that *Rhodobacter capsulatus* lives photoheterotrophically (Caiazza et al. 2007).

The assumption of ferric iron deposition by anoxygenic phototrophic bacteria has far reaching consequences for the evolution of life. It is assumed in geology that the occurrence of BIF necessarily indicates the production of free oxygen. With the possible deposition of ferric iron by anoxygenic photosynthesis (Widdel et al. 1993), BIF might have been formed already during the long lasting anoxygenic conditions. The occurrence of free oxygen in the atmosphere can, therefore, no longer be deduced from the occurrence of BIF and might have happened much later in earth history. The predominance of anaerobic life on earth might have been much longer than assumed. Free oxygen as electron acceptor in aerobic respiration results in a higher energy yield and is a prerequisite of the development of higher multicellular organisms. The later presence of free oxygen in the atmosphere – assuming that BIF originates from anoxygenic photosynthesis – would much closer coincide with the development of multicellular higher organism.

The existence of anaerobic life during long periods of planetary evolution also has implications for astrobiology. Use of oxygen as a signature for life on extraterrestrial planets has been a major concept in exobiology. With the knowledge of anaerobic life in mind, this concept has to be altered to include signatures for detection of anaerobic life forms.

1.2 Iron-Depositing Microorganisms in History

During historic times, deposition of iron by microorganisms had an impact on the development of cultures and civilizations. Iron ore formations resulting from iron-depositing microorganisms (Raseneisenstein, bog iron) are found in all temperate regions and were used as an important resource. Such formations develop where water containing soluble iron is continuously penetrating through sandy or clay soil formations and the iron is deposited and, therefore, locally enriched, when oxidized zones are reached. Bog iron ores, for example are formed worldwide at the margins of swamps and bogs. Ferric iron is mobilized in the anoxic zones within the bog and deposited as ferrous iron when the water reaches the aerobic zones at the margin of the bog. Similar reactions may occur in wet lowlands where ground water containing soluble iron is ascending and ferrous iron is precipitated in the aerobic zone typically 20–60 cm below the soil surface (Raseneisenstein). Depending on the amount of iron, climate and soil type, solidified iron-containing sediments may form with time. In Northern Germany, the main period of iron ore formation in lowlands was 9.000 to 4.500 years ago but the process is still ongoing today.

These solidified iron-containing sediments have been used, on the one hand, as building material. Many stone buildings in Northern Germany and other regions in Central Europe are constructed to a large extent of bog iron blocks (Fig. 1). On the other hand, bog iron has been an important resource for the production of iron and steel since the Iron Age until early industrialization. Bog iron has been retrieved from wet lowland deposits and harvested from bogs. Under optimal climatic and environmental conditions, iron deposition at the margin of bogs continued after harvesting and a new harvesting could be started after some decades. Steel production from bog iron has left remnants in the landscape that can still be seen today (Fig. 2a, b).

1.3 Iron-Depositing Microorganisms in Water Treatment

Another consequence of the activity of iron-depositing microorganisms is the formation of ochrous depositions (oxidized iron compounds) in groundwater wells and drinking water distribution systems (Cullimore and McCann 1978; Ridgway et al. 1981).

Fig. 1 *Left*: Bog iron used as building material for a house, Wörlitzer Park, Germany. *Right*: Exhibit of a large specimen of bog iron, Wörlitzer Park, Germany

Fig. 2 Medieval remnants from iron production in a forest in Norway, close to Koppang. *Left*: Pit used for the production of charcoal. *Right*: deposit of slag (see *arrow*) from iron production from bog iron

Anoxic ground waters may contain high concentrations of reduced iron and manganese compounds. As soon as this anoxic water gets in contact with oxygen, ochrous depositions are formed. These depositions may already occur in ground water extraction wells and are responsible for the so-called aging of groundwater wells, which might lead to well clogging with time (Hässelbarth and Lüdemann 1967a, b, 1972; Ralph and Stevenson 1995; Barbic et al. 2000; de Mendonca et al. 2003). The main morphotypes of iron-depositing bacteria described in these habitats are *Gallionella*, *Leptothrix* and *Siderocapsa*. The reduced efficiency of extraction wells

Fig. 3 Ochrous depositions in a drinking water pipe made from cast iron (diameter 25 cm), which had been in use for more than 80 years in Berlin, Germany

due to ochrous depositions results in the necessity of expensive and time consuming well regeneration measures. The same microbial process is used intentionally during drinking water treatment for the removal of iron and manganese in filter systems under micro-aerophilic conditions. This process has been proven to work very reliably and energy efficiently (Mouchet 1992). Many water works still use chemical oxidation with high oxygen concentrations, which is very energy consuming.

Iron bacteria living in biofilms in the drinking water distribution system may use very low concentrations of dissolved iron that pass the iron filters during treatment to form iron depositions at the wall of the pipes that will result in a substantial reduction of diameter with time (Fig. 3). Another consequence of this deposition is the formation of micro-habitats where different bacteria may hide and survive even in the presence of disinfectants. Such depositions may even support survival of hygienically relevant microorganisms that subsequently appear sporadically in the free water phase.

Unwanted iron depositions also occur in technical installations in different industries whenever anoxic ground water is extracted either for drainage, or production of process water. Iron-depositing bacteria (*Siderocapsa*) have also been described in bottled mineral water, where deposits have been observed on the walls and bottoms of the bottles (Svorcova 1975).

In activated sludge in waste water treatment systems, filamentous iron bacteria identified by molecular methods have been described and partly being made responsible for bulking problems (Kämpfer 1997). Iron deposition, however, does rarely occur in these habitats.

2 Microbiology of Iron-Depositing Microorganisms

2.1 Microscopic Discovery of Iron-Depositing Microorganisms

The first descriptions of iron bacteria date back to the middle and end of the nineteenth century, when reports on iron bacteria were published mostly in connection with problems occurring due to ochrous depositions (Ehrenberg 1836; Kützing 1843; Mettenheimer 1856–1858; Giard 1882; Winogradsky 1888). Several monographs on iron bacteria have been published during the next decades, all being based on microscopic and morphological descriptions (Molisch 1910; Ellis 1919; Cholodny 1926; Naumann 1921, 1929, 1930; Dorff 1934; Skuja 1948, 1956; Beger 1949). In these publications, the occurrence of bacteria and other organisms depositing iron has been described in many natural water and soil habitats as well as in technical environments such as extraction wells for drinking water production. The bacteria have been classified due to the morphological appearance of the cells and the iron deposits and many different "genera" were described such as *Gallionella, Leptothrix, Sphaerothrix, Clonothrix, Mycothrix, Siderocapsa, Siderococcus, Planctomyces,* and *Naumanniella* (Cholodny 1924). The description of the genera and species was based exclusively on morphological features and ecological aspects. Many attempts have been made to isolate iron-depositing bacteria in pure culture (Smith 1982). Classification of the organisms was difficult and disputed among the various microbiologists resulting in many synonyms and continuous renaming and re-assigning of species and genera. Conclusions on metabolic and physiological capabilities of the iron bacteria originating from this time have to be regarded with caution because early reports on pure culture studies probably refer to enriched mixed cultures. With the techniques available today, a few genera of iron-depositing bacteria such as *Leptothrix* are easy to be obtained in pure cultures. For *Gallionella*, more sophisticated cultivation conditions have to be applied (Hanert 1968; Hallbeck and Pedersen 1991). Pure cultures of many other iron bacteria are still to be achieved by use of more complex cultivation techniques or by newly to be developed methods (see below).

Deposition of ferric iron by bacteria results in arrangement of the iron on and in various types of bacterial EPS leading consequently to different morphologies of the cells and colonies. Depending on the species, but also on the growth conditions, morphology and deposition patterns vary. The main morpho-types being described in the first publications include rods and cocci with ferric iron deposited around each individual cell (Naumanniella-type), or around colonies of these cells (e.g., Siderocapsa-type) as well as cells with diffuse deposition of ferric iron around cells, (e.g., Mycothrix-type). In addition, bacteria have been described forming sheaths out of their EPS, and depositing ferric iron on the outer side of these sheaths (e.g., Leptothrix-type) or depositing iron on a twisted band of organic fibers, excreted by the cells (*Gallionella* sp.). Even various Eukaryotes (fungi, protozoa, mosses) have been described depositing iron on excreted EPS or cell wall materials. Further unusual morphotypes of iron bacteria have been described in the second part of the twentieth century such as the budding, net-forming bacteria *Pedomicrobium* and *Hyphomicrobium* (Aristovskaya 1961; Tyler and Marshall 1967).

The same morphotypes can still be seen nowadays in natural and technical environments (Fig. 4). However, it cannot be deduced from the microscopic studies if the morphological variations observed within one morphotype represent different species or phenotypic variations within one species (Fig. 5).

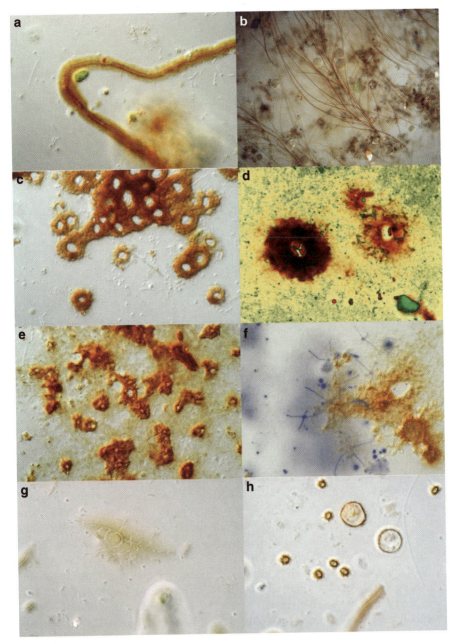

Fig. 4 Morphotypes of iron-depositing bacteria from Nationalpark Unteres Odertal (**a–f, h**) and from the river Atna, Rondane, Norway (**g**). (**a**) Leptothrix sp., (**b**) Clonothrix, (**c**) "Siderocapsa", (**d**) Siderocapsa – CLSM picture after staining with propidium iodide and FiTC-labeled lectin, (**e, f**) channels due to hyphae forming bacteria (**f**) after staining with DAPI, (**g**) unknown morphotype, (**h**) Naumanniella-type. (**b**) 40 × objective; (**a, c–h**) 100 × objective

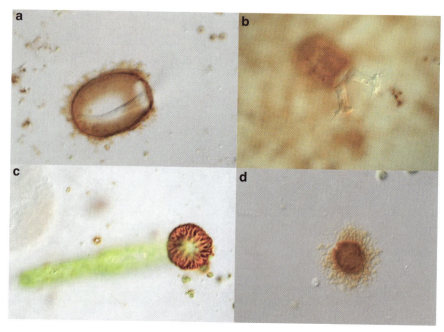

Fig. 5 Eukaryotes from backwaters and channels in the Nationalpark Unteres Odertal, depositing iron on their shells and/or on their holdfast. (**a**, **b**) peritrichous ciliates, (**c**) green algae with encrusted holdfast, (**d**) choanoflagellate. (**a**, **b**) 40 × objective; (**c**, **d**) 100 × objective

2.2 First Descriptions of Pure Cultures

A *Leptothrix* species was the first iron-depositing bacterium to be obtained in pure culture. *Leptothrix cholodnii* syn. *Leptothrix discophora*,(Mulder and van Veen 1963; Spring et al. 1996), *L. mobilis* (Spring et al. 1996) and *L. discophora* (Emerson and Ghiorse 1992) are available as pure cultures from culture collections. Only few other genera and species of neutrophilic iron bacteria have been described based on pure cultures: *Gallionella* (Hallbeck and Pedersen 1990, 1991, 1995; Luttersczekalla 1990), *Hyphomicrobium* (Hirsch and Conti 1964a, b), and *Pedomicrobium* (Gebers 1981).

Description and assignment of most other species remains vague, due to missing pure cultures. Genera such as *Siderocapsa*, *Sideromonas*, *Naumanniella* and others have frequently been found in environmental samples and described based on morphological traits. Some pure cultures have been obtained and it was claimed that these cultures represented members of the Siderocapsa-group. For instance, Dubinina and Zhdanov (1975) reported the isolation of a bacterium, which deposited iron bacterium-depositing iron in a Siderocapsa-like manner in pure culture. This bacterium was first assigned to the genus *Arthrobacter* (Dubinina and Zhdanov 1975) and later, based on 16s rDNA sequences, to the genus *Pseudomonas* (Chun et al. 2001). It was, however, not possible to demonstrate that these bacteria are identical to the bacteria that are responsible for the Siderocapsa-structures observed in the environment. The problem is that the deposition of oxidized iron among bacteria and other microorganisms is only a morphological and physiological characteristic, which bacteria from many different phylogenetic lineages might share. The question remains whether *Siderocapsa*, as it is found in nature, is really a *Pseudomonas*, or whether *Pseudomonas* is just an easily culturable organism, while the "real Siderocapsa organism" is still unknown. Results from FISH analyses of natural biofilms with dense populations of *Siderocapsa* indicate that the bacteria in the ring-shaped structures belong to the betaproteobacteria. Since *Pseudomonas putida* belongs to the gammaproteobacteria, these observations support the second hypothesis.

It has been observed in many enrichment cultures and pure cultures that bacteria which, based on phylogenetic parameters belong to different genera or even families, express similar morphotypes, for example, ring-shaped iron deposits around cells (Siderocapsa- or Naumanniella-type). It can be assumed that pure cultures of bacteria exist that are able to deposit iron but have not been recognized as iron-depositing bacteria because they have never been grown under conditions where this feature would be expressed. It remains to conclusively demonstrate the phylogenetic identity of morphotypes found in nature and correlate them, if possible, with pure cultures available.

The taxonomic and phylogenetic assignment of most morphologically described iron bacteria remains unsolved until today.

2.3 *Physiology*

The common and striking feature of neutrophilic iron bacteria is of course their ability to deposit ferric iron. The mechanisms of iron deposition remain so far unclear. Deposition of ferric iron occurs under aerobic conditions and under anaerobic conditions in the presence of nitrate or sulfate as electron acceptor (Straub et al. 1996; Kappler et al. 2005b). Oxidized manganese compounds are often found to be deposited together with ferric iron. In addition, further metals, for example, aluminum have been detected in colonies of *Siderocapsa* in river Elbe biofilms (Lünsdorf et al. 1997).

There are several theories on the physiological role of iron deposition among nonphototrophic, neutrophilic bacteria. These basic ideas have already been proposed several decades ago (Naumann 1921):

- Energy conservation
  Oxidation of ferrous iron to ferric iron is an exogenic reaction. It has, therefore, been assumed that reduced iron is used as electron donor for energy production through a respiratory chain with oxygen or nitrate as electron acceptor. With the exception of *Gallionella*, which have been shown to grow autotrophically, no definite experimental proof for this hypothesis has yet been found. In some

studies, growth of iron bacteria has been observed after addition of reduced iron compounds, which were oxidized by the bacteria (Präve 1957; Emerson and Revsbech 1994b). This was interpreted as indication of energy conservation from iron oxidation. It might, however, also be explained by the production of easily degradable compounds from complex molecules such as agar or humic substances, due to the action of radicals produced during iron oxidation, which then support microbial growth.

For iron bacteria depositing the ferric iron in a ring some distance away from the cells energy conservation from iron oxidation would require an extended electron transport through the exopolymeric structures.

- Release of iron from chelators
  Iron compounds are chelated by organic molecules such as humic substances and thereby transported in water systems. Microbial degradation of humic substances or other organic compounds which act as chelators for iron leads to a release and subsequent deposition of the ferric iron. In this case, the iron deposition would just be waste material disposed off around the cells.
- Formation of microhabitats
  Oxidation of iron may not be coupled to energy production and still be beneficial for the cells. Microhabitats are created by unspecific oxidation and deposition of ferric iron around cells and give protection from environmental stresses.
- Oxygen scavenging
  Iron-depositing bacteria live in micro-aerophilic habitats. Oxidation of iron may act as a mechanism to scavenge oxygen and reduce risk of oxidative damage to the cells (Hallbeck and Pedersen 1995).

Information on the physiology of iron-depositing bacteria is scarce due to a lack of pure culture studies. Originally it has been assumed that iron-depositing bacteria grow autotrophically (Winogradsky 1888). This view was already criticized by Molisch (1910); Lieske (1919) and Cholodny (1926) who considered only some species of *Gallionella* to be autotrophic. Based on the examinations with pure cultures of *L. discophora* (Emerson and Ghiorse 1992), it is now known, that this organism is heterotrophic and iron deposition occurs as result of proteins on the outer side of the sheath, without connection to any energy metabolism (Boogerd and de Vrind 1987; Emerson and Ghiorse 1993). Isolated proteins have been shown to initiate deposition of ferric iron and oxidized manganese compounds without bacteria being present. *Gallionella* has been shown to grow autotrophically and heterotrophically (Hallbeck and Pedersen 1991). In addition, *Gallionella* is able to grow lithoautotrophically with thiosulfate or sulfide as electron donor (Luttersczekalla 1990).

Assumptions on the degradative potential of iron-depositing bacteria were made on the availability of organic substances in their natural habitat and few examinations in pure cultures. The hypothesis that these bacteria might use humic substances as carbon source and consequently should possess the respective degradation enzymes and pathways, led to the conclusion, that these bacteria might be very important even for degradation of organic pollutants. Recent examinations in our working group demonstrate that degradation of pharmaceutical residues occurs in the zone of iron deposition during treatment of groundwater in sand filters.

3 Diversity and Phylogeny of Iron-Depositing Microorganisms

Recent investigations in our working group using a combination of different cultivation techniques and molecular methods (see below) revealed a large diversity of iron-depositing bacteria in natural habitats as well as in drinking water systems.

3.1 New Pure Cultures

Pure cultures were obtained from the following habitats:

- Nationalpark Unteres Odertal, a river flood plain at the German–Polish border, which is flooded during winter time and falling dry during summer time leaving only the river Oder and some canals with iron enriched water as ideal habitats for iron bacteria
- Tierra del Fuego, Argentina, where iron bacteria develop in thick mats in creeks at the margin of extended swamp areas (Fig. 6).

Fig. 6 Microbial mat depositing iron in a small creek (about 50 cm broad) at the outlet of a bog on Tierra del Fuego, near Estanzia Harberton, Argentina

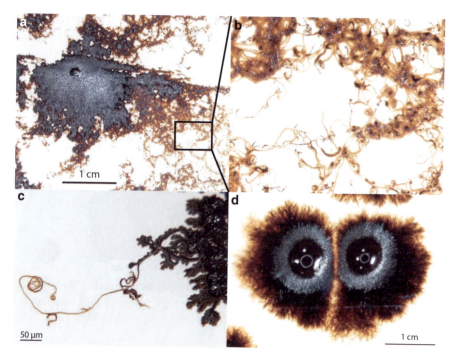

Fig. 7 Colonies of pure cultures of newly isolated iron-depositing bacteria. (**a–c**) *Leptothrix* species, (**d**) colony of an actinobacterium

- Various groundwater wells in Berlin where anaerobic ground water is extracted.

A wide variety of media were used for isolation of pure cultures. On the one hand, this included well-established media for the cultivation of *Leptothrix* and other classical iron bacteria. On the other hand, media were developed simulating conditions as they are encountered at the sampling sites. This strategy resulted in a large collection of pure cultures of iron-depositing bacteria covering well-known genera such as *Leptothrix* (Fig. 7), *Pedomicrobium*, *Pseudomonas*, and *Hyphomicrobium*, as well as many isolates which according to phylogenetic analysis (see below) are not closely related to known iron bacteria.

3.2 Molecular Phylogeny

Pure cultures were studied for their phylogenetic affiliation by amplification of 16S rRNA genes and subsequent sequencing. Phylogenetic trees were constructed using the neighbor-joining method, following distance analysis calculated by using the Jukes and Cantor correction. Besides known iron-depositing bacteria such as *Leptothrix*, *Pedomicrobium*, *Pseudomonas*, and *Hyphomicrobium*, many isolates

Fig. 8 Phylogenetic tree based on 16S rRNA gene sequences showing the relationship of several isolates of iron-depositing bacteria. The tree was constructed using the neighbor-joining method, following distance analysis calculated by using the Jukes and Cantor correction. Bootstraps based on 1,000 replications are indicated at branching nodes. *Streptomyces bellus* is used as the out-group

were affiliated to taxa for which iron deposition has not been described so far (Fig. 8). Several isolates were even assigned to actinobacteria (Fig. 7d).

These observations, together with the findings of other examiners, clearly demonstrate that representatives of very different phylogenetic lineages are able to deposit ferric iron. Consequently, this means that the diversity of iron-depositing bacteria has largely been underestimated, but correlates with the description of many morphotypes from natural habitats.

Clone libraries of 16S rDNA were obtained from natural biofilm communities from river Oder in the Nationalpark Unteres Odertal. Many of the clones obtained clustered close to bacteria, which are known to deposit iron. Other clones were distributed in many other taxa without relation to known iron-depositing bacteria. Further investigations, for example, by FISH are necessary to show whether these taxa play a role in iron deposition in natural habitats. Probes for some of the sheathed iron-depositing bacteria (Siering and Ghiorse 1997) and for *Pedomicrobium* (Braun et al. 2009) have been developed.

With the isolation of pure cultures representing many morphotypes of iron-depositing bacteria and the molecular tools available, we will now be able to solve some of the mysteries around the phylogeny and physiology of iron bacteria which have fascinated scientists since more than a century.

Acknowledgments The study on the iron-depositing bacteria in the National Park "Unteres Odertal" was funded by a grant of the German Bundesministerium für Bildung und Forschung (BMBF), 02WU0715.
The investigations in Tierra del Fuego are conducted in cooperation with Prof. Sineriz, University of Tucuman, and are financially supported by the Humboldt foundation.
We would like to thank Arne Espelund, Trondheim, Norway, for the introduction in ancient iron production in Norway.

References

Aristovskaya TV (1961) Accumulation of iron in breakdown of organomineral humus complexes by microorganisms (in Russian). Dokl Akad Nauk S.S.S.R. 136:954–957

Barbic F, Comic L, Pljakic E (2000) Iron and manganese bacteria populations in groundwater sources. Eur Water Manage 3:26–30

Beger H (1949) Beiträge zur Systematik und geographischen Verbreitung der Eisenbakterien. Berichte der deutschen botanischen Gesellschaft 62:7–13

Beger H (1952) Leitfaden der Trink- und Brauchwasserhygiene, Schriftenreihe des Vereins für Wasser-, Boden- und Lufthygiene 5. Piscator, Stuttgart

Boogerd FC, de Vrind JPM (1987) Manganese oxidation by Leptothrix discophora. J Bact 169:489–494

Braun B, Richert I, Szewzyk U (2009) Detection of iron-depositing *Pedomicrobium* species in native biofilms from the Odertal National Park by a new, specific FISH probe. J Microbiol Meth 79:37–43

Caiazza NC, Lies DP, Newman DK (2007) Phototrophic Fe(II) oxidation promotes organic carbon acquisition by *Rhodobacter capsulatus* SB1003. Appl Environ Microbiol 73:6150–6158

Cholodny N (1924) Zur Morphologie der Eisenbakterien *Gallionella* und *Spirophyllum*. Ber Dtsch Bot Ges 42:35–44

Cholodny N (1926) Die Eisenbakterien: Beiträge zu einer Monographie, Pflanzenforschung 4

Chun J, Rhee M-S, Han J-I, Bae KS (2001) *Arthrobacter siderocapsulatus* Dubinina and Zhdanov 1975AL is a later subjective synonym of *Pseudomonas putida* (Trevisan 1889) Migula 1895AL. Int J Syst Evol Microb 51:169–170

Cullimore DR, McCann AE (1978) The identification, cultivation and control of iron bacteria in ground water. In: Skinner FA, Shewan JM (eds) Aquatic microbiology. Academic, New York, pp 1–32

de Mendonca MB, Ehrlich M, Cammarota MC (2003) Conditioning factors of iron ochre biofilm formation on geotextile filters. Can Geotech J 40:1225–1234

Dorff P (1934) Die Eisenorganismen, Systematik und Morphologie. Pflanzenforschung, Jena 16:1–62

Drabkova VG (1971) Iron bacteria in some lakes of the Karelian isthmus. Hydrobiol J 7:21–27

Dubinina GA (1978) Mechanism of oxidation of divalent iron and manganese by iron bacteria growing in neutral medium. Mikrobiologiya 47:591–599

Dubinina G, Zhdanov AV (1975) Recognition of iron bacteria "*Siderocapsa*" as Arthrobacters and description of *Arthrobacter siderocapsulatus* sp-nov. Int J Syst Bacteriol 25:340–350

Ehrenberg CG (1836) Vorläufige Mitteilungen über das Vorkommen fossiler Infusorien und ihre große Verbreitung. In: Poggendorf's Annalen der Physik und Chemie 38:213-227

Ellis D (1919) Iron bacteria. J Soc Chem Ind 38:486

Emerson D, Ghiorse WC (1992) Isolation, cultural maintenance, and taxonomy of a sheath-forming strain of *Leptothrix discophora* and characterization of manganese-oxidizing activity associated with the sheath. Appl Environ Microbiol 58:4001–4010

Emerson D, Ghiorse WC (1993) Ultrastructure and chemical composition of the sheath of Leptothrix discophora SP-6. J Bact 175:7808–7818

Emerson D, Revsbech NP (1994a) Investigation of an iron-oxidizing microbial mat community located near Aarhus, Denmark: field studies. Appl Environ Microbiol 60:4022–4031

Emerson D, Revsbech NP (1994b) Investigation of an iron-oxidizing microbial mat community located near Aarhus, Denmark: laboratory studies. Appl Environ Microbiol 60:4032–4038

Gebers R (1981) Enrichment, isolation, and emended description of *Pedomicrobium ferrugineum* Aristovskaya and *Pedomicrobium manganicum* Aristovskaya. Int J Syst Bact 31:302–316

Giard A (1882) Sur le Crenothris Kühniana (Rabenhorst) cause de l'infection des eaux de Lille. Comptes rendu Acad. d. So, XCV, pp 247–249

Glathe H, Ottow JCG (1972) Ecological and physiological aspects of the mechanism of iron oxidation and ochreous deposit formation: a review. Zbl Bakteriol 127:749–769

Hallbeck L, Pedersen K (1990) Culture parameters regulating stalk formation and growth rate of *Gallionella ferruginea*. J Gen Microbiol 136:1675–1680

Hallbeck L, Pedersen K (1991) Autotrophic and mixotrophic growth of *Gallionella ferruginea*. J Gen Microbiol 137:2657–2661

Hallbeck L, Pedersen K (1995) Benefits associated with the stalk of *Gallionella ferruginea* evaluated by comparison of a stalk-forming and a non-stalk-forming strain and biofilm studies in situ. Microb Ecol 30:257–268

Hanert H (1968) Untersuchungen zur Isolierung, Stoffwechselphysiologie und Morphologie *von Gallionella ferruginea* Ehrenberg. Arch Mikrobiol 60:348–376

Hässelbarth U, Lüdemann D (1967a) Die biologische Verockerung von Brunnen durch Massenentwicklung von Eisen- und Manganbakterien. Bohrtechnik Brunnenbau Rohrleitungsbau 18:363–368

Hässelbarth U, Lüdemann D (1967b) Die biologische Verockerung von Brunnen durch Massenentwicklung von Eisen- und Manganbakterien (II). Bohrtechnik Brunnenbau Rohrleitungsbau 18:401–406

Hässelbarth U, Lüdemann D (1972) Biological incrustation of wells due to mass development of iron and manganese bacteria. Wat Treatm Exam 21:20–29

Hirsch P, Conti SF (1964a) Biology of budding bacteria II. Growth and nutrition of *Hyphomicrobium* spp. Arch Microbiol 48:358–367

Hirsch P, Conti SF (1964b) Biology of budding bacteria. I. Enrichment, isolation and morphology of *Hyphomicrobium* spp. Arch Mikrobiol 48:339–357

Jones JG (1981) The population ecology of iron bacteria (genus *Ochrobium*) in a stratified eutrophic lake. J Gen Microbiol 125:85–93

Kämpfer P (1997) Detection and cultivation of filamentous bacteria from activated sludge. FEMS Microbiol Ecol 23:169–181

Kappler A, Newman DK (2004) Formation of Fe(III)-minerals by Fe(II)-oxidizing photoautotrophic bacteria. Geochim Cosmochim Acta 68:1217–1226

Kappler A, Pasquero C, Konhauser KO, Newman DK (2005a) Deposition of banded iron formations by anoxygenic phototrophic Fe(II)-oxidizing bacteria. Geology 33:865–868

Kappler A, Schink B, Newman DK (2005b) Fe(III) mineral formation and cell encrustation by the nitrate-dependent Fe(II)-oxidizer strain BoFeN1. Geobiology 3:235–245

Kützing FT (1843) Phytologia generalis oder Anatomie. Physiologie uns Systemkunde der Tange, Leipzig, FA Brockhaus

Lieske R (1919) Zur Ernährungsphysiologie der Eisenbakterien. Zbl Bakteriol 39:369

Lünsdorf H, Brümmer I, Timmis KN, Wagner-Döbler I (1997) Metal selectivity of in situ microcolonies in biofilms of the Elbe river. J Bact 179:31–40

Luttersczekalla S (1990) Lithoautotrophic growth of the iron bacterium *Gallionella ferruginea* with thiosulfate or sulfide as energy source. Arch Microbiol 154:417–421

Mettenheimer C (1856–1858) Ueber Leptothrix ochracea u. ihre Beziehungen zu Gallionella ferruginea. Abhandl. der Senckenberg Naturforsch. Gesellsch. 10

Molisch H (1910) Die Eisenbakterien. Gustav Fischer, Jena

Mouchet P (1992) From conventional to biological removal of iron and manganese in France. JAWWA 84:158–167

Mulder EG, van Veen WL (1963) Investigations on the Sphaerotilus-Leptothrix group. Ant V Leeuwenhoek 29:121–153

Naumann E (1921) Untersuchungen über die Eisenorganismen Schwedens. I. Die Erscheinungen der Sideroplastie in den Gewässern des Teichgebiets Aneboda. Kungl Svenska Vetenskapsakademiens Handlingar 62(4)

Naumann E (1929) Die eisenspeichernden Bakterien. Kritische Übersicht der bisher bekannten Formen. Zbl Bakteriol 78:512–515

Naumann E (1930) Die Eisenorganismen. Grundlinien der limnologischen Fragestellung. Int Revue d ges Hydrobiol u Hydrographie 24:81–96

Präve P (1957) Untersuchungen über die Stoffwechselphysiologie des Eisenbakteriums Leptothrix ochracea Kützing. Arch Mikrobiol 27:33–62

Ralph DE, Stevenson JM (1995) The role of bacteria in well clogging. Wat Res 29:365–369

Ridgway HF, Means EG, Olson BH (1981) Iron bacteria in drinking water distribution system: elemental analysis of *Gallionella* stalks, using X-ray energy-dispersive microanalysis. Appl Environ Microbiol 41:288–297

Schmidt WD, Overbeck J (1984) Studies of "iron bacteria" from Lake Pluss I. Morphology, finestructure and distribution of *Metallogenium* sp. and *Siderocapsa geminata*. Z Allg Mikrobiol 24:329–339

Siering PL, Ghiorse WC (1997) Development and application of 16S rRNA-targeted probes for detection of iron- and manganese-oxidizing sheathed bacteria in environmental samples. Appl Environ Microbiol 63:644–651

Skuja H (1948) Taxonomie des Phytoplanktons einiger Seen in Uppland, Schweden. Symb Bot Ups 9(3):1–399

Skuja H (1956) Taxonomische und biologische Studien über das Phytoplankton schwedischer Binnengewässer. Nova Acta Reg Soc Sci Uppsala IV 16:1–404

Smith S (1982) Culture methods for the enumeration of iron bacteria from water well samples: a critical literature-review. Ground Water 20:482–485

Spring S, Kämpfer P, Ludwig W, Schleifer KH (1996) Polyphasic characterization of the genus *Leptothrix*: New descriptions of *Leptothrix mobilis* sp. nov. and *Leptothrix discophora* sp. nov. nom. rev. and emended description of *Leptothrix cholodnii* emend. Syst Appl Microbiol 19:634–643

Straub KL, Benz M, Schink B, Widdel F (1996) Anaerobic nitrate-depended microbial oxidation of ferrous iron. Appl Environ Microbiol 62:1458–1460

Svorcova L (1975) Iron bacteria of the genus *Siderocapsa* in mineral waters. Z Allg Mikrobiol 15:553–557

Tyler PA, Marshall KC (1967) Hyphomicrobia: a significant factor in manganese problems. J Am Water Works Assoc 59:1043–1048

Widdel F, Schnell S, Heising S, Ehrenreich A, Assmus B, Schink B (1993) Ferrous iron oxidation by anoxygenic phototrophic bacteria. Nature 362:834–835

Winogradsky S (1888) Üeber Eisenbakterien. Bot Zeitung 46:261–270

Microbial Biofouling: Unsolved Problems, Insufficient Approaches, and Possible Solutions

Hans-Curt Flemming

Abstract Microbial biofouling is a very costly problem, keeping busy a billion dollar industry providing biocides, cleaners, and antifouling materials worldwide. Basically, five general reasons can be identified, which continuously compromise the efficacy of antifouling strategies:

1. Biofouling is detected by its effect on process performance or product quality and quantity. Early warning systems are very rare, although they could save costly countermeasures necessary for removing established fouling.
2. Usually, biofouling is diagnosed only indirectly, when other explanations fail. The common practice is to take water samples, which give no information about site and extent of biofouling deposits.
3. When finally the diagnosis "biofouling" is established, biocides are used which, in many cases, for the best kill microorganisms but do not really remove them. Killing, however, is not cleaning while frequently the presence of biomass and not its physiological activity is the problem.
4. Biofouling is a biofilm phenomenon and based on the fact that biofilms grow at the expense of nutrients; oxidizing biocides can make things even worse by breaking recalcitrant molecules down into biodegradable fragments. Nutrients have to be considered as potential biomass.
5. Efficacy control is performed again by process performance or product quality and not optimized by meaningful biofilm monitoring, verifying successful removal.

Thus, further biofouling is predictable. To overcome this vicious circle, an integrated strategy is suggested, which does not rely on one type of countermeasure,

---

H.-C. Flemming (✉)
Biofilm Centre, University of Duisburg-Essen, Universitätsstraße 5, 47141 Essen, Germany
and
IWW Water Centre, Moritzstraße 26, 45476 Muelheim, Germany
e-mail: hc.flemming@uni-due.de

and which acknowledges that antifouling effects are essentially time dependent: long-term claims have to meet different (and more difficult) goals than short-term ones. An appropriate strategy includes the selection of low-adhesion, easy-to-clean surfaces, good housekeeping, early warning systems, limitation of nutrients, improvement of cleaners, strategic cleaning and monitoring of deposits. The goal is: to learn how to live with biofilms and keep their effects below the level of interference in the most efficient way.

1 Introduction

"Biofouling refers to the undesirable accumulation of a biotic deposit on a surface" (Characklis 1990). This definition is borrowed from heat exchanger technology (Epstein 1981) and applies both to the deposition of macroscopic organisms such as barnacles or mussels ("macrofouling") and to microorganisms ("microbial biofouling"). This chapter is focused on microbial biofouling.

In contrast to abiotic kinds of fouling (scaling, organic and particle fouling), biofouling is a special case because the foulant, that is the microorganisms, can grow at the expense of biodegradable substances from the water phase, turning them into metabolic products and biomass. Therefore, microorganisms are particles which can multiply. They produce extracellular polymeric substances (EPS), which keep them together and glue them to the surface and also add to the fouling. Biofouling is not only a problem in technical environments but also equally in health and medical contexts (Costerton et al. 1987). Contamination of drinking water frequently originates from biofilms and biofilm development on implants, and wounds is a common cause of serious illness and sometimes death (Gilbert et al. 2003). However, medical aspects will not be further considered in this chapter, which focuses on technical systems.

"Biofilm" is an expression for a wide variety of manifestations of microbial aggregates. Biofilms are the oldest and most successful form of live on Earth with fossils dating back 3.5 billion years and represent the first signs of life on Earth (Schopf et al. 1983). Aggregation and the association to surfaces offer substantial ecological advantages for microorganisms (Flemming 2008). Practically, all surfaces in nonsterile environments which offer sufficient amounts of water are colonized by biofilms, even at extreme pH values, high temperatures, high salt concentrations, radiation intensities and pressure (O'Toole et al. 2000; Flemming 2008). Biofilms are involved in the biogeochemical cycles of virtually all elements and are carriers of the environmental "self-purification" processes. The process is always the same: microorganisms on surfaces convert dissolved or particulate nutrients from the water phase and/or from their support into metabolites and new biomass. This is the principle of biofiltration systems used in drinking water and wastewater purification as well as many other biotechnological applications (Flemming and Wingender 2003). Of all forms of life, microbial biofilms certainly are the most ubiquitous and successful, with the highest survival potential.

Biofilms, however, can occur in the wrong place and at the wrong time. In that case, they are addressed as biofouling. It is observed in many different fields ranging from ship hulls, oil, automobile, steel, and paper production, food, beverage industries to water desalination and drinking water treatment, storage, and distribution (Flemming 2002; Henderson 2010).

In antifouling efforts, it is worthwhile to keep in mind that biofilm organisms have developed effective, versatile, and multiple defence strategies over billions of years against a multitude of stresses, including those caused, for example, by heavy metals, irradiation, biocides, antibiotics, and host immune systems. Therefore, an easy and lasting victory over biofouling cannot be expected.

Furthermore, it has to be considered that antifouling success is time dependent and not permanent. Sooner or later, all surfaces will be colonized by microbial biofilms. Kevin Marshall, one of the key researchers of early biofilm research once commented antifouling efforts: "The organism always wins" and he is right. The question is only how long the time span of nonfouling can be extended. The temporal requirements vary from hours to days (e.g., removable catheters, food and beverage industry, pharmaceutical industry) to months and years (e.g., desalination membranes, ship hulls, and environmental sensors). This makes it difficult to extrapolate short-term experiment results to long-term success.

2 The Costs of Biofouling

Biofouling is a costly problem. Although it is a common phenomenon, there is little quantitative data about the caused costs. Admittedly, it is very difficult to assess such costs as they are composed by a number of various factors: from interference with process performance, decrease of product quality and quantity, to material damage by microbial attack which even can include minerals (Sand and Gehrke 2006) or metals (Little and Lee 2007), preventive overdosing of biocides and cleaners, and finally, most expensive, interruptions of production processes and shortened life-time of plant components due to extended cleaning. An additional matter of expense is represented by treatment of wastewater contaminated by antifouling chemicals.

Collectively, biofouling causes considerable damage and supports an economically healthy antifouling industry offering everything from antifouling surfaces and materials, biocides, cleaners, and consulting services – this market is worth billions of dollars annually worldwide, considering the volume of the biocide divisions of major chemical companies. There are two reasons for this (1) the dimension of the problem with so many industrial areas concerned, and (2) the poor efficacy of many antifouling efforts which requires frequent and ongoing countermeasures.

As an example for a cost assessment, Flemming et al. (1994) estimated the costs of biofouling in a membrane application at Water Factory 21, Orange County, to 30% of the operating costs, at that time about $750.000 per year – such a rate has not much changed since. The estimate considered not only on the costs for membrane cleaning itself and labor costs but also down-time during cleaning, pretreatment costs, including biocides and other additives, an increased energy demand due to higher transmembrane and tangential hydrodynamic resistance, and shortened lifetime of the membranes. In a case study, the author is currently involved in treatment of seawater for injection into oilfields for replacement of the oil was performed by nanofiltration membranes. The client reports: "Due to biofouling, membrane life is reduced from 3 to 1 year, so over the life of the plant the cost of membrane replacement will be increased by 3. If it is taken into account that each membrane costs 2500 € and each plant has around 700 membranes, one can easily calculate a yearly investment cost of 1.75 million instead of 0.58 million. That means an extra cost of 1.17 million a year just for membrane replacement, but this can easily increase significantly if man hours involved in replacements, filters and piping replacement cost, fees paid to the client for downtimes and low quality water and further factors are taken into account." Such cost assessments, even if crude, reflect how complex and essentially arbitrary any numbers are, but they show one thing for certain: that they are high. In particular, downtime caused by biofouling amounts to surprisingly high overall costs: Azis et al. (2001) estimated the costs for biofouling in desalination to 15 billion US$ yearly worldwide.

In heat exchangers, the decrease of efficacy of heat transfer is the first aspect of biofouling-related costs and contributes to the "fouling factor" (Characklis et al. 1990; Zhao et al. 2002; Hillman and Anson 1985; Flemming and Cloete 2010). Biofouling – very conservatively assumed – accounts for about 20% of overall fouling in energy generation. To match the fouling factor, preventive extended dimensioning of heat exchanger plants is a common practice. Thus, biofouling directly increases the capital costs of, for example, a power plant (Murthy and Venkatesan 2009). In power plants around the world, thousands of tons of chlorine are spent each day to combat biofilms, which amounts to high values in terms of biocide and wastewater treatment costs (Cloete 2003). Again, down-time for cleaning causing loss of production and labor costs contribute a much larger share of costs.

Treatment of wastewater contaminated with antifouling additives represents an emerging cost factor as the release of biocides is increasingly restricted and will cause more effort for removal – a problem which will come further into focus in Europe when new EU guidelines which limit the biocide content in effluents come into action (Flemming and Greenhalgh 2009; Cheyne 2010). What clearly makes more sense is putting more effort in prevention of biofouling by advanced strategies.

In marine environments, the primary cost factor of biofouling is the increase in drag resistance, for example, on ship hulls (Schultz 2007; Edyvean 2010) or on heat exchanger surfaces (Andrewartha et al. 2010). Marine biofouling begins with the adhesion of microorganisms which form a microbial biofilm ("slime") to which other organisms may adhere, settle, and grow (Stancak 2004). Characklis (1990) calculated that on ship hulls, a biofilm with a thickness as small as 25 μm can increase drag by 8% and a roughness element of 50 μm will increase drag by as much as 22%.

3 Biofouling: An Operationally Defined Parameter

As indicated earlier, the natural phenomenon underlying biofouling is biofilms. The term "Biofouling" is operationally defined and is not determined by any objective scientific reason and standard, but only on process efficiency considerations. All nonsterile technical water systems bear biofilms, but not all of them suffer from biofouling. Therefore, a threshold level must exist above, which biofouling begins. This "level of interference" (a.k.a. as "pain threshold") is defined mostly by economical considerations defined by the extent to which biofilm effects can be tolerated without inacceptable losses in process performance or product quality and quantity (Flemming 2002). Beyond this point, which can be quite different in various industries, biofouling begins. This can be illustrated by the well-known logistic curve as shown in Fig. 1.

This threshold of interference is a felt limit, which reflects the fouling tolerance of an operator. Although it may be felt differently in different technical fields, it is safe to assume that eventually a 30% loss of productivity, product quality loss, or process efficacy will alert any operator who will try to identify and eliminate the reason. Then, usually a vicious circle begins which is sketched in Fig. 2. The main stations in that circle are (1) indirect detection of biofouling by product or process quality loss, (2) indirect and not very robust verification of biofouling, (3) no nutrient limitation although nutrients are potential biomass, (4) more or less blind use of biocides instead of cleaning, and (5) no proper verification of remedial action.

Virtually, every industrial field which has to struggle with biofouling has developed its own "culture" to handle this problem, and there is not much lateral

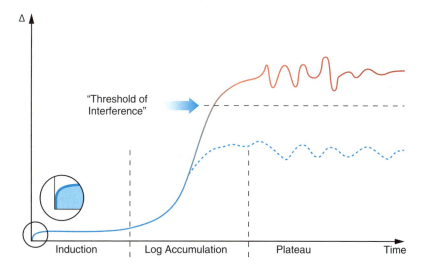

Fig. 1 Development of biofilms below and above the "threshold of interference". Δ = Parameter for biofilm effect, for example, friction resistance, hydraulic resistance, thickness, etc. (after Flemming and Ridgway 2009) *Inset*: Primary adhesion

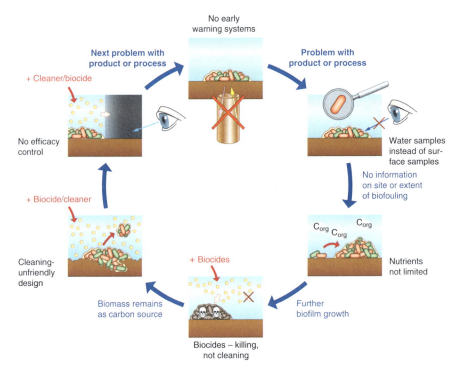

Fig. 2 The vicious circle of conventional anti-fouling efforts

learning from other fields. For example, antifouling systematics in food, beverage, pharmaceutical and microelectronics industries (Cole 1998; Verran and Jones 2000; Wirtanen and Salo 2003) are much more advanced compared to the state of the art in other biofouling-concerned technologies such as power generation (Henderson 2010), membrane treatment (Flemming and Ridgway 2009), or process water use in automobile, paint or cosmetics and medical products manufacture.

4 An Integrated Antifouling Strategy

Breaking this vicious circle is not possible with one-shot solutions but rather by proper process analysis and integrated, holistic approaches – which still represent rare exercises in the field. This chapter is intended to contribute to such an approach.

4.1 Detection, Sampling, and Analysis of Biofouling

As already pointed out, usually, biofouling is diagnosed only if problems in product quality or process performance occur, which cannot be explained by conventional technical or chemical reasons. Then, it is a common practice to take water samples at points of use of the water and determine the number of planktonic bacteria. The result may be quite misleading as numbers of planktonic bacteria neither indicate location nor extent of biofilms in a system. This leads to the generation of piles of useless data, which is frequently observed in practice. Early warning systems (see Sect. 4.5) are usually missing.

However, if taken systematically, water samples still can indicate hot spots of biofouling in a technical system. An example from practice was a water purification system in which microbial counts were determined from water samples after intake reservoir, flocculation, and filtration units were low but after the ion exchanger unit they increased for three orders of magnitude. This revealed the ion exchanger as origin of the biofouling problem. Thus, systematically upstream water sampling can lead to foci of biofouling.

Then, it is a good idea to take surface samples, preferably from defined surface areas, and to analyze them in the laboratory (Schaule et al. 2000). Quantification of microorganisms is usually performed by determination of viable counts ("colony-forming units," cfu). However, they reveal only the "tip of the iceberg" because the proportion of cultivable organisms in environmental and technical microbial populations is usually less than 1% of the actual total number of bacteria present in the sample (Rompré et al. 2002). Most cells, particularly in the depth of biofilms, do not multiply on the commonly employed nutrient agars and, thus, will not be detected by cultivation methods. Nevertheless, they are part of total biomass and in cases where the physical properties of biomass cause the problem, it makes perfect sense to determine the total microbial cell numbers by fluorescence microscopy (Schaule et al. 2000). If the proportion of cultivable bacteria (cfu) is high, for example, 10–100%, it indicates the presence of nutrients and a high biofouling potential (Wingender and Flemming 2004).

Another point to be considered is the frequent coincidence of biofouling with nonbiological fouling. Figure 3 shows an example of mixed fouling on the feed side of a reverse osmosis membrane, eventually blocking it beyond cleaning. In hindsight, it is always difficult to tell what came first. But it is well known that biofilms promote the precipitation of minerals (Arp et al. 2001; van Gulck et al. 2003). EPS components such as polysaccharides seem to play a crucial role in such processes (Braissant et al. 2003; Flemming and Wingender 2010).

4.2 Low-Fouling Surfaces

One of the most obvious targets in antifouling strategies is the selection or development of surfaces, which are not readily colonized by microorganisms and, ideally, easy to clean. Such surfaces are referred to as "low fouling." Clearly, rough surfaces are more prone to microbial colonization than smooth surfaces. This has been confirmed with stainless steel surfaces: even on the smoothest surface, bacteria can attach.

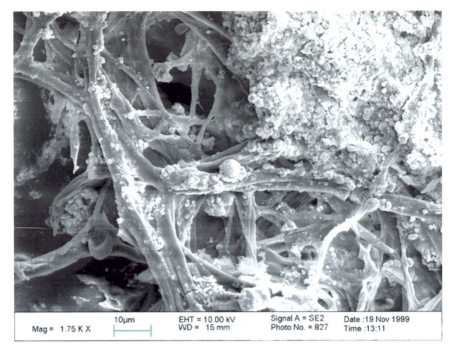

Fig. 3 Mixed microbial and abiotic deposit on a terminally fouled reverse osmosis membrane feed water surface (courtesy of G. Schaule, IWW Mülheim)

This is the result of unsuccessful approaches to prevent biofouling in heat exchangers by electropolishing (Characklis 1990; Jullien et al. 2003).

It is worth to take a closer look into the scenario of a microorganism approaching a surface prior to adhesion, which is schematically depicted in Fig. 4. A surface submerged in water will first be covered by a conditioning film, a long known phenomenon (Baier 1982). This is the result of the "race to the surface" of all molecules and particles present in the water phase, even at very low concentrations. Biopolymers meet surfaces prior to bacteria. This is due to the fact that bacteria simply do not move as fast as molecules in the water phase. Once in contact, they tend to be kept to the surface by a multiple hook-and-loop mechanism, provided by weak physicochemical interactions such as hydrogen bonds, van der Waals and weak electrostatic interactions at contact with the surface. Biopolymers have a very high number of possible binding sites, for example, polarized bonds, OH groups, or charged groups. If only 1% of them interact with a surface, the overall binding energy can exceed that of single covalent bonds by far. The molecules may migrate and spread out on the surface and attach irreversibly. Eventually, the conditioning can partially mask original surface properties. Cells in suspension usually are surrounded by a more or less thick layer of EPS and, for some organisms, cellular appendages such as fimbriae and pili. These molecules and appendages are sticky and make first contact to surfaces, interacting with both conditioning film and the

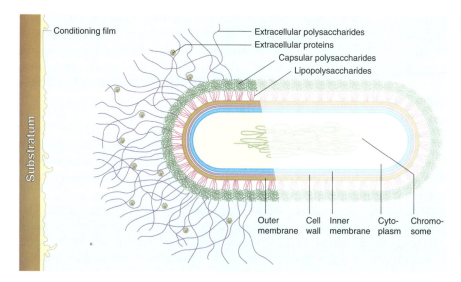

Fig. 4 Schematical depiction of a Gram-negative bacterium approaching a submerged surface

surface itself. The cells do not need to be viable for adhesion as the phenomena is controlled by physicochemical interactions.

Already in 1971, Marshall et al. tried to understand microbial primary adhesion as the interaction between "living colloids" and surfaces, applying the theory of Derjaguin, Landau, Vervey, and Overbeck (DLVO), a concept which was further evaluated for a long time (Hermansson 2000), but it clearly did not allow for realistic predictions as recent research confirmed (Schaule et al. 2008). Obviously, this approach does not acknowledge all factors involved in primary adhesion, in particularly not the role of EPS, or cellular appendices.

Many approaches have been pursued to prevent biofilm formation, some of which may be critically discussed here:

1. *Tributyl tin antifouling compounds.* They are extremely successful in biofouling prevention and have been widely used in antifouling paints for ships (Howell and Behrends 2010; ten Hallers-Tjabbes and Walmsley 2010). However, they are so toxic to marine organisms that they have been widely banned from use, although they were considered as "wonder weapons" for quite some time. It perfectly fulfilled its purpose from an antifouling point of view but it caused inacceptable economical and environmental damage (Maguire 2000; van der Oost et al. 2003).
2. *Natural antifouling compounds.* Such compounds have been isolated mainly from marine plants, which are practically not colonized by bacteria (Terlezzi et al. 2000). De Nys et al. (2006, 2010) have isolated signalling molecules from an Australian seaweed, exhibiting activity against bacterial colonization. More marine antifouling products have been investigated by Armstrong et al. (2000). Turley et al. (2005) used pyrithiones as antifoulants. Dobretsov (2009) concentrated on quorum sensing molecules as targets to prevent biofouling,

suggesting using quorum sensing inhibitors for biofouling control. In their review, they present an overview on the wide variety of quorum sensing molecules. Unfortunately, not all biofilm organisms can be addressed by one single quorum sensing inhibitor. The problem with natural antifouling compounds is that (1) most of them are only scarcely available, (2) that they are difficult to apply on a constant basis on a surface, (3) they do not completely prevent biofilm formation on inanimate surfaces, (4) that they will select for organisms which can overcome the effect, and (5) they are inherently biodegradable and, thus, their effect can be short-lasting. Apart from that, they will have to undergo the EU biocide guideline procedure, which is assessed to cost about 5–10 million € per substance (see Flemming and Greenalgh 2009; Cheyne 2010).

3. *Surfaces with lotus effect* (Nienhuis and Barthlott 1997; Marmur 2004). This effect relates to the "purity of the sacred lotus" (which is shared by less sacred cabbage leaves as well) based on the particular structure of the wax layers on the leaf surface. A highly hydrophobic pattern of needles in micrometer distances will prevent water from moistening the surface due to the physicochemical interactions of three phases: solid, liquid, and gaseous. By nature, this effect is not possible with immersed surfaces. Also, as soon as surface active substances cover the hydrophobic pattern, surface tension decreases and water is no longer repelled. Thus, the lotus effect can be taken advantage of only on solid–air interfaces and only if no surfactants are used.

4. *Silver-coated surfaces*. These are presently very much favoured with many reports on decreased adhesion (e.g., Gu et al. 2001; Gray et al. 2003; de Prijck et al. 2007). The efficacy of silver deserves critical considerations. It is still unclear whether this is an antideposition or antimicrobial effect. With regard to the question of microbial response, the Ag^+-ion is generally accepted as the effective agent (Silver 2003). The most important problem with silver efficacy is that the organisms will develop resistance after extended exposure. In environmental systems, occurrence of resistance has to be expected rather sooner than later, that is in terms of some weeks or a few months (Flemming 1982; Silver 2003). Furthermore, reactions of the silver ion with abiotic compounds have to be considered in open systems, which decrease their active concentration. Thus, silver resistance seems to be widely underestimated while its efficacy is even more overestimated.

5. *Surface-bound biocides*. Biocides attached to surfaces, such as already suggested by Hüttinger et al. (1982) and continue to be investigated and patented in many versions (see Table 1), all are suspected to draw their efficacy from biocides leaching into the water phase. If this is excluded, some basic questions have to be considered:

   – What happens with bacteria which are killed by contact and essentially will cover the surface?
   – How such biocides may act, because by concept, they do not enter the cytoplasm;

Table 1 Examples for approaches to minimize primary biofilm formation by surface modifications

Principle	References
Smoothing of surfaces	Whitehead and Verran (2009)
Superhydrophilic surfaces	Vladkova (2009)
Superhydrophobic surfaces	Schackenraad et al. (1992), Marmur (2004), Genzer and Efimenko (2006), Whitehead and Verran (2009)
Microstructured surfaces	Bers and Wahl (2004), Carman et al. (2006)
Si- and N-doped carbon coatings	Zhao et al. (2002)
UV-activated TiO_2 coatings, Ag nanoparticles	Sunada et al. (2003)
Pulsed surface polarization, pulsed electrical fields	Perez-Roa et al. (2006), Schaule et al. (2008), Giladi et al. (2008)
Polyether-polyamide copolymer (PEBAX) coating	Louie et al. (2006)
Low-surface energy coatings	Vladkova (2009)
Surface conditioning	Marshall and Blainey (1990)
Incorporation of antimicrobials	Whitehead and Verran (2009)
Biocides directly generated on surfaces	Wood et al. (1996, 1998)
Surface-bound biocides and antimicrobial peptides	Hüttinger et al. (1982), Tiller et al. (2002), Milovic et al. (2005), Madkour et al. (2008), Zasloff (2002), Leeming et al. (2002), Parvici et al. (2007), Lewis and Klibanov (2005), Park et al. (2006), Klibanov (2007)
Combined, multiple approaches	Majumdar et al. (2008)

- Do they also act on cells which attach but are physiologically inactive (e.g., in the viable-but-noncultivable state? Oliver 2005, 2010)
- How can deposition of abiotic foulants on these surfaces be prevented? (Webster and Chisholm 2010)

6. *UV irradiation.* The use of UV irradiation is also discussed for fouling control (e.g., Patil et al. 2007). However, it has to be taken into consideration that UV irradiation can only kill those cells, which are exposed to the UV rays. They can be shielded by particles while passing the irradiation chamber. Furthermore, the efficacy of UV light against biofilms is more than doubtful, in particular, if the biofilm has trapped particles. And even if the biofilm bacteria are killed, they provide nutrients for others and they are not removed from any surface.

There is a vast variety of further approaches to prevent adhesion or at least to minimize it and slow down biofilm formation. Table 1 is listing only a few of them – it is essentially incomplete because this field is one of the most innovative and under continuous development, generating hundreds of publications every year. Among those, the review of Meseguer Yebra et al. (2004) is particularly interesting as it is dedicated to environmentally friendly antifouling coatings. Many innovations come from the field of marine technology (Finnie and Williams 2010). Concerns over the environmental impact of antifouling biocides (Howell and Behrends 2010; ten Hallers-Tjabbes and Walmsley 2010) have led to interest in the development of biocide-free control solutions (Finnie and Williams 2010).

There are some general problems, which have prevented the expected breakthrough of such approaches so far, in spite of the sometimes enthusiastic and advertising character of some of the publications. There are some sobering general problems:

1. The duration of the effect. In many cases, the tests are carried out only for a few hours or days. If such approaches are applied to surfaces which are to be protected only for a short time, it may be sufficient, but mostly not for long time applications. It has to be taken into account that fouling protection is a matter of time – extrapolation from short periods to longer ones is usually not valid.
2. The test system itself. In many cases, *E. coli, P. aeruginosa* or other standard organisms are used, mostly as single strains and after washing, which leads to very unrealistic conditions with the actual sticky components at least partially removed by the washing process. Then, the number of colony-forming units (cfu) per surface area usually is the parameter to determine success, although it might be taken into account that the cells may react to contact to some of these surfaces, becoming noncultivable (Flemming 2010). Therefore, cfu numbers will be lower than the numbers of actually present cells and success is overestimated.
3. A further problem with all antimicrobial coatings in technical or environmental systems will be the covering by abiotic compounds such as humic substances, oil etc., and by inactivated cells sooner or later, masking the original effect. An example: copper-plating of ship hulls (Howell and Behrends 2010) only extents the phase until copper-tolerant microorganisms completely cover the surface and allow less copper-tolerant organisms to settle on top, eventually leading to biofouling. A longer lag phase of biofouling, however, may be very valuable, as long as the time limitation of this phase is taken into account, because time is a crucial factor in efficacy of such surfaces.

An interesting novel approach may be the employment of environmentally responsive polymers (Ista et al. 1999). An example is the use of pulsed polarized surfaces. This is also in its early experimental development but may open an interesting window in mitigation of biofouling. Schaule et al. (2008) used surfaces coated with indium-tinoxide (ITO) and polarized them at ± 600 mV under potentiostatic conditions in a pulsing routine of 1 min. Originally, this approach was intended to prevent microbial adhesion but failed to do so. However, unexpectedly, it significantly inhibited biofilm development after primary adhesion of microorganisms (Fig. 5). Of course, all arguments as mentioned above are valid for these approaches equally.

4.3 Limitation of Biofilm Growth

Given the fact that it is very difficult to prevent biofilm formation on a long term, the next plausible approach is to limit the extent of biofilm growth to limit its growth-related effects. The most obvious approach is to keep the nutrient content as low as possible, both in the water phase or leaching from the substratum. Carbon

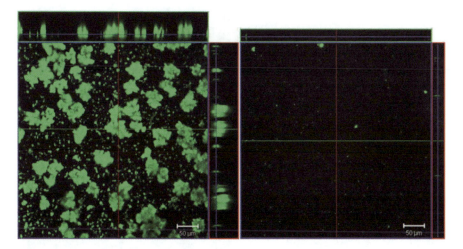

Fig. 5 Biofilm development after 164 h under potentiostatic conditions (−/+600 mV, pulse frequency 60 s), control (left), polarized biofilm (right) (Schaule et al. 2008)

Table 2 Effect of sand filtration on biofilm development on a flat cell membrane (Griebe and Flemming 1998)

Parameter	Unit	Before filter	After filter
Total cell count	[cells/cm^2]	1.0×10^8	5.5×10^6
Colony count	[cfu/cm^2]	1.0×10^7	1.2×10^6
Protein	[µg/cm^2]	78	4
Carbohydrates	[µg/cm^2]	26	3
Uronic acids	[µg/cm^2]	11	2
Humic substances	[µg/cm^2]	41	12
Biofilm thickness	[µm]	27	3
Flux decline	[%]	35	<2

sources have to be considered primarily, although lifting of nitrogen and phosphorus limitation may also be a reason for increased biofilm formation. If biofouling can be considered as a "biofilm reactor in the wrong place" as pointed out earlier, it is logical to use a "biofilm reactor in the right place." The "right place" is ahead of any system to be protected. For the case of nutrients in water, this has been successfully implemented to prevent biofouling in membrane systems (Griebe and Flemming 1998, Table 2) and is increasingly applied now in membrane technology and also in the protection of heat exchangers against biofouling. It does not completely eliminate biofilm growth but allows for keeping it below the threshold of interference as depicted in Fig. 1.

Figure 6 shows thin cuts of the membrane and biofilms (a) before and (b) after the sand filter in the above cited study. Clearly, the protected membrane was not free of biofilm but the effect of this much thinner biofilm was below the threshold of interference. It is not necessary to kill or remove such biofilms but well possible to live with them.

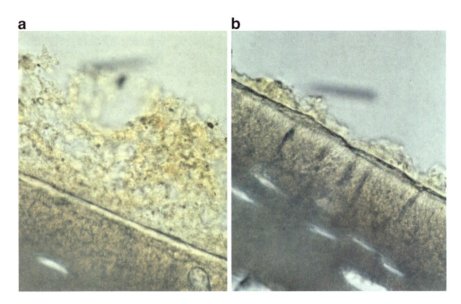

Fig. 6 (**a**) Biofilm on a reverse osmosis membrane before sand filter, (**b**) after sand filter. Magnification: 400 fold. (Griebe and Flemming, unpublished)

Of course, in this context it has to be taken into account that some additives (e.g., antiscalants, flocculants, phosphate, biodegradable biocides and components of synthetic polymeric materials such as plasticizers, anti-oxidants and flame retardants) can unintentionally contribute to nutrient supply and support biofilm growth. This has been observed in biofouling case histories of water distribution systems (Kilb et al. 2003).

Maintenance of high shear forces also helps to limit the extent of biofilm growth; however, it may lead to thinner but mechanically more stable biofilms (Characklis 1990) because higher shear forces will select for EPS with higher cohesion forces and wash away those polymers which cannot stick to the matrix.

Limitation of the extent of biofilm development seems to be an equally obvious as neglected aspect in antifouling strategies. However, it may be one of the most pragmatic approaches and can be applied creatively, adapted to the system to be protected, if taken into account. Of course, this is not generally applicable but if it is applied where it is possible, it leads to considerable success. It simply requires a shift of thinking from the "medical paradigm" toward the use of understanding of biofilm dynamics (Flemming 2002).

4.4 Biocides Versus Cleaning

A reason for frequent antifouling failures is the "medical paradigm" on which current common antifouling measures are based: biofouling is considered as a kind of a "technical disease," caused by microorganisms and can be "healed" by killing these, using biocides. This is usually called "disinfection," although it does not fit into the proper definition of this word. However, biofouling is commonly not the result of a sudden invasion of microorganisms but of a more or less rapid deposition and growth. In many cases, that is due to an increase of accessible nutrients and can occur quite unexpectedly, even after application of oxidizing biocides. A consequence of the medical paradigm is the expectation that killing of the organisms will solve the problem. However, in the first place, it is surprisingly difficult to kill biofilm organisms (Schulte et al. 2005). They have developed many ways to tolerate biocide concentrations, which would kill suspended organisms easily (Gilbert et al. 2003). The common way to determine the success of biocide application by cultivation methods will not reveal if biofilm has been removed or if it still remains as dead biomass on biofouled surfaces. This can only be determined by parameters, which actually reflect biomass, for example by direct enumeration of cells using fluorescence staining of nucleic acids. The difference is illustrated in biocide application experiments inspired by practice in heat exchanger. In an annular rotating reactor as described by Lawrence et al. (2000), a biofilm was grown from a drinking water population and run with drinking water enriched by 0.1% CASO broth as nutrient source. It was repeatedly treated with a combination of hydrogen peroxide and peracetic acid (28 ppm H_2O_2, 1.2 ppm peracetic acid for 1 h). The results are shown in Fig. 7 (Schulte 2003).

The colony counts indicate a significant reduction in "living" bacteria, which usually would have been interpreted as a substantial biomass removal. But microscopic enumeration of total cell numbers revealed that most of the biomass still remained on the surface. The cells simply did not multiply in the cultivation assay

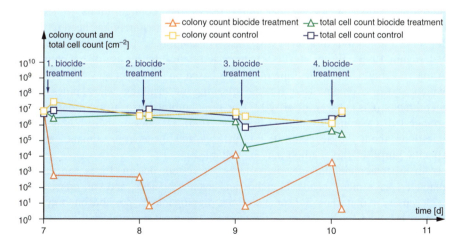

Fig. 7 Treatment of biofilm grown in an annular rotating reactor with combined hydrogen peroxide and peracetic acid (28 ppm/1.2 ppm) for 1 h. Quantification of microorganisms by cultivation (cfu) and total cell determination by fluorescence staining with DAPI (Schulte 2003)

used for their quantification. And obviously, they quickly recovered after every biocide treatment step.

The quantities of biocides and/or cleaners applied to control biofouling are mostly based on "gut feeling" instead of indicative data and specifically targeted measures. It is a disconcerting fact that the state of art in cleaning still is more an art than a science. Although serious research is performed in terms of surface cleaning, the application of this research has not yet gravitated to practice. Arbitrary mixtures of complexing substances, enzymes, shock dosages of oxidizing and nonoxidizing biocides, pH-shocks and others are common practice.

Killing is not cleaning – this has to be taken into account, because in technical processes such as cooling systems or water filtration, biomass itself, regardless if alive or dead, still causes the problems. It does not help to only inactivate a part of the population – which will soon recover after the biocide is rinsed out. The most advanced cleaning concepts have been developed in the food industry (Wirtanen and Salo 2003; Whitehead and Verran 2009). Here, surfaces continuously become contaminated by food components, which have to be removed. Much of the work is dedicated to protein adsorption and desorption (Vladkova 2009).

4.4.1 Mechanical Cleaning

Mechanical cleaning of biofouled surfaces is probably the oldest and most successful. Brushing teeth may serve as a metaphor, which also illuminates the fact that it is not possible to mechanically clean surfaces once and forever. But the analogy also illustrates that timely, properly carried out cleaning is quite efficient. In technical environments, it is used in many different applications, including "pigs" as mechanical plugs of various configurations to remove deposits in pipelines. In the food industry, ultrasonic treatment has been reported for some applications as a successful means of keeping surfaces clean (Boulangé-Petermann 1996). Wu et al. (2008) suggested defouling by use of nanobubbles. A macro version of this cleaning method is based on the use of air–water flushing which is for example used for the cleaning of drinking water pipes or in membranes (Cornelissen et al. 2007) or the combined use of air scouring and sponge ball cleaning (Psoch and Schwier 2006). Of course, there remain open questions about the accessibility of surfaces and the energy demand, but for specific purposes this method may be suitable.

An aspect crucial for cleaning success is system design. Pipes can vary in diameter, have many bends, branches and even dead legs, and can consist of a wide variety of materials. For some sections, pigging may be possible but only if the system is suited for by design – for the rest, chemical cleaning will be the only choice (Cloete 2003). Part of the problem with cleaning is the fact that biofouling is not always homogeneously distributed in a system and can be focused at certain sites, for example, fittings, valves and especially air–water–solid interfaces. Therefore, information about the actual location of biofouling foci is required but usually missing and mostly not even aimed for. To pinpoint the location, a systematic sampling approach is required. Access to surfaces is a great advantage and should be considered in construction from the very beginning. Usually, this implicitly leads also to a less fouling-prone system, avoiding dead legs, tortuous piping with various diameters and rough surfaces.

However, in spite of the efficacy of mechanical cleaning methods, it selects for biofilms with strong cohesive and adhesive properties. This has been shown in early work on ocean thermal energy recovery (OTEC) when repeated mechanical cleaning led to very sticky biofilms which (Nickels et al. 1981), which were resistant to any chemical cleaning and could only be removed mechanically.

4.4.2 Chemical and Biochemical Cleaning

The main requirement of a cleaner is to overcome the adhesion of the biofilm to the surface and the cohesion forces, which keep the biofilm together to disperse it. The mechanical stability of biofilms is mainly attributed to weak physicochemical forces such as hydrogen bonding, weak ionic interactions, hydrophobic and van der Waals interactions and entanglement (Mayer et al. 1999; Flemming and Wingender 2010 a). While surface active substances mainly address hydrophobic and van der Waals interactions, complexing substances act on ionic bonds. Hydrogen bonds can be addressed by so-called chaotropic agents such as urea, tetramethyl urea and others which interfere with the shell of water molecules surrounding biopolymers.

Matrix stability provided by entanglement of the biopolymers (Wloka et al. 2006) can be weakened by either oxidizing biocides or by enzymes, both shortening the chain length of the polymers. Enzyme applications in the food industry have been extensively studied (Lequette et al. 2010). In another very recent publication, Kolodkin-Gal et al.(2010) reported biofilm disassembly in *B. subtilis*, *P. aeruginosa*, and *S. aureus* by a mixture of D-amino acids, releasing amyloid fibers that linked the cells together. Bacteriophages induce a wide range of polysaccharide-degrading enzymes in their hosts. Dispersion by induction of a prophage, followed by cell death and subsequent cell cluster disaggregation has been observed (Webb et al. 2003). However, phage enzymes are very specific and rarely act on more than a few closely related polysaccharide structures. Phages and bacteria can coexist symbiotically within biofilms, suggesting that they would make poor tools for the control of biofilm formation. Combinations of phage enzymes and disinfectants have been recommended as possible control strategies under certain conditions (Tait et al. 2002) with the phage added before addition of disinfectant being more effective than either of these alone. Mixtures of enzymes are commonly used, composed on arbitrary base. However, Brisou (1995) already showed that there were a vast variety of target structures that enzymes had to interact with, indicating that there is no single enzyme or enzyme mixture to effectively remove biofilms. Klahre et al. (1998) report poor performance of enzymes alone in antifouling efforts in paper mills, particularly in long-term applications. The enzymes themselves are rapidly degraded by extracellular proteases.

An interesting, but possibly overestimated approach to get rid of biofilms is the employment of signalling molecules regulating biofilm development are prime targets for biological biofilm removal (Webb et al. 2003). In technical systems, however, there are no reports of successful applications so far. Recently, a substituted fatty acid, *cis*-11-methyl-2-docecenoic acid, called "diffusible signal factor" (DSF), was recovered from *Xanthomonas campestris*, which was responsible for virulence as well as for the induction of release of endo-β-1,4-mannanase which degrades mannose containing polysaccharides (Dow et al. 2003). Davies and Marques (2009) suggested the activation of stress regulons, which may be involved in biofilm dispersion. They reported *cis*-2-decenoic acid as a fatty acid messenger, produced by *P. aeruginosa*, capable of inducing dispersion of biofilms formed by *E. coli*, *K. pneumoniae*, *Proteus mirabilis*, *S. pyrogenes*, *S. aureus*, *B. subtilis*, and the yeast *C. albicans*. Such a "universal biofilm disperser" is of great interest in medical and technical systems. Equally recently, the role of nitric oxide has been revealed as signal for biofilm dispersion (Barraud et al. 2009). However, all these signalling molecules can only influence bacteria in their vegetative state. When they are dormant, they cannot respond to the signals. And it is a fact that in biofilms, a large proportion of the cells is in a resting, dormant or viable but nonculturable (VBNC) state and not affected by these molecules. Furthermore, none of them eradicates existing even fully viable biofilms completely. Therefore, their use as antifoulants in technical systems appears very limited, particularly if one considers that it is not easy and by no means cheap to produce them in the required amounts and to apply them in the required sites. Due to their biological nature, they have limited life spans and may be degraded before they have completed their function. It is the great variability of EPS, which protects biofilms and, in turn, limits the success of enzymatic antifouling strategies.

4.5 Surface Monitoring

A big problem for the implementation of timely countermeasures is the already mentioned fact that the surfaces of technical systems usually are poorly or not at all accessible. The response in practice is frequently preventive overdosing of cleaning and biocidal chemicals, which in turn can damage industrial equipment. Also, an environmental burden is generated when the employed substances are released into wastewater, which will have to undergo specific treatment for removal of these substances. Efficiency control of cleaning is usually performed in the reverse way as biofouling is detected: by improvement of process parameters or product quality. This means that it is very indirect and vague and can easily lead into a new round in the vicious circle as depicted in Fig. 2.

Conventional monitoring methods employ sampling of defined surface areas or on exposure of test surfaces ("coupons") with subsequent analysis in the laboratory. A classical example is the "Robbins device" (Ruseska et al. 1982), which consists of plugs smoothly inserted into pipe walls, experiencing the same shear stress as the wall itself. After given periods of time, they are removed and analyzed in the laboratory for all biofilm-relevant parameters. The disadvantage of such systems is the time lag between analysis and result.

What is lacking is information about site and extent of fouling deposit and, thus, effective and down-to-the-point countermeasures. Therefore, it would be very useful to have "eyes in the system," which allow for the detection of biofilm growth before it turns into biofouling. This is the case for advanced surface monitoring (Flemming 2003; Janknecht and Melo 2003). Systems are needed which have to provide information about the presence of a deposit, its quantity, thickness and distribution, the nature and composition of the deposit, and the kinetics of formation and removal to assess the fouling potential and cleaning success. Some criteria and characteristics for early warning systems are:

- Continuous and automated detection of fouling-relevant parameters;
- Reliable and fast response;
- Easy handling combined with little maintenance;
- Feasible in controlled conditions in the laboratory and in the field situation;
- Easy to interpret output software.

This information should be available online, *in situ*, in real-time, nondestructively, and suitable for data processing and automatization of response.

Such systems will be mainly based on physical principles, some of which have been addressed earlier (Nivens et al. 1995). To meet the demands as listed above, a technique must

- Function in an aqueous system;
- Not require sample removal;
- Provide real-time data;
- Be specific for the surface, that is minimize signal from organisms or contaminants in the water phase.

As early as 1985, Hillman and Anson gave the most comprehensive overview until today on physical measuring principles related to fouling, which meet quite a few of requirements above. Strangely, almost none of them made it to the market, although they were very sophisticated and original. They included already changes in ultrasound, heat transfer resistance, and Nivens et al. (1995) have given an excellent overview on continuous nondestructive biofilm monitoring techniques, including Fourier transform-infrared (FT-IR) spectroscopy, microscopic, electrochemical and piezoelectric techniques. An important aspect has to be taken into account: physical methods mostly respond to nonspecific effects of biofilms. Three levels of information provided by these methods can be identified (Flemming 2003):

4.5.1 Level 1 Monitoring

Level 1 monitoring devices can be classified as systems which detect the kinetics of deposition of material and changes of thickness of deposit layer but cannot differentiate between microorganisms and abiotic deposit components.

They provide information nondestructively, on line, *in situ*, continuously, in real time, and with a realistic potential as a signal for automatic countermeasures. Some examples of this category have already been successfully applied to biofilms (Table 3). It is surprising that they are only very reluctantly accepted in the market an developed further to practical application, although they might save considerable values – blind dumping of biocides still remains the common practice.

An interesting approach for biofouling monitoring in separation membranes has been developed by Vrouwenvelder et al. (2006), using a membrane fouling simulator. This device both determines friction resistance and allows for optical inspection of the surfaces. In an exciting study, it was combined with NMR imaging, revealing the actual hydrodynamics in the device and the dominant role of the spacer for biofouling in membrane modules (Von der Schulenburg et al. 2008).

Table 3 Examples of level 1 monitoring devices

Device	Principle	References
Rotoscope	Determination of light absorption in response to deposit formation	Cloete and Maluleke (2005)
Differential turbidity measurement (DTM)	Two turbidity measurement devices, one of them cleaned, determination of difference between both caused by surface deposit on measurement window	Klahre and Flemming (2000) and Wetegrove and Banks (1993)
Hot wire, heat transfer resistance	Determination of heat transfer changes in response to deposit	Hillman and Anson (1985), Fillaudeau (2003)
Fiberoptical device (FOS)	Determination of light backscattered from deposit on top of a light fiber	Tamachkiarow and Flemming (2003)
Electrochemical measurement device	Determination of changes in electric conductivity caused by material deposited on the surface	Nivens et al. (1995), Bruijs et al. (2000), Mollica and Cristiani (2003)
Acoustic fouling detector	Acoustic backscattering by deposit	Hillman and Anson (1985)
Quartz crystal microbalance	Quenching of frequency of quartz crystal by deposit	Nivens et al. (1995), White et al. (1996), Helle et al. (2000)
Mechatronic surface sensor	Sonic actuator and detector on surface, determination of vibration response	Pereira et al. (2007)
Surface acoustic waves	Determination of difference between speed of acoustic waves with and without deposit	Ballantine and Wohltien (1989)
Photoacoustic spectroscopy sensor	Absorption of eletromagnetic radiation inside a sample, where nonradiative relaxation processes convert the absorbed energy into heat. Due to thermal expansion of medium, a pressure wave is generated, which can be detected by microphones or piezoelectronic transducers	Schmid et al. (2003, 2004)
Friction resistance measurement	Determination of pressure drop due to biofilm roughness	Hillman and Anson 1985, Eguia et al. (2008)

4.5.2 Level 2 Monitoring

In level 2, systems can be categorized, which can distinguish between biotic and abiotic components of a given deposit. A suitable way to accomplish this is the specific detection of signals of biomolecules. Examples are:

Use of autofluorescence of biomolecules such as amino acids, for example, tryptophane or other biomolecules (Angell et al. 1993; Nivens et al. 1995; Zinn et al. 1999; Kerr et al. 1998; Wetegrove 1998). Such molecules are considered as representative for the presence of biological material. Again, this is only true for systems which normally do not contain biomass, for example, heat exchangers, membrane systems for water treatment, or process water systems. However, it seems that the discrimination of the fluorescence signals of such molecules is difficult to identify, in particular in presence of quenching substances.

FTIR-ATR-spectroscopy specific for amid bands. This approach is suitable for systems, which usually do not contain biological molecules, for example, cooling or process water systems. One way to follow this approach is a bypass pipe with IR transparent windows. For measurement, the water is drained transiently and the measurement is performed (White et al. 1996; Flemming et al. 1998). A very elegant system has been developed by Wetegrove and Banks (1993) and is called the "rotating disk device." This device is based on a disc which is mounted eccentrically on an axis. The lower part is immersed into a water system. After given intervals, the disk turns upside and is analyzed by IR spectroscopy. Strictly spoken, these systems are not completely continuous but they still fulfil fundamental demands of monitoring systems.

Microscopical observation of biofilm formation in a bypass flow chamber and morphological identification of microorganisms (Nivens et al. 1995). Microscopically, however, it may be difficult to distinguish microorganisms from agglomerated abiotic material without application of a dye. Also, microscopic observation requires either a microscopist who more or less continuously carries out the work or a powerful image analysis system, which encounters the same problems as the microscopist when complex deposits accumulate.

4.5.3 Level 3 Monitoring

Systems, which provide detailed information about the chemical composition of the deposit or directly address microorganisms. An example:

FTIR-ATR-spectroscopy in a flow-through cell. In such an approach, not only the amide bands are considered but also the entire spectrum of medium infra red, which has proven to be the most indicative for biological material. The system is composed of an IR-transparent crystal of zinc selenide or germanium which is fixed in a flow-through cell (White et al. 1996; Flemming et al. 1998). The attenuated total reflection (ATR) spectroscopy mode allows to specifically receive the signals of material depositing on the ATR crystal because the IR beam penetrates the medium it is embedded in, that is, water, only into a maximum depth of 1–2 µm. Such systems allow to distinguish and identify abiotic and biotic material, which attaches to the crystal surface. Raman spectroscopy and microscopy may also provide such information (Wagner et al. 2009).

NMR imaging of deposits in pipes or porous media. Nuclear magnetic resonance imaging (NMRI) techniques were employed to identify and selectively image biofilms growing in aqueous systems (Hoskins et al. 1999). This technique can give information about the extent and spatial distribution of the biofilm and/or deposit and information about the chemical nature of certain components.

5 Conclusions

From the considerations as outlined in this review, it is obvious that successful strategies against biofouling should be based on integrated approaches, which consider the entire system to be protected. It is not possible to eradicate biofilms once and forever. Requests like this from practice remind to the well-known situation when children hate to brush their teeth. Until now, there is no way to brush teeth once and forever. If this is acknowledged, it is possible to learn how to live with biofilms and minimize biofouling problems – which requires just some attention.

The most promising approaches include technical hygiene ("good housekeeping") to minimize the fouling potential of the water phase, for example, by keeping bacterial numbers as well as nutrients as low as possible. Easy-to-clean surfaces and materials which do not support microbial adhesion and growth are another element of integrated antifouling strategies; here, one can learn a lot from food industry, in particular, the implementation and application of the HACCP concept (hazard analysis critical control point; Mortimore, 2001). That includes cleaning-friendly design of the systems and material surfaces, which are smooth and do not leach biodegradable substances such as plasticizers and other additives to plastics.

Establishment of early warning capacity is important to initiate timely countermeasures. Surface monitoring will help a lot, either performed by regular sampling of accessible surfaces or by following fouling-related parameters using

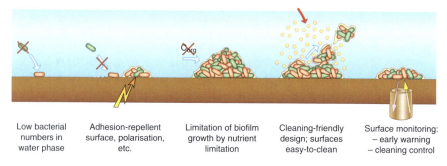

Fig. 8 Key elements of an integrated antifouling strategy

devices as presented earlier. Integrated approaches are, in a nutshell, depicted in Fig. 8:

Practically, all components required for integrated solutions are already available and only have to be assembled and adapted to their particular application. However, such solutions require a shift of paradigms and of the point of view, away from the so much desired one-shot solutions. Tailored solutions can be applied right now and only have to be selected and adapted from the big range of tools, many of which have been mentioned in this review. And although optimal solutions not always exist, the already present ones would proof very effective if applied in the context of holistic approaches. This is where further research should be dedicated – in particular, for longer-term, sustainable solutions. The benefit would be much more success in antifouling and much less environmental damage by biocides, disinfectants and other components, which we do not want to further pollute our waters.

References

Andrewartha J, Perkins K, Sargison J, Osborn J, Walker G, Henderson A, Hallegraeff G (2010) Drag force and surface roughness measurements on freshwater biofouled surfaces. Biofouling 26:487–496

Angell P, Arrage AA, Mittelmann MW, White DC (1993) Online, non-destructive biomass determination of bacterial biofilms by fluorimetry. J Microbiol Meth 18:317–327

Armstrong E, Boyd KG, Burgess JG (2000) Prevention of marine biofouling using natural: compounds from marine organisms. Biotechnol Annu Rev 6:221–241

Arp G, Reimer A, Reitner J (2001) Photosynthesis-induced biofilm calcifcation and calcium concentrations in Phanerozoic Oceans. Science 292:1701–1704

Azis PKA, Al-Tisan I, Sasikumar N (2001) Biofouling potential and environmental factors of seawater at a desalination plant intake. Desalination 135:69–82

Baier RE (1982) Conditioning surfaces to suit the biomedical environment: recent progress. J Biomec Engg 104:257–271

Ballantine DS, Wohltien H (1989) Surface acoustic devices for chemical analysis. Anal Chem 61:188–193

Barraud N, Hasset DJ, Hwang S-H, Rice SA, Kjelleberg S, Webb JS (2009) Involvement of nitric oxide in dispersion of *Pseudomonas aeruginosa*. J Bact 188:7344–7353

Bers AV, Wahl M (2004) The influence of natural surface microtopographies on fouling. Biofouling 20:43–51

Boulangé-Petermann L (1996) Processes of bioadhesion on stainless steel surfaces and cleanability: a review with special reference to food industry. Biofouling 10:275–300

Braissant O, Cailleau G, Dupraz C, Verreccia EP (2003) Bacterially induced mineralization of calcium carbonate in terrestrial environments: the role of exopolysaccharides and amino acids. J Sed Res 73:485–490

Brisou JF (1995) Biofilms: methods for enzymatic release of microorganisms. CRC, Boca Raton, New York, London, Tokyo, p 204

Bruijs MCM, Venhuis LP, Jenner HA, Daniels DG, Licina GJ (2000) Cooling water biocide optimisation using an on-line biofilm monitor. KEMA Tech. Op. Serv. PO Box 9035, Arnhem, Gelderland; 6800 ET Netherlands

Carman ML, Estes TG, Feinberg AW, Schumacher JF, Wilkerson W, Wilson LH, Callow ME, Callow JA, Brenan AB (2006) Engineered antifouling microtopographies: correlating wettability with cell attachment. Biofouling 22:11–21

Characklis WG (1990) Microbial biofouling. In: Characklis WR, Marshall KC (eds) Biofilms. Wiley, New York, pp 523–584

Characklis WG, Turakhia MH, Zelver N (1990) Transport and interfacial transfer phenomena. In: Characklis WG, Marshall KC (eds) Biofilms. Wiley, New York, pp 265–340

Cheyne I (2010) Regulation of marine antifouling in international and EC law. In: Dürr S, Thomason JC (eds) Biofouling. Wiley-Blackwell, Chichester, pp 306–318

Cloete TE (2003) Biofouling: what we know and what we should know. Mat Corr 54:520–526

Cloete ET, Maluleke M (2005) The use of the rotoscope as an on-line, real-time, non-destructive biofilm monitor. Wat Sci Technol 52:211–216

Cole GC (1998) Pharmaceutical production facilities. Design and applications. CRC, Boca Raton

Cornelissen ER, Vrouwenvelder JS, Heijman SGJ, Viallefont XD, van der Kooij D, Wessels LP (2007) Periodic air/water cleaning for control of biofouling in spiral wound membrane elements. J Mem Sci 287:94–101

Costerton JW et al (1987) Bacterial biofilms in nature and disease. Ann Rev Microbiol 41:435–464

Davies DG, Marques CNH (2009) A fatty acid messenger is responsible for inducing dispersion in microbial biofilms. J Bacteriol 191:1393–1403

de Nys R, Givskov M, Kumar N, Kjelleberg S, Steinberg P (2006) Furanones. Prog Mol Subcell Biol 42:55–86

de Nys R, Guenther J, Uriz MJ (2010) Natural control of fouling. In: Dürr S, Thomason JC (eds) Biofouling. Wiley-Blackwell, Chichester, pp 109–120

de Prijck K, Nelis H, Coenye T (2007) Efficacy of silver-releasing rubber for the prevention of *Pseudomonas aeruginosa* biofilm formation in water. Biofouling 23:405–411

Dobretsov S (2009) Inhibition of marine biofouling by biofilms. In: Flemming H-C, Murthy RS, Venkatesan R, Cooksey KE (eds) Marine and industrial biofouling. Springer, Heidelberg, pp 293–314

Dow JM, Crossman L, Findlay K, He Y-Q, Feng J-X, Tang J-L (2003) Biofilm dispersal in *Xanthomonas campestris* is controlled by cell-cell signalling and is required for full virulence to plants. Proc Natl Acad Sci USA 100:10995–11000

Edyvean R (2010) Consequences of fouling on shipping. In: Dürr S, Thomason JC (eds) Biofouling. Wiley-Blackwell, Chichester, pp 217–225

Eguia E, Truebo A, Rio-Calogne B, Giron A, Amieva JJ, Bielva C (2008) Combined monitor for direct and indirect measurement of biofouling. Biofouling 24:75–86

Epstein N (1981) Fouling: technical aspects. In: Somerscales EFC, Knudsen JG (eds) Fouling of heat transfer equipment. Hemisphere, Washington, pp 31–53

Fillaudeau L (2003) Fouling phenomena using hot wire methods. In: Heldman D (ed) Encycl Agric Food Biol Eng 56:315–324

Finnie AA, Williams DN (2010) Paint and coatings technology for the control of marine fouling. In: Dürr S, Thomason JC (eds) Biofouling. Wiley-Blackwell, Chichester, pp 185–206

Flemming H-C (1982) Bacterial growth on ion exchanger resin – investigations with a strong acidic cation exchanger. Part II: Efficacy of silver against aftergrowth during non-operation periods. Z Wasser Abwasser Forsch 15:259–266

Flemming H-C (2002) Biofouling in water systems: cases, causes, countermeasures. Appl Envir Biotechnol 59:629–640

Flemming HC (2003) Role and levels of real time monitoring for successful anti-fouling strategies. Wat Sci Technol 47(5):1–8

Flemming H-C, Cloete TE (2010) Environmental impact of controlling biofouling and biocorrosion in cooling water systems. In: Rajagopal S, Jenner HA, Venugopalan VP (eds) Operational and Environmental Consequences of Large Industrial Cooling Water Systems 365–380

Flemming H-C, Greenalgh M (2009) Concept and consequences of EU biocide guideline. In: Flemming H-C, Venkatesan R, Murthy PS, Cooksey KC (eds) Marine and industrial biofouling. Springer, Heidelberg, pp 189–200

Flemming H-C, Ridgway HF (2009) Biofilm control: conventional and alternative approaches. In: Flemming H-C, Venkatesan R, Murthy PS, Cooksey KC (eds) Marine and industrial biofouling. Springer, Heidelberg, pp 103–118

Flemming, H.-C. (2008) Biofilms. In: Encyclopedia of life sciences. John Wiley, Chichester http://http://www.els.net/ [DOI: 10.1002/9780470015902.a0000342]

Flemming H-C, Wingender J (2003) Biofilms. In: Steinbüchel A (ed) Biopolymers, vol 10. VCH Wiley, Weinheim, pp 209–245

Flemming H-C, Wingender J (2010) The biofilm matrix: key for the biofilm mode of life. Nat Rev Microbiol 8:623–633

Flemming H-C, Schaule G, McDonogh R, Ridgway HF (1994) Mechanism and extent of membrane biofouling. In: Geesey GG, Lewandowski Z, Flemming H-C (eds) Biofouling and biocorrosion in industrial water systems. Lewis, Chelsea, MI, pp 63–89

Flemming H-C, Tamachkiarowa A, Klahre J, Schmitt J (1998) Monitoring of fouling and biofouling in technical systems. Wat Sci Technol 38:291–298

Genzer J, Efimenko K (2006) Recent developments in superhydrophobic surfaces and their relevance to marine fouling: a review. Biofouling 22:339–360

Giladi M, Porat Y, Blatt A, Wasserman Y, Kirson ED, Dekel E, Palti Y (2008) Microbial growth inhibition by alternating electric fields. Antimicrob Agents Chemother 52:3517–3522

Gilbert P, McBain AJ, Rickard AH (2003) Formation of microbial biofilm in hygienic situations: a problem of control. Int Biodet Biodegr 51:245–248

Gray JE, Norton PR, Alnounu R, Marolda CL, Valvano MA, Griffiths K (2003) Biological efficacy of electroless-deposited silver on plasma activated polyurethane. Biomaterials 24:2759–2765

Griebe T, Flemming H-C (1998) Biocide-free antifouling strategy to protect RO membranes from biofouling. Desalination 118:153–156

Gu J-D, Belay B, Mitchell R (2001) Protection of catheter surfaces from adhesins of *Pseudomonas aeruginosa* by a combination of silver ions and lectins. World J Microbiol Biotechnol 17:173–179

Helle H, Vuoriranta P, Välimäki H, Lekkala J, Aaltonen V (2000) Monitoring of biofilm growth with thickness-shear mode quartz resonators in different flow and nutrition conditions. Sens Actuators B Chem 71:47–54

Henderson P (2010) Fouling and antifouling in other industries: power stations, desalination plants, drinking water supplies and sensors. In: Dürr S, Thomason JC (eds) Biofouling. Wiley-Blackwell, Chichester, pp 288–305

Hermansson M (2000) The DLVO theory in microbial adhesion. Coll Surf B Biointerfaces 14:105–119

Hillman RE, Anson D (1985) Biofouling detection monitoring devices: status assessment. New England Marine Research Laboratory, Duxbury, MA, p 119, Ordering address: Research Report Center, P.O. Box 50490, Palo Alto, CA 94303, USA

Hoskins BC, Fevang L, Majors PD, Sharma MM, Georgiou G (1999) Selective imaging of biofilms in porous media by NMR relaxation. J Magn Res 139(1):67–73

Howell D, Behrends B (2010) Consequences of antifouling coatings: the chemist's perspective. In: Dürr S, Thomason JC (eds) Biofouling. Wiley-Blackwell, Chichester, pp 226–242

Hüttinger KJ, Müller H, Bomar MR (1982) Prevention of biodeterioration of cellulose by chemically bound preservatives. Mater Org 17:285–298

Ista LK, Pérez-Luna VH, López GP (1999) Surface-grafted, environmentally sensitive polymers for biofilm release. Appl Envir Microbiol 65:1603–1609

Janknecht P, Melo L (2003) Online biofilm monitoring. Rev Envir Sci BioTech 2:269–283

Jullien C, Bénézech T, Carpentier B, Lebret V, Faille C (2003) Identification of surface characteristics relevant to the hygienic status of stainless steel for the food industry. J Food Eng 56:77–87

Kerr MJ, Cowling CM, Beveridge MJ, Parr ACS, Head DM, Davenport J, Hodkiess T (1998) The early staes of marine biofouling and ist effect on two types of optical sensors. Envir Int 24:331–343

Kilb B, Lange B, Schaule G, Wingender J, Flemming H-C (2003) Contamination of drinking water by coliforms from biofilms grown on rubber-coated valves. Int J Hyg Envir Health 206(6):563–573

Klahre J, Flemming H-C (2000) Monitoring of biofouling in papermill water systems. Wat Res 34:3657–3665

Klahre J, Lustenberger M, Flemming H-C (1998) Mikrobielle Probleme bei der Papierfabrikation. Teil III: Monitoring. Papier 52:590–596

Klibanov AM (2007) Permantly microbial materials coatings. J Mat Chem 17:2479–2482

Kolodkin-Gal I, Romero D, Cao S, Clardy J, Kolter R, Losick R (2010) D-amino acids trigger biofilm disassembly. Science 328:627–629

Lawrence JR, Swerhone GDW, Neu TR (2000) A simple rotating annular reactor for replicated biofilm studies. J Microb Meth 42:215–224

Leeming K, Moore CP, Denyer SP (2002) The use of immobilized biocides for process water decontamination. Int Biodet Biodegr 49:39–43

Lequette Y, Boels G, Clarisse M, Faille C (2010) Using enzymes to remove biofilms of bacterial isolates sampled in the food-industry. Biofouling 26:421–431

Lewis K, Klibanov AM (2005) Surpassing nature: rational design of sterile-surface materials. Trends Biotechnol 23:343–348

Little BJ, Lee JS (2007) Microbiologically influenced corrosion. Wiley, Hoboken, NJ

Louie JS, Pinnau I, Ciobanu I, Ishida KP, Ng A, Reinhard M (2006) Effects of polyether–polyamide block copolymer coating on performance and fouling of reverse osmosis membranes. J Membr Sci 280:762–770

Madkour M, Ahmad E, Tew GN (2008) Towards self-sterilizing medical devices: controlling infection. Polym Int 57:6–10

Maguire RJ (2000) Review of the persistence, bioaccumulation and toxicity of tributyltin in aquatic environments in relation to Canada's toxic substances management policy. Water Qual Res J Can 35:633–675

Majumdar P, Lee E, Patel N, Ward K, Stafslien SJ, Daniels J, Chisholm BJ, Boudjouk P, Callow M, Callow J, Thompson S (2008) Combinatorial materials research applied to the development of new surface coatings IX: an investigation of novel antifouling/fouling-release coatings containing quaternary ammonium salt groups. Biofouling 24:185–200

Marmur A (2004) The lotus effect: superhydrophobicity and metastability. Langmuir 20:3517–3519

Marshall KC, Blainey B (1990) Role of bacterial adhesion in biofilm formation and biocorrosion. In: Flemming HC, Geesey GG (eds) Biofouling and biocorrosion in industrial water systems. Springer, Heidelberg, pp 29–46

Marshall KC, Stout R, Mitchell R (1971) Mechanism of the initial events in the sorption of marine bacteria to surfaces. J Gen Microbiol 68:337–348

Mayer C, Moritz R, Kirschner C, Borchard W, Maibaum R, Wingender J, Flemming H-C (1999) The role of intermolecular interactions: studies on model systems for bacterial biofilms. Int J Biol Macromol 26:3–16

Meseguer Yebra D, Kiil S, Dam-Johansen K (2004) Antifouling technology: past, present and future steps towards efficient and environmentally friendly antifouling coatings. Prog Org Coat 50:75–104

Milovic NM, Wang J, Lewis K, Klibanov A (2005) Immobilized N-alkylated polyethylenimine avidly kills bacteria by rupturing cell membranes with no resistance developed. Biotech Bioeng 90:715–722

Mollica A, Cristiani P (2003) On-line biofilm monitoring by "BIOX" electrochemical probe. Wat Sci Tech 47:45–49

Mortimore S (2001) How to make HACCP work in practice. Food Contr 12:209–215

Murthy PS, Venkatesan R (2009) Industrial biofilms and their control. In: Flemming H-C, Murthy PS, Venkatesan R, Cooksey KC (eds) Marine and industrial biofouling. Springer, Heidelberg, pp 65–101

Nienhuis C, Barthlott W (1997) Characterization and distribution of water-repellent, self-cleaning plant surfaces. Ann Bot 79:667–677

Nickels J, Bobbie RJ, Lott DF, Maritz RF, Benson PH, White DC (1981) Effect of manual brush cleaning on biomass and community structure of microfouling film formed on aluminium and titanium surfaces exposed to rapidly flowing seawater. Appl Environ Microbiol 41:1442–1453

Nivens DE, Palmer RJ, White DC (1995) Continuous nondestructive monitoring of microbial biofilms: a review of analytical techniques. J Ind Microbiol 15:263–276

O'Toole G, Kaplan HB, Kolter R (2000) Biofilm formation as microbial development. Ann Rev Microbiol 54:49–79

Oliver JD (2005) The viable but nonculturable state in bacteria. J Microbiol 43:93–100

Oliver, J.D. (2010) Recent findings on the viable but nonculturable state in pathogenic bacteria. FEMS Microbiol Rev 34:415–425

Park D, Wang J, Klibanov AM (2006) One-step painting-like coating procedures to make surfaces highly and permanently biocidal. Biotechnol Prog 22:584–589

Parvici J, Antoci V, Hickok NJ, Shapiro IM (2007) Self protective smart orthopaedic implants. Expert Rev Med Dev 4:55–64

Patil JS, Kimoto H, Kimoto T, Saino T (2007) Ultraviolet radiation (UV-C): a potential tool for the control of biofouling on marine optical instruments. Biofouling 23:215–230

Pereira A, Mendes J, Melo L (2007) Using nanovibrations to monitor biofouling. Biotech Bioeng 99:1407–1414

Perez-Roa RE, Tompkins DT, Paulose M, Grimes CA, Anderson MA, Noguera DR (2006) Effects of localized, low-voltage pulsed electric fields on the development and inhibition of *Pseudomonas aeruginosa* biofilms. Biofouling 22:383–390

Psoch C, Schwier S (2006) Direct filtration of natural and simulated river water with air sparging and sponge ball application for fouling control. Desalination 197:190–204

Rao TS, Kora AJ, Chandramohan P, Panigrahi BS, Narasimhan SV (2009) Biofouling and microbial corrosion problem in the thermo-fluid heat exchanger and cooling water system of a nuclear test reactor. Biofouling 25:581–591

Rompré A, Servais P, Baudart J, de-Roubin M-R, Laurent P (2002) Detection and numeration of coliforms in drinking water: current methods and emerging approaches. J Microbiol Meth 49:31–54

Ruseska I, Robbins J, Lashen ES, Costerton JW (1982) Biocide testing against corrosion-causing oilfield bacteria helps control plugging. Oil Gas J 80:253–264

Sand W, Gehrke T (2006) Extracellular polymeric substances mediate bioleaching/biocorrosion via interfacial processes involving iron(III) ions and acidophilic bacteria. Res Microbiol 157:49–56

Schackenraad JM, Stokroos I, Bartels H, Busscher HJ (1992) Patency of small caliber, superhydrophobic E-PTFE vascular grafts: a pilot study in rabbit carotid artery. Cells Mater 2:193–199

Schaule G, Griebe T, Flemming H-C (2000) Steps in biofilm sampling and characterization in biofouling cases. In: Flemming H-C, Szewzyk U, Griebe T (eds) Biofilms. Technomic, Lancaster, pp 1–21

Schaule G, Rumpf A, Weidlich C, Mangold K-M, Flemming H-C (2008) The effect of pulsed electric polarization of indium tin oxide (ITO) and polypyrrole on biofilm formation. Wat Sci Technol 58:2165–2172

Schmid T, Helmbrecht C, Panne U, Haisch C, Niessner R (2003) Process analysis of biofilms by photoacoustic spectroscopy. Anal Bioanal Chem 375:1124–1129

Schmid T, Panne U, Adams J, Niessner R (2004) Investigation of biocide efficacy by photoacoustic spectroscopy. Wat Res 38:1189–1196

Schopf JW, Hayes JM, Walter MR (1983) Evolution on earth's earliest ecosystems: recent progress and unsolved problems. In: Schopf JW (ed) Earth's earliest biosphere. Princeton University Press, New Jersey, pp 361–384

Schulte S (2003) Wirksamkeit von Wasserstoffperoxid gegenüber Biofilmen. Ph.D. dissertation, University of Duisburg-Essen, Germany

Schulte S, Wingender J, Flemming H-C (2005) Efficacy of biocides against biofilms. In: Paulus W (ed) Directory of microbicides for the protection of materials and processes. Kluwer Academic, Doordrecht, The Netherlands, pp 90–120, Chapter 6

Schultz MP (2007) Effects of coating roughness and biofouling on ship resistance and powering. Biofouling 23:331–341

Silver S (2003) Bacterial silver resistance: molecular biology and uses and misuses of silver compounds. FEMS Microbiol Rev 27:341–353

Stancak M (2004) Biofouling: it's not just barnacles any more. http://www.csa.com/discoveryguides/biofoul/overview.php

Sunada K, Watanabe T, Hashimoto K (2003) Studies on photokilling of bacteria by TiO_2 thin film. J Photochem Photobiol A 156:227–233

Tamachkiarow A, Flemming H-C (2003) On-line monitoring of biofilm formation in a brewery water pipeline system with a fibre optical device (FOS). Wat Sci Tech 47(5):19–24

Tait K, Skillman LC, Sutherland IW (2002) The efficacy of bacteriophate as a method of biofilm eradication. Biofouling 18:305–311

Ten Hallers-Tjabbes CC, Walmsley S (2010) Consequences of antifouling systems: an environmental perspective. In: Dürr S, Thomason JC (eds) Biofouling. Wiley-Blackwell, Chichester, pp 243–251

Terlezzi A, Conte E, Zupo V, Mazzella L (2000) Biological succession on silicone fouling-release surfaces: long term exposure tests in the harbour of Ischia, Italy. Biofouling 15:327–342

Tiller JC, Lee SB, Lewis K, Klibanov AM (2002) Polymer surfaces derivatized with poly (vinyl-N-hexylpyridinium) kill airborne and waterborne bacteria. Biotechnol Bioeng 79:465–471

Turley PA, Fenn RJ, Ritter JC, Callow ME (2005) Pyrithiones as antifoulants: environmental fate and loss of toxicity. Biofouling 21:31–40

Van der Oost R, Beyer J, Vermeulen NPE (2003) Fish bioaccumulation and biomarkers in environmental risk assessment: a review. Environ Toxicol Pharmacol 13:57–149

Van Gulck JF, Rowe RK, Rittmann BE, Cooke AJ (2003) Predicting biogeochemical calcium precipitation in landfill leachate collection systems. Biodegradation 14:331–346

Verran J, Jones M (2000) Problems of biofilms in the food and beverage industry. In: Walker J, Surmann S, Jass J (eds) Industrial biofouling detection, prevention and control. Wiley, Chichester, UK, pp 145–173

Vladkova T (2009) Surface modification approach to control biofouling. In: Flemming HC, Murthy PS, Venkatesan R, Cooksey KC (eds) Industrial and marine biofouling. Springer, Heidelberg, pp 135–163

Von der Schulenburg DA, Akpa BS, Gladden LF, Johns ML (2008) Non-invasive mass transfer measurements in complex biofilm-coated structures. Biotech Bioeng 101:602–608

Vrouwenvelder JS, Bakker SM, Wessels LP, van Paassen JAM (2006) The membrane fouling simulator as a new tool for biofouling control of spiral wound membranes. Desalination 204:170–174

Wagner M, Ivleva NP, Haisch C, Niessner R, Horn H (2009) Combined use of confocal laser scanning microscopy (CLSM) and Raman microscopy (RM): investigations on EPS-matrix. Wat Res 43:63–76

Webb J, Thompson LS, James S, Charlton T, Tolker-Nielsen T, Koch B, Givskov M, Kjelleberg S (2003) Cell death in *Pseudomonas aeruginosa* biofilm development. J Bacteriol 185:4585–4592

Webster DC, Chisholm BJ (2010) New directions in antifouling technology. In: Dürr S, Thomason JC (eds) Biofouling. Wiley-Blackwell, Chichester, pp 366–387

Wetegrove RL (1998) Monitoring of film forming living deposits. US Patent No. 5,796,478

Wetegrove RL, Banks R (1993) Monitoring film fouling in a process stream with a transparent shunt and a light detecting means. US Patent No. 5,185,533

White DC, Arrage AA, Nivens DE, Palmer RJ, Rice JF, Sayler GS (1996) Biofilm ecology: on-line methods bring new insights into MIC and microbial biofouling. Biofouling 10:3–16

Whitehead KA, Verran J (2009) The effect of substratum properties on the survival of attached microorganisms on inert surfaces. In: Flemming H-C, Murthy PS, Venkatesan R, Cooksey KC (eds) Marine and industrial biofouling, Springer series on biofilms. Springer, Heidelberg, pp 13–33

Wingender J, Flemming H-C (2004) Contamination potential of drinking water distribution network biofilms. Wat Sci Tech 49:277–285

Wirtanen G, Salo S (2003) Disinfection in food processing: efficacy testing of disinfectants. Rev Environ Sci Biotechnol 2:293–306

Wloka M, Rehage H, Flemming H-C, Wingender J (2006) Structure and rheological behaviour of the extracellular polymeric substance network of mucoid *Pseumonas aeruginosa* biofilms. Biofilms 2:275–283

Wood P, Jones M, Bhako M, Gilbert P (1996) A novel strategy for control of microbial biofilms through generation of biocide at the biofilm-surface interface. Appl Envir Microbiol 62:2598–2602

Wu Z, Chen H, Dong Y, Mao H, Sun J, Chen S, Craig VSJ, Hu J (2008) Cleaning using nanobubbles: defouling by electrochemical generation of bubbles. Coll Mat 328:10–14

Zasloff M (2002) Antimicrobial peptides of multicellular organisms. Nature 415:389–395

Zhao Q, Liu Y, Müller-Steinhagen HM (2002) Effects of interaction energy on biofouling adhesion. In: Proceedings of the conference on fouling, cleaning and desinfection in food processing, Cambridge University, pp 41–47

Zinn MS, Kirkegaard DR, Palmer RJ, White DC (1999) Laminar flow chamber for continuous monitoring of biofilm formation and succession. Biofilms 310:224–232

Advances in Biofilm Mechanics

Thomas Guélon, Jean-Denis Mathias, and Paul Stoodley

Abstract A knowledge of the mechanical properties of bacterial biofilms is required to more fully understand how a biofilm will physically respond, and adapt, to the physical forces, such as those caused by fluid flow or particle or bubble impingement, acting upon it. This is particularly important since biofilms are problematic in a wide diversity of scenarios and spatial and temporal scales and many control strategies designed to remove biofilms include a mechanical component such as fluid flow, particle or bubble impingement or a physical contact with the surface generated by scraping or brushing. Knowing when, and how, a biofilm might fail (through adhesive or cohesive failure) will allow better prediction of accumulation and biomass detachment, key processes required in the understanding of the structure and function of biofilm systems. However, the measurements of mechanical properties are challenging. Biofilms are living systems and they readily desiccate if removed from the liquid medium, it is not clear how quickly their mechanical properties might change when removed from their indigenous environment into a testing environment. They are also very thin and are inherently attached to a surface. They cannot be formed into standard test coupons such as plastics or solids, and cannot readily be poured or placed into conventional viscometers or rheometers, such as liquids and gels. Measured parameters such as the elastic and shear modulus, adhesive strength or tensile strength are sparse but are increasingly appearing in the literature. There is a large range of reported values for these properties, although there is general agreement that biofilms are viscoelastic. Biofilms have been assessed with various experimental methods depending on the desired characteristic and available equipment. The aforementioned challenges and

---

T. Guélon • J.-D. Mathias
Cemagref – LISC (Laboratory of engineering for complex systems), 24, avenue des Landais, 50 085-63 172, Aubière Cedex 1, France

P. Stoodley (✉)
National Centre for Advanced Tribology, University of Southampton, University RoadSouthampton, SO17 1BJ, UK
e-mail: pstoodley@gmail.com

lack of standard methods or equipment for testing attached biofilms have led to the development of many creative methods to tease out aspects of biofilm mechanical properties. In this paper, we review some of the more common techniques and highlight some recent results.

1 Introduction

Bacteria in natural, industrial and clinical settings predominantly live in surface-associated communities called biofilms (Costerton et al. 1995). They develop at any interface with conditions compatible with microbial life. Important examples where biofilms occur are the teeth and gums (dental plaque) (Socransky and Haffajee 2002; Bradshaw and Marsh (1999), riverbed and marine sediments, plant leaves, soil, and engineered systems such as waste water treatment systems (Woolard and Irvine 1994; Wagner and Loy 2002) industrial process plants and marine structures (Beech et al. 2005; Coetser and Cloete 2005), and sewage pipelines and bioreactors (Godon et al. 1997).

Biofilms can be beneficially used in industrials bioreactors for the production of useful chemicals such as acetic acid, ethanol or hydrogen (Bartowsky and Henschke 2008, Liao et al. 2010), or in the waste water treatment process, where they convert organic carbon and ammonia waste into utilizable biomass (for fertilizer, or fuel) and gas (Woolard and Irvine 1994; Wagner and Loy 2002). Indeed, microbial biofilms play a major role in the global recycling of elements vital to life. On the contrary, biofilms can be problematic in many industrial processes such as water distribution pipelines, ship hulls or biofouling in the food industry, where they can cause drag, corrosion and contamination. Pathogens in such biofilms can represent a public health risk. On medical devices biofilms are almost universally undesirable and often result in chronic infection, with acute outbreaks. To sum up, for applications where biofilms have a beneficial purpose, it is desirable to retain biofilms in the system, however, the removal, or prevention, of biofilms is of prime importance when biofilms are harmful and have a negative impact on industrial processes, or human health. The efficacy of any removal process which involves degradation of the matrix by chemical or enzymatic digestion or the application of a physical force such as natural fluid flow or mechanical scraping will be determined in part by the mechanical properties of the biofilm.

Understanding and characterizing the mechanical properties of bacterial biofilms constitutes an important scientific and economic issue. Indeed, the physical properties of biofilms will determine the shape and mechanical stability of the biofilm structure and consequently affect both mass transfer and detachment processes. A primary interest in determining the mechanical properties of bacterial biofilms is to predict and manage biofilm accumulation and detachment. These processes represent key-issues to many applications in the context of the control of bacterial biofilms. Active natural biofilm detachment and dispersal of cells from biofilm structures (i.e. initiated by the bacterial communities themselves) is important as it allows the micro-organisms to escape from the biofilm matrix, where they can contaminate the bulk fluid and colonize downstream surfaces. The recent discovery of active dispersal also provides the opportunity to exploit the natural mechanisms for biofilm control through exogenous manipulation using natural dispersal molecules (Barraud et al. 2006; Davies and Marques 2009). Hence, it is important to understand all of the factors affecting the detachment of biofilm in any given system. However, it is difficult to test and calculate the mechanical properties of bacterial biofilms because of a number of factors. (1) Biofilms are living materials and are difficult to control – they do not form convenient testing coupons, which are typically used for tensile and compression tests of solid materials. (2) Biofilms are highly hydrated, with the most prolific and problematic biofilms, such as those encountered in marine biofouling, growing in aqueous environments and will quickly desiccate when removed from their liquid medium, thus necessitating the desire to perform tests in submersion, or at least at high humidity. (3) Biofilms are usually thin (microns) with very small volumes and so it is difficult to harvest enough material to perform conventional tests designed to measure fluid properties. (4) Biofilms are inherently attached to surfaces and the process of removal from a surface may compromise the very properties which are trying to be measured. (5) Biofilms are generally very soft and require sensitive equipment capable of imparting and measuring low stresses and strains. (6) Biofilms of medical and dental interest contain pathogens and therefore require special biological handling conditions, not normally found in conventional mechanical testing facilities.

There are various studies in the literature that use different and imaginative techniques to assess the material characteristics of bacterial biofilms. However, with mechanical testing, the testing method employed will often influence the interpretation. In this paper, we review some of the most common methods ranging from cone and plate rheometry (Picologlou et al. 1980) and the later pioneering work by Ohashi and Harada (1994), who proposed a centrifugation method to assess the tensile force required to separate biofilm fragments to more recent techniques like microcantilever (Aggarwal et al. 2009) or the flow cell method (Mathias and Stoodley 2009). We broadly classify these methods into two categories (1) methods which use a directly applied and controlled loading force and (2) methods which use hydrodynamic loadings. The direct applied force means that the bacterial biofilm is affected directly by a mechanical force, usually without hydrodynamic flow, such as micro-cantilevers (Poppele and Hozalski 2003; Aggarwal et al. 2009; Aggarwal and Hozalski 2010), indenters (Cense et al. 2006b) or T-shaped probes (Chen et al. 1998, 2005), which are used to pull (tensile testing under a normal load) or push (compression testing under a normal load) the biofilm. Spinning disk rheometry has been used to load biofilms in shear (Vinogradov et al. 2004). With regards to methods using hydrodynamic loading, bacterial biofilms are subjected to a fluid flow such as in flow cells (Stoodley et al. 1999a, b, 2001a, 2002) or Couette–Taylor type reactors (Coufort et al. 2007; Rochex et al. 2008).

The structural behaviours of the biofilm subjected to a defined applied load are measured to determine the mechanical properties of the biofilm. Several behaviour laws exist, which link the various materials parameters. In several studies, bacterial biofilms have been reported to behave as viscoelastic materials with a relaxation time equal to few minutes (Klapper et al. 2002; Cense et al. 2006a, b; Lau et al. 2009), while in other studies they are described as elastic (Ohashi et al. 1999; Aravas and Laspidou 2008). The choice of either measurement method depends on which mechanical property is being investigated, and conversely the method employed dictates the sorts of properties that can be characterized. In the first part of the review, we investigate the different behaviours of biofilms and present the main basis of constitutive laws. Then, we detail methods using either direct applied forces or hydrodynamics loading. A summary of results of mechanical properties specifically relating to the particular methodology used is presented. Finally, we discuss future developments in the domain of biofilm mechanics.

2 Mechanical Background

Here, we present the mechanical basis and language (Table 1) required to discuss, and characterize, bacterial biofilms from a materials standpoint. As we explain above, a variety of different methods can be used to assess different mechanical properties. The determination of the rheological properties of biofilms and the interpretation of the findings is complicated by the structural heterogeneity of biofilms, which is a common characteristic of biofilms. The term "structural heterogeneity" refers to the observation that in many biofilms the EPS matrix is interspersed with water pores and channels (i.e. Costerton et al. 1995), (Fig. 1). The situation is further complicated by local density variations in the biofilm. Biofilms tend to be denser at the base and there is evidence that some of the water channels are devoid of bacterial cells but may contain low density strands of EPS (de Beer et al. 1994).

In our discussions, we simplify the situation by considering the biofilm as a homogeneous material and only the macroscopic parameters are explained and detailed. Various mechanical behaviours have been highlighted in the literature including elasticity, viscoelasticity and plasticity. Furthermore, the failure of bacterial biofilms has been studied in (Cense et al. 2006a, b; Aggarwal and Hozalski 2010) to assess biofilm strength. Mechanical definitions are recalled hereafter.

Some of these terms will be defined in greater detail below.

2.1 Elasticity

A number of studies report a linear behaviour of bacterial biofilm (Körstgens et al. 2001a, b; Mathias and Stoodley 2009) when they are subjected to rapid low loadings. In this case, the elasticity framework can be used to describe the mechanical behaviour of bacterial biofilm. In the general framework, anisotropic

Table 1 Quantifiable and descriptive terms of relevance to biofilm mechanics

Term	Description	Units SI and base units (mass, length, time)
Adhesive or interfacial failure	Failure at the plane of attachment between two materials, in this case the biofilm and the attachment surface	N.A.
Cohesive failure	Failure (i.e. the breaking of bonds) within the bulk material, in this case the biofilm	N.A.
Constitutive Law	A set of equations which describe the deformational response of any material to a mechanical stress	N.A.
Elasticity	The property of a material that allows it to returns to its original shape after the removal of an applied load	Pa (kg m^{-1} s^{-2})
Failure	Permanent damage of a material caused by an applied load	N.A.
Force	An entity which has the ability to do work or cause physical change, expressed as F=ma in Newton's second law	N (kg m s^{-2})
Hydraulics	The scientific study of water and other liquids	N.A.
Hydrodynamics	The characteristics of a flowing fluid	N.A.
Linear Range	A range of applied stress over which the degree of deformation (elastic and/or viscous) is directly proportional to the stress	N.A.
Load	The magnitude and manner in which a force is applied	N.A.
Plasticity	The property of a solid, which allows it to undergo a permanent deformation (change its shape) under an applied load	N.A.
Strain	The deformation of a material in response to an applied stress. Simple strain is the ratio of the change in length caused by loading to the original length	Dimensionless (−)
Stress	In mechanics, stress refers to the magnitude of a force applied over a given area. Stresses can be *normal*, i.e. perpendicular to the surface of a material, or *shear*, i.e. parallel or tangential to a surface	Pascals (Pa), N m^{-2}, kg m^{-1} s^{-2}
Ultimate strength	The maximum stress a material can withstand before permanent failure	Pa (kg m^{-1} s^{-2})
Viscoelasticity	The properties of a material, which allow it to display both solid (elastic) and fluid (viscous) deformation when subjected to a load	N.A.
Viscosity	The resistance of a fluid to deformation (flow) by an applied load	Pa s (kg s^{-1} m^{-1})
Yield strength	The stress at which a material is irreversibly deformed	Pa (kg m^{-1} s^{-2})

N.A. Not applicable

behaviours can be defined. An anisotropic behaviour can be used if bacterial biofilms respond differently to forces which are applied from different directions. For the sake of simplicity, we only consider the linear isotropic elastic behaviour,

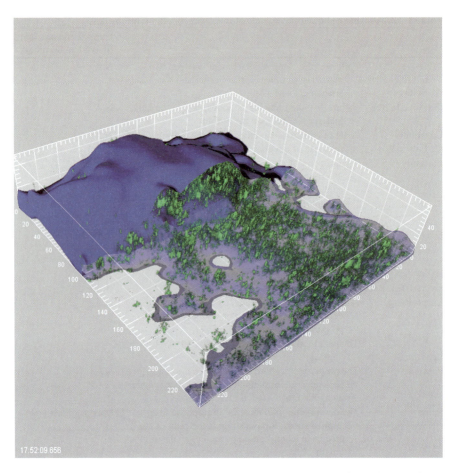

Fig. 1 *Streptococcus mutans* biofilm exhibiting heterogeneous structure of cell clusters with an undulating surface. Individual bacterial cells were stained green (live) or red (dead), and the EPS slime matrix surrounding the cells was stained blue. The EPS surface is made transparent in the "cutaway" in the foreground to show cells within the biofilm. *Scale*: major divisions = 10 μm

that is to say bacterial biofilms exhibit the same mechanical behaviour in all directions. It corresponds to the following relation between the stress σ (expressed in Pa) and the strain ε (dimensionless):

$$\begin{pmatrix} \sigma_{xx} \\ \sigma_{yy} \\ \sigma_{zz} \\ \sigma_{xy} \\ \sigma_{xz} \\ \sigma_{yz} \end{pmatrix} = \frac{E}{(1+v)(1-2v)} \begin{bmatrix} 1-v & v & v & 0 & 0 & 0 \\ v & 1-v & v & 0 & 0 & 0 \\ v & v & 1-v & 0 & 0 & 0 \\ 0 & 0 & 0 & 1-2v & 0 & 0 \\ 0 & 0 & 0 & 0 & 1-2v & 0 \\ 0 & 0 & 0 & 0 & 0 & 1-2v \end{bmatrix} \begin{pmatrix} \varepsilon_{xx} \\ \varepsilon_{yy} \\ \varepsilon_{zz} \\ \varepsilon_{xy} \\ \varepsilon_{xz} \\ \varepsilon_{yz} \end{pmatrix}. \quad (1)$$

In this equation, E represents Young's modulus (expressed in Pa) which characterizes the rigidity of the biofilm. If E is high, the biofilm displacements are low and vice versa. The coefficient ν represents the Poisson's ratio and characterize the perpendicular strain caused by a normal loading. In many studies, utilizing hydrodynamic loading or rheometry, biofilms are subjected to a wall shear stress. In this case, the shear modulus G is used:

$$G = \frac{E}{2(1+\nu)}. \qquad (2)$$

Young's modulus has been evaluated in (Stoodley et al. 1999a, b; Körstgens et al. 2001a, b; Mathias and Stoodley 2009; Aggarwal and Hozalski 2010) and the shear modulus in (Stoodley et al. 1999a, b; Towler et al. 2003).

2.2 Viscoelasticity

Microrheological *in-situ* measurements of biofilm "streamers" based on tracking particles embedded in a biofilm in response to changes over time in response to elevations in liquid shear stress clearly demonstrated the viscoelastic properties of biofilms (Stoodley et al. 1999a, b, 2002). This behaviour is time dependent, and it is important to not confuse viscoelasticity with plasticity. In the case of viscoelastic solids, the theory of viscoelasticity allows us to take into account completely reversible deformation over time. However, data from flow cell experiments (Stoodley et al. 2001a, 2002), and viscosity measurements on detached biofilms (Ohashi and Harada 1994) suggest that pure culture biofilms behaved like viscoelastic fluids, in which biofilms exhibit both irreversible viscous deformation and a reversible elastic recoil response. These experiments show that biofilms, in response to stress, exhibited behaviour consistent with those of both an elastic solid and a viscous liquid (see Fig. 2b). Note that in the case of ideal fluids the strain is completely irreversible. Figure 2 shows the idealized structural strains for various classes of materials when the material is exposed to an applied load. When a constant stress is applied to a viscoelastic material (e.g. bacterial biofilm), creep is observed over time with the strain progressively increasing (Fig. 2b). Conversely when a viscoelastic material is subjected to a constant strain, a progressive relaxation of the induced stress is observed over time (Fig. 2c). These two phenomena describe the time dependent behaviour of the material which may be analyzed in the frame of the theory of linear viscoelasticity (Lemaitre and Chaboche 1988; Couarraze and Grossiord 1991; Nowacki 1965; Mandel 1958). According to this theory, the stress $\sigma(t)$ resulting at any time $t > 0$ from an imposed history of strain $\varepsilon(t)$ is given by Boltzmann's equation (Nowacki 1965):

$$\sigma(t) = \int_0^t R(t,\tau)\, \dot{\varepsilon}(\tau)\, d\tau + \sum_i H(t - t_i)\, R(t, t_i)\, \Delta\varepsilon(t_i), \qquad (3)$$

where $\dot{\varepsilon}(t)$ is the time derivative of the strain loading function; $\varepsilon(t)$, $\Delta\varepsilon(t_i)$ denotes a set of strain discontinuities occurring at specified times t_i and $H(t)$ corresponds to the Heavyside function.

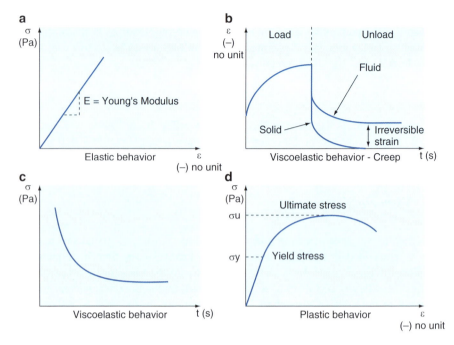

Fig. 2 Different kinds of biofilms behaviour. (**a**) Elastic: when a stress is applied the material immediately "stretches" in proportion to the magnitude of the stress. If the material is not stressed to failure (i.e. breaking) when the stress is removed the material immediately contracts to its original dimensions. (**b**) Viscoelastic under applied stress: the strain response of a viscoelastic material to an applied stress. When a stress is applied first there is an immediate elastic response and then the material slowly stretches over time. (**c**) Viscoelastic under applied strain: when a viscoelastic material is stretched the initial imparted stress dissipates over time. (**d**) Plastic behaviour: when the material is stressed it behaves elastically until the yield point is reached when the stress–strain relationship becomes non-linear

In this fundamental expression, $R(t, t_i)$ is called "relaxation function" which depends on the mechanical properties of the material and on the period of time $t - t_i$ elapsed since the application of a constant imposed strain. Owing to the principles of thermodynamics applied to the mechanics of viscoelastic media (Mandel 1958), this function can be developed as a sum of decaying exponential terms in the form of a Dirichlet's series with t_0 the time at which a constant strain is imposed:

$$R(t, t_0) = \sum_{i=0}^{r} E_i \, e^{-\alpha_i (t-t_0)}, \qquad (4)$$

where E_i and α_i are material parameters. This relationship can be represented by an analogous generalized Maxwell's chain with $r + 1$ branches, where branches consist of a spring (elastic) connected to a dashpot (viscous). $\alpha_i^{-1} = \eta_i/E_i$ is called relaxation time of branch i. Note that in the case of solid materials, the relaxation time of branch 0 is equal to 0.

For further reference, viscoelastic behaviour in biofilms has been highlighted in (Stoodley et al. 1999a, b, 2002; Cense et al. 2006a, b; Hohne et al. 2009; Lau et al. 2009). An experimentally derived creep curve for a *Staphylococcus aureus* biofilm is shown in Fig. 3 exhibiting classical viscoelastic features. The response was

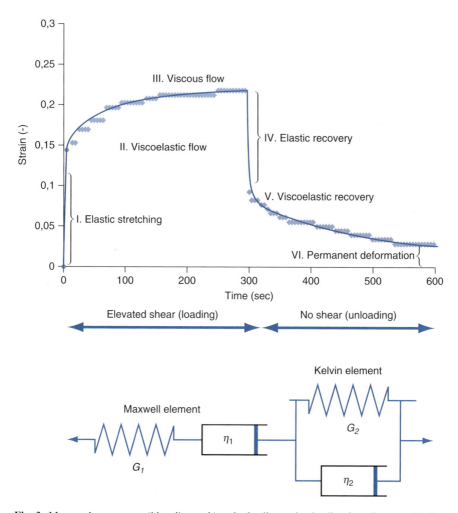

Fig. 3 Measured creep curve (*blue diamonds*) under loading and unloading for a *S. aureus* biofilm grown in a flow cell. The response was modelled (*solid blue line*) using a *Burger spring* (elastic constants) and *dashpot* (viscous constants) model shown underneath. When the shear was elevated (loading), there was an immediate elastic response (I) represented in the model as G_1. This was followed by a time-dependent viscoelastic response (II) represented as a second elastic constant G_2 and viscosity constants η_1 and η_2. Once the spring G_2 was "fully extended" the biofilm continued to deform by viscous flow alone (III), represented by viscosity η_1. When the load was removed (i.e. flow was turned off), there was an immediate elastic recoil (IV) represented by G_1. Finally, there was a viscoelastic recovery (V) governed by the Kelvin element in the model. The permanent deformation (VI) was due to unrecovered fluid flow suggesting that the biofilm behaved as a viscoelastic fluid

modeled using a Burger's model of springs (elastic constants) and dashpots (representing viscosity), see caption for details. In this case, the Burger model incorporates both reversible behaviour due to the springs in the model and irreversible behaviour due to the in series dashpot in the model. In this case, the load was induced through hydrodynamic shear in a flow cell, although spinning disk rheometry testing showed similar mechanical behaviour.

In some cases, it is convenient to use dynamic tests in order to analyse viscoelasticity. Viscoelasticity in the linear range can be analysed using dynamic mechanical tests by applying a small oscillatory strain on the biofilm and measure the resulting stress, or vice versa. In purely elastic materials, the stress and strain are exactly in phase since an applied stress results in an immediate deformation (i.e. strain). However, if a material also exhibits viscous behaviour there will be a delay between the applied stress and the induced strain resulting in a phase shift between the sinusoidal curves. The shear modulus (G) represents the relationship between the stress and the strain. It can be decomposed in a complex expression as follows:

$$G = G' + iG'',$$

where $i^2 = -1$. G' is the storage modulus and G'' is the loss modulus:

$$G' = \frac{\sigma_0}{\varepsilon_0} * \cos \delta, \qquad (5)$$

$$G'' = \frac{\sigma_0}{\varepsilon_0} * \sin \delta, \qquad (6)$$

and where σ_0 and ε_0 are the amplitudes of stress and strain and δ is the phase shift between them. G' enables us to quantify the elastic behaviour of the biofilm and G'' the viscosity of the biofilm.

2.3 Plasticity

In some instances, biofilm testing exhibits an irreversible feature, corresponding to permanent strain. In this case, the plasticity framework can be used. Note that the study of plastic behaviour of biofilms has been seldom addressed in the literature. In most cases, an elastic behaviour precedes the plastic behaviour (see Fig. 2d). The yield strength can be defined where the plastic behaviour starts (see Fig. 2d). This yield strength is denoted σ_Y. This yield strength must be then defined. The most popular criterion can be used, called "Von Mises criterion". For this purpose, an equivalent stress is calculated as follows:

$$\sigma_{EQ} = \frac{1}{2}[(\sigma_{xx} - \sigma_{yy})^2 + (\sigma_{zz} - \sigma_{yy})^2 + (\sigma_{xx} - \sigma_{zz})^2 + 6(\sigma_{xy} + \sigma_{xz} + \sigma_{yz})^2]. \qquad (7)$$

Then, the plasticity occurs when $\sigma_{EQ} > \sigma_Y$. Many plasticity models are described in the literature (i.e., Chaboche 2008). In the case of perfect-plastic behaviour, the evolution of the plastic strain ε^P can be defined as follows:

$$d\varepsilon^P = \frac{3}{2} d\lambda \frac{\sigma^D}{\sigma_{EQ}}, \tag{8}$$

where σ^D corresponds to the deviatory part of the stress tensor and λ denotes the plastic multiplicator. When the stress increases, the material reaches a point where it finally fails (see Fig. 2d). This point corresponds to the "tensile strength" of the material in the case of a tensile test. The term "ultimate strength" is also used in the literature. Plastic behaviour was discussed in (Körstgens et al. 2001a, b) and the yield strength and the ultimate strength have been evaluated in Körstgens et al. (2001a, b), Aggarwal and Hozalski (2010), however, few studies have been designed specifically to study plasticity.

2.4 Failure

How, and when, a biofilm fails is arguably the most important mechanical parameter from a practical perspective and must be clearly defined. Two types of failure, cohesive and adhesive, are commonly described based on observation. Cohesive failure occurs within the biofilm and leads to the detachment of parts of the biofilm, while some underlying biofilm or bacterial cells remain attached. Cohesive failure occurs when the loading is higher than the ultimate strength of the biofilm. In the case of fluid flow, the load is caused by a combination of skin friction, pressure drag and lift. This is a failure caused by the intrinsic properties of the material. In biofilms, the situation is complicated by the fact that biofilms are not homogeneous materials and can be mechanically stronger at some locations than others. Also, because of the heterogeneous structure of biofilms loads are not evenly distributed thus failure can be site specific. For example, a two-dimensional fluid–structure interaction model which coupled fluid and biofilm interactions utilized by Taherzadeh et al. (2010) predicted that biofilm streamers would fail where stresses were concentrated, at the location where the streamer was attached to the upstream cell cluster, or at a bend in the streamer tail as it waved in the flow. Another complication in measuring biofilm mechanical properties is that biofilms are not static entities but are alive and can grow, respond and adapt to mechanical shear (and as such can be considered as "self-repairing" materials) so that tests to failure should be conducted over relatively short time scales.

The second type of failure is interfacial, or adhesive, fracture. It occurs at the interface of the biofilm and the surface on which the biofilm grows. Generally, adhesive failure leads to the detachment (or sloughing) of large pieces, or even the whole biofilm, and can therefore be more desirable for biofilm removal strategies than localized cohesive failures, which can leave underlying biofilm to rapidly regrow. Adhesive failure depends on the property of the interface between the biofilm and the surface. Several models have been developed in the literature to evaluate this de-adhesion in a mechanical point of view (Mignot et al. 1980; Xu and Needleman 1994), for example. The most popular model used to simulate interfacial decohesion is the cohesive zone model (CZM). A review of CZM is available in (Xu and Needleman 1994). Briefly, the CZM has been used to simulate fracture as a gradual phenomenon. The separation of two elements takes place across an extended crack "tip", or cohesive zone. This zone increases proportionally to cohesive tractions (Ortiz and Pandolfi 1999). This type of model can be useful to simulate interfacial failure in biofilms.

3 Methods for Mechanically Testing Biofilms

The framework for describing the mechanical properties of biofilms has been presented as elasticity, viscoelasticity, plasticity and strength. Our intent is to present in the following section the main methods which enable us to characterize the mechanical behaviour of biofilms. The identified mechanical parameters and the interest of each method are presented.

3.1 Hydrodynamic Loading

A variety of methods have been used to estimate mechanical properties of bacterial biofilms. In this section, we discuss methods using hydrodynamic loading. To impose a hydrodynamic load, there are two possibilities. The first is to impose a movement to the bacterial biofilm (an imposed rotation of the biofilm for example). The second is to impose a hydrodynamic flow by controlling the speed of the flow. The following section aims at distinguishing methods using hydrodynamic loading following these two possibilities.

3.2 Imposed Hydrodynamics Flow

3.2.1 Flow-Cell

Stoodley et al. (1999a, b) have developed a method for conducting simple stress–strain and creep tests on mixed and pure culture biofilms in situ by observing the structural deformations caused by changes in hydrodynamic shear stress. The principle of the flow cell method is to grow a bacterial biofilm under steady flow and then to impose a hydrodynamic "load" by changing the flow. The flow cell method is used with a microscopy to observe biofilm structural deformation and detachment. The resulting biofilms were heterogeneous and consisted of filamentous streamers that were readily deformed by changing the wall shear stress. The flow cells have the advantage that they are relatively simple and cheap to operate and can be used to perform simple stress–strain and creep tests on attached biofilms by using the flow rate to vary the shear stress. In Fig. 4, we can see the experimental set up of the flow cell method. The biofilm is grown in a flow cell. The system is composed of two parts: one for the observation of the displacements of the biofilm with digital video imaging controlled by a computer, and the second part is a recirculation loop attached to a mixing chamber, which has a waste overflow. A pump is used upstream of the biofilm to regulate fluid flow. Figure 5 shows a biofilm grown in a flow cell with high flow.

Several articles used the method developed by Stoodley (Klapper et al. 2002; Aravas and Laspidou 2008; Mathias and Stoodley 2009). Young's modulus has been approximated using this technique. Viscoelastic effects have been documented (Stoodley et al. 1999a, 2002; Mathias and Stoodley 2009) and some

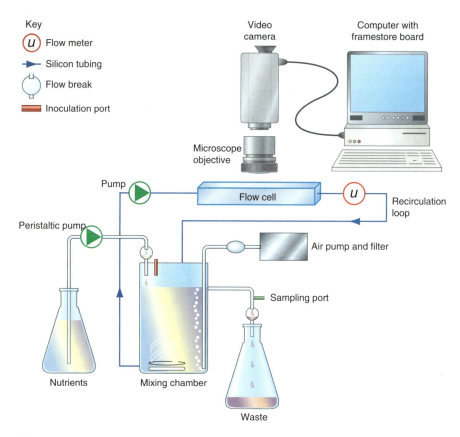

Fig. 4 Experimental set-up for growing and monitoring biofilms during transient changes in fluid shear stress

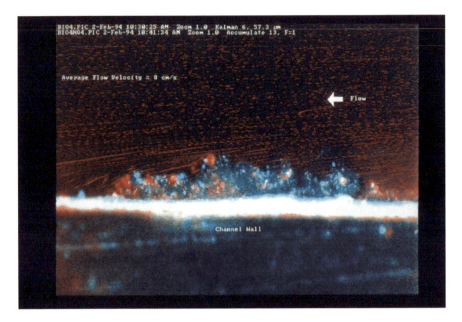

Fig. 5 Confocal side view of a biofilm using particle image velocimetry to reveal the fluid pathlines around the biofilm. The average flow velocity was 8 cm s^{-1}. The flow over and around the biofilm imparts a shear stress over the surface, downstream drag and the aerofoil effect creates lift

strength values have been estimated (see Table 1). Mathias and Stoodley (2009) used Digital Images Correlation to characterize the mechanical behaviour of biofilms in response to wall shear stress using digital video micrographs taken from biofilm flow cells (Fig. 6). DIC is a more sophisticated use of digital microscopy, which has been developed by materials scientists to quantify material behaviour by computing the displacement field in response to an applied force (Sutton et al. 1983). While this method is useful for precisely measuring structural deformations in real time, the deformations could only be related to the average wall shear stress since precise measurements of local forces acting on the biofilm were not possible. However, coupled fluids–solids models such as that described by Taherzadeh et al. (2010) might be used to calculate the predicted local shear stresses and drag forces from biofilm structure, flow cell dimensions and bulk fluid flow rate.

3.2.2 Flow Dynamics Gauging

Fluid dynamic gauging (FDG) is a technique for the measurement of the thickness of soft deposit layers on solid surfaces immersed in a liquid environment (Chew et al. 2004a, b). Möhle et al. (2007) used this method to measure the mechanical strength of biofilms by applying a shear stress by aspirating fluid from above the

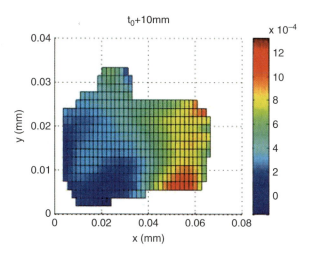

Fig. 6 Example of displacement field over a biofilm cell cluster calculated with DIC. The magnitude of longitudinal strain caused by increasing the flow is indicated by the heat map. The greatest strains occurred at the downstream part of the biofilm (to the right)

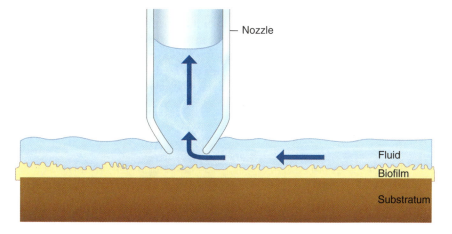

Fig. 7 Fluid dynamic gauging experiment: the fluid flows over the biofilm and is sucked by the gauge creating a shear stress

biofilm (Fig. 7). The adhesive strength was estimated by monitoring when the biofilm was completely removed from the substratum. Moreover, the FDG offers the possibility to calculate the wall shear stress in the rotating disk reactor. Möhle et al. (2007) have shown that the biofilm thickness was dependent on substrate concentration and shear stress. The main advantage of this method is that it is easier to perform a test than with the flow cell technique. This method also provides the advantage that forces acting on the biofilm derive from fluid flow tangential to substratum surface. This is rather typical for many technical as well as natural biofilm systems. FDG has application in detecting the type of fouling of filtration membranes.

3.3 Imposed Displacement on Bacterial Biofilm

3.3.1 Rotating Disk

The rotating disk biofilm reactor (RDBR) is a cylindrical reactor in which a disk rotates in the culture medium (Vinogradov et al. 2004) (Fig. 8). The rotational speed of the disk is controlled by a magnetic stirrer. In addition to imparting a shear stress on the biofilm the rotation also ensures that the reactor is completely mixed. Coupons or rheometer disks can be held within the rotating disk to grow replicate biofilms. Biofilms grown in the RDBR can be removed for testing with techniques such as FDG.(Möhle et al. 2007) or spinning disk rheometry.

3.3.2 Couette-Taylor

The Couette–Taylor reactor is composed of two concentric cylinders, the inner cylinder is fixed and the outer cylinder is rotated by a motor drive. The measurement of torque required to prevent rotation of the inner cylinder under the effect of the viscosity force of the fluid contained between two cylinders can be used to measure drag. Several teams have adapted the Couette–Taylor reactor to assess mechanical parameters of bacterial biofilms. This reactor can used to evaluate the influence of fluid dynamic shear on the extent of biofilm and mode of detachment of bacteria from

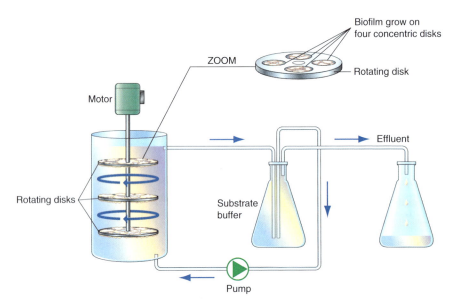

Fig. 8 Rotating disk technique: the disks are arranged on the axis of the rotating motor, which creates a shear force on biofilms developed on the surface. There is a recirculation loop with a substrate bluffer, which feeds the bioreactor

biofilms. By changing the rotation speed of the inner cylinder the wall shear stress can be varied. Couette–Taylor type reactors provide a constant wall shear stress distribution on surfaces and are suitable for the cultivation of biofilms in turbulent flows (Fig. 9). Recently, (Willcock et al. 2000 and Rickard et al. 2004) developed a concentric cylinder reactor, consisting of four cylindrical sections contained within concentric chambers, thus creating a variety of wall shear stresses in the same reactor, since the rotating speeds of the surface are proportional to their radius. This reactor allows the simultaneous generation of different shear rates on the same inoculating population (Willcock et al. 2000) but the four chambers were independently fed. In conclusion, with these types of reactors, the shear modulus and the cohesion strength can be calculated. The detachment phenomenon and the potential biodiversity impact can also be investigated (Coufort et al. 2007; Rochex et al. 2008; Derlon et al. 2008).

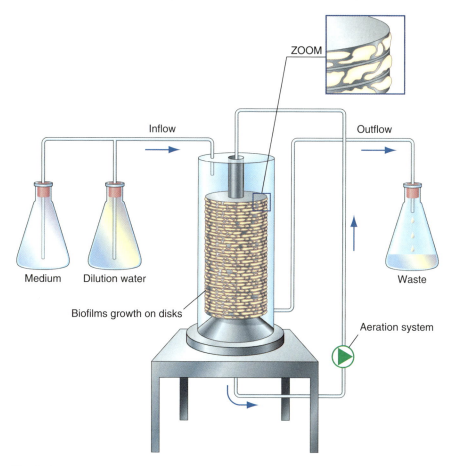

Fig. 9 Couette–Taylor reactor. The experimental set-up with a recirculation loop air feed and a medium reservoir with a pump used to feed the bioreactor on the *left*. A zoomed schematic of the two cylinders is shown on the *right*. The internal cylinder is fixed and the rotation of the external cylinder creates the shear stress. The *arrows* indicate the flow direction of the air and dilution water

Table 2 Summary of reported biofilm mechanical properties with hydrodynamics loads techniques

Parameter measured	Method used	Range of value	References
Young's modulus	– Flow cell	17–353 Pa	Stoodley et al. (1999a, b), Mathias and Stoodley (2009), Dunsmore et al. (2002)
Shear modulus	– Flow cell	1.1–27 Pa	Stoodley et al. (1999a, b), Mathias and Stoodley (2009), Towler et al. (2003)
	– Rotating disk	0.2–24 Pa	
Cohesive shear strength	– Couette–Taylor reactor	2–13 Pa	
	– Rotating disk and fluid dynamic gauging	6–7 Pa	Möhle et al. (2007), Coufort et al. (2007)

Some of the main results of mechanical properties measured using various methods of hydrodynamic loading are shown in Table 2.

4 Method Using a Direct Applied Force

In this section, we discuss methods which test biofilms by using a directly applied force. We broadly classify these techniques in three categories. The differences are due to the type of effort that is sustained by the bacterial biofilm. Indeed, with a direct applied force, the bacterial biofilm can be subjected to a tensile force, a compressive force or a shear force.

4.1 Tensile Testing

4.1.1 Centrifugation

Ohashi and Harada (1994, 1996) grew bacterial biofilms on circular plates in a traditional biofilm reactor. Then, the plates were perpendicularly fixed on a rotary table. The detachment of the biofilm was investigated by imposing various magnitudes of centrifugation forces, which generated a tensile force. The adhesion strength was defined as the force required to resist the tensile test stress immediately before failure (detachment). Data from these methods are difficult to interpret because the applied forces depend on the mechanism (rotary table and biofilm-attached plates) and geometry of the separation events, which were not evaluated. Moreover, biofilm strength could be obtained as a range but not as a single quantity.

4.1.2 Tensile Test Device Using Biofilm-Attached Tubes

This technique consists in performing a tensile test on pair of tubes (Ohashi et al. 1999). The tubes were placed one on top of the other and the bacterial biofilm developed at the interface of the tubes. Then, each pair of tubes was tested in tension by pulling them apart. The lower tube was pin-fixed to a supporting rod and the upper tube was pulled by until tubes separated. This technique was used primarily to investigate biofilm detachment; however, it could be readily adapted so that the displacements are registered by a video camera and the tensile force is measured, thus allowing the stress–strain curve to be constructed to determine parameters such as Young's modulus, and tensile strength.

4.1.3 Microcantilever Technique

The microcantilever method is a micromechanical technique used to measure the tensile strength of bacterial aggregates (Poppele and Hozalski 2003). This method is based on a method described by Yeung and Pelton (1996) for estimating the tensile strength of abiotic flocs and fractal dimensions. The method consists in picking up single floc particles and pulling them apart using glass suction pipettes, during which the required rupture forces is measured. Poppele introduced a microcantilever technique by which a tensile test is performed on a biofilm fragment immobilized by suction between a straight micropipette and a cantilevered micropipette (Fig. 10). Microbial aggregates were tested by moving the microscope stage and attached cantilever away from the capture pipette. The increasing deflection of the cantilever produces an increasing force that continues until either the sample separates or the suction between the sample and one of the pipettes is broken. The trials were recorded with a microscope equipped with a video capture device. Deflection of the cantilever arm was determined by measuring the distance the cantilever moved as the aggregate

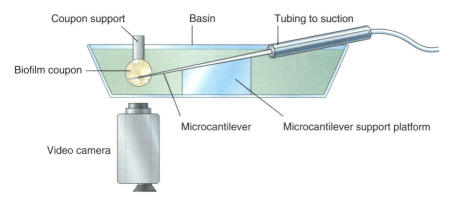

Fig. 10 Microcantilever device, the biofilm grows on the coupon, thanks to the suction transmitted in the microcantilever by the suction tube, we can measure the failure strain. The camera allows us to follow the displacement of the biofilm during the test

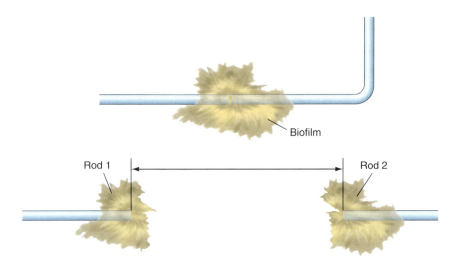

Fig. 11 Microcantilever technique separation of floc redrawn after Poppele and Hozalski (2003). *Top panel*: frame immediately prior to separation. The floc is fixed by the capture pipette (*left*) and is under stress from the force applied through the cantilever. *Bottom panel*: floc and cantilever after separation

separated. The stress applied to the aggregate was determined by dividing the applied force by the area over which the separation occurred.

The separation area was estimated by assuming a circular separation area pattern ith a diameter equal to the observed width of the floc at the point of separation. The manipulation is showed Fig. 11.

The method can be used to determine the strength of biofilm fragments sampled from any system. The main drawback of this method is that the tests are conducted on fragments taken from bacterial biofilms. The tests are not performed with intact biofilms and thus the location and orientation of the fragment in the biofilm are unknown.

To solve this problem, Aggarwal et al. (2009) proposed a modification of the microcantilever method that permits the testing of mechanical properties of intact biofilms. An acrylic platform is built to conduct the strength testing under a microscope. The platform is designed to keep the biofilm sample disk submerged

4.2 Compression Testing

4.2.1 Uniaxial Compressive Test

As for the tensile tests, compression tests can also be used to measure mechanical properties of the biofilm. The compression tests are usually conducted using indenters but can also be performed using rheometers, which are capable of measuring force as a function of gap thickness as biofilm is "squeezed" between

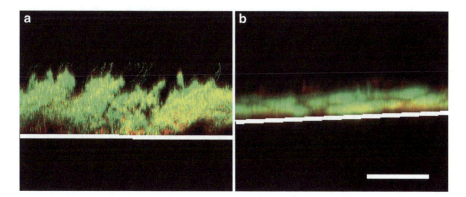

Fig. 12 Confocal side view of a biofilm formed from the dental pathogen *Streptococcus mutans* (**a**) before and (**b**) after 10 min compression with a force of 0.1 N. The biofilm was highly compressible. Image taken from Vinogradov et al. (2004) reprinted with permission from Cambridge University Press

two plates (Fig. 12). Uniaxial compression experiments have been used to obtain the stress–strain curve and consequently determine the "apparent" Young's modulus, the shear modulus or the yield strength (Körstgens et al. 2001a, b). The term "apparent" is used due to the fact that the biofilm is a viscoelastic material. This means that during the timescale of the deformation, some relaxation processes take place, which are based on conformational changes of polymer segments, thus the slower the indentation test the greater the contribution of viscous flow to the deformation. The advantage of this technique is that it is relatively quickly and easy to perform.

4.2.2 Micro-indentation

To perform a compression test, the microindentation method (also called microindentation hardness test) can be used to determine the hardness of a material as its resistance to deformation. The principle of this method is to impress an indenter of specific geometry into the surface of the test specimen using a known applied force (Fig. 13) and measuring the corresponding force on the biofilm. To adapt this method for use with bacterial biofilms, biofilms were grown on a round glass coupon, positioned in a cell chamber of the microindentation device and compression tests were carried out by pressing the indenter into the sample. Relaxation tests can also be performing with this technique (Cense et al. 2006a, b). However, a general concern with uniaxial testing is that it is not necessarily the most relevant loading regime which biofilms experience in the real world. In many cases, particularly in flow situations, shear is arguably the most relevant stress. Nevertheless, normal testing (compression or tensile) can be useful in determining the relative influence of particulates and solutes on biofilm mechanical properties.

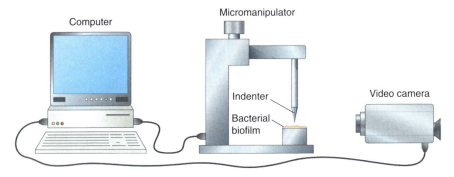

Fig. 13 Micro-indentation technique. The micro-indenter compresses the biofilm and the video camera helps visualize the displacement of biofilm to produce a force displacement curve

4.3 Shear Tests

4.3.1 Spinning Disk Rheometry

Rheometers are usually used to measure the mechanical properties of viscous liquids and gels. Conventionally, a few millilitres of sample is placed on a bottom plate or in a cup and a cone or flat disk is brought down on the sample such that under ideal conditions the sample completely fills the gap. The disk can then be rotated at varying speeds in the same direction or oscillated back or forth at a controlled or measured rate and displacement. In some rheometers, there the displacement (strain) is controlled and the shear stress measured, while in others the shear stress is controlled and the resulting strain is measured. Creep and relaxation shear tests can be performed, respectively, by (1) applying a shear stress and monitoring the strain (i.e. rotation of the disk) over time, or (2) by applying a strain and monitoring the dissipation of stress over time. Responses can also be measured when the sample is unloaded. Dynamic testing can also be performed by oscillating the rotation and measuring the viscoelastic parameters. While rheometry is useful for measuring biofilm mechanical properties, there are practical problems since biofilms are very thin and it is desirable to perform measurements in the attached state. In our lab, we developed a technique to use a TA Instruments AR1000 rheometer to measure viscoelasticity in biofilms by growing the biofilms directly on flat rheometer plates and performing creep and compression tests in submersion (Vinogradov et al. 2004; Towler et al. 2003). Because the thickness of the biofilms was highly variable, we did not test with a defined gap thickness, as is conventional, but rather used a defined normal compressive force of 0.1 N and let the gap thickness vary over the course of the test. We also built a reservoir on the bottom plate of the rheometer using modelling clay however, commercially available geometries for measuring in submersion or high humidity are available with most instruments.

4.3.2 Pulling Testing Shear

Chen et al. (1998) designed a micromanipulation technique, which was similar to a scratch test in which they used a specially designed T-shaped probe to measure the adhesive strength of biofilms formed in pipes or on the surface of glass test studs (Chen et al. 2005) (Fig. 14). The principle is to pull a whole biofilm from the glass test stud surface using the T-shaped probe, thus creating a shear stress. The applied force on the probe for pulling the biofilm is recorded in a computer. Results showed that the adhesive strength was found to depend on the condition under which they were grown.

The main results of mechanical properties using a method with direct force applied are showed in Table 3.

5 Discussion, Perspectives

5.1 Identification of Mechanical Parameters

Over the last two decades, several creative techniques to assess mechanical properties of biofilms have been developed depending on the desired mechanical

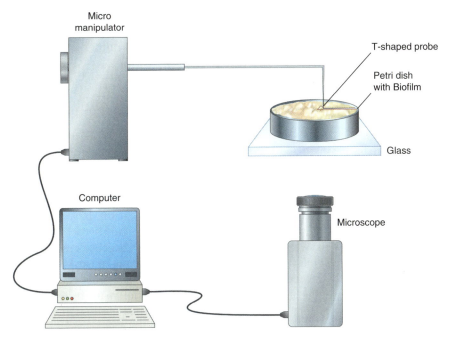

Fig. 14 Pulling test. The T-shaped probe pulls the biofilm developed on the surface of the tube. The microscope is used to quantify the displacement and the computer controls the force transmitted by the micro manipulator to pull the biofilm

Table 3 Summary of biofilm mechanical properties with direct applied force techniques

Parameter measured	Method used	Range of value	References
Adhesive strength	– Centrifugation – Micromanipulation technique	5–50 Pa 0.05–0.2 J m^{-2}	Ohashi and Harada (1994), Chen et al. (1998)
Tensile strength	– Centrifugation – Tensile test with tubes – Microcantilever method	0–8 Pa 500–1,000 Pa 61–5,842 Pa	Ohashi and Harada (1996), Ohashi et al. (1999), Aggarwal et al. (2009), Aggarwal and Hozalski (2010)
Young modulus	– Uniaxial compression – Microcantilever method	6,000–50,000 Pa 990–1,550 Pa	Körstgens et al. (2001a, b), Aggarwal et al. (2009), Aggarwal and Hozalski (2010)
Yield stress	– Uniaxial compression	900–2,000 Pa	Körstgens et al. (2001a, b)
Adhesion shear stress	– Micromanipulation technique	0.12–0.65 J m^{-2}	Chen et al. (2005)
Storage modulus Loss modulus	– Microindentation and confocal microscopy	1,000–8,000 Pa 5,000–10,000 Pa	Cense et al. (2006a, b)
Cohesive energy	– Atomic force microscopy	0.1–2.05 nJ μm^{-3}	Ahimou et al. (2007)
Failure strain	– Tensile test – Microcantilever method	150–320% 224–256%	Ohashi et al. (1999), Aggarwal and Hozalski (2010)
Effective shear modulus	– Disk rheometry	80–200 Pa	Vinogradov et al. (2004)

parameters to be measured, the desired loading regime and equipment and materials to hand. The characterization of the mechanical behaviour of bacterial biofilms remains challenging. While there appears to be consensus that biofilms are viscoelastic, there is a very wide range (many orders of magnitude) of reported elastic moduli. It is not clear what the main contributing factors are to this large variability but it is likely a combination of the bacteria strains used, the growth conditions, and the testing methods themselves. The assessment of the viscoelastic and plastic behaviour remains complicated because experiments on living, intact biofilms are very hard to conduct. Indeed, ideally the loading must be precisely controlled as well as the geometry of the biofilm. As a living entity, the biofilm might also change over the duration of testing, particularly if tests are carried out over time scales longer than a few minutes. Another difficulty lies in the fact that bacterial biofilms are heterogeneous materials and the behaviour of the biofilm may depend on different parameters such as bacteria density for example. Applying the full-field measurement method can help meet the challenge. Full-field measurement methods enable us to calculate local rigidities and to obtain a rigidity field. A first application to biofilms has been addressed recently (Mathias and Stoodley 2009) on 2D images from biofilms grown in flow cells. However, the response of the bacterial biofilm depends on the local thickness of the biofilm as well as the local wall shear stress. In future work, it should be possible to perform 3D measurement of bacterial biofilms using time-lapse 3D confocal microscopy to assess to the geometry of the biofilm and its 3D displacement in response to wall shear stress. Wall shear stresses acting on individual clusters could be better estimated using a pressure drop measurement rather than using calculations based on the flow rate through a clean tube; however, they would still remain averaged over the entire flow cell. Once the experimental fields are obtained, inverse identification methods, used by mechanical engineers, can be used. For this purpose, two techniques seem to be promising. The first consist in using a finite element model to identify the mechanical parameters. Indeed, we can vary the elastic moduli in the numerical model to match the numerical displacement fields with those from experiments to estimate mechanical properties. The second method consists in using the virtual field method (VFM) (Grédiac et al. 2006). This method is based on the choice of pertinent virtual displacement fields which enable us to simplify the mechanical equilibrium equations and to deduce the mechanical properties.

5.2 Biofilm Failure

Characterizing the failure of bacterial biofilms is of prime interest from a practical point of view and hence constitutes an important economic key issue. However, the failure of biofilm must be clearly defined. Indeed, when some pieces break off due to an applied stress other pieces might remain so, unlike a structural material, failure does not necessarily compromise the entire structure. Moreover, failure can be of benefit for biofilm populations in terms of improving overall mechanical stability (i.e. be shedding weaker parts, or optimizing structure for drag reduction) and for biological dispersal. The different types of fracture must be clearly distinguished.

Cohesive failure occurs within the biofilm matrix. It leads to the detachment of parts of the biofilm, while some cells or biofilm remains attached to the surface. This type of failure can be advantageous for the biofilm because the biofilm can rapidly grow back (a process termed regrowth), thus replacing weaker parts with stronger parts. The second type of fracture is interfacial or adhesive fracture. It occurs at the interface of the biofilm with a substratum. Generally, it leads to the detachment of the whole biofilm or segments of whole biofilm from a surface, and may therefore be considered undesirable for the biofilm. This type of failure is due to some interfacial mechanism, which bonds the base layer of cells to the substratum. In this case, interfacial models might be used to characterize the adhesion features of the biofilm.

Both of these types of failures can be provoked by wall shear stress. However, some mechanisms can increase the effect of the wall shear stress on failure. There are two main mechanisms. First, the failure can be due to fatigue through some type of hydrodynamic cycling. Fatigue is the progressive and localized structural damage that occurs when a material is subjected to cyclic loading. For example, a high value of cycles with a low value of the wall shear stress can lead to the failure of biofilm. This phenomenon has been seldom addressed in the literature but is likely an important factor in the study of biofilm failure. This failure mechanism might become more significant with ageing. Material ageing is commonly understood as changes of material properties with time, due to some chemical reaction for example. This leads to a decrease of the strength and biofilm failures may occur with a low value of the loading.

It is clear that many different phenomena play a role in biofilm failure and conventional models of material failure are not adequate for direct application to biofilms. Biofilms are heterogeneous and the theory of continuum mechanics may be of limited use in describing or predicting how biofilms respond to mechanical forces. Moreover, the material behaviour of the biofilm must be clearly identified before we can characterize failure or even define what we mean by mechanical failure when it comes to bacterial biofilms. The future of biofilm mechanics certainly lies in the development and the enhancement of new experimental and modelling techniques. It is also clear that the interest in biofilm mechanics is growing rapidly and researchers from various disciplines are turning their attention to it, and coming up with highly creative and innovative ways in modifying conventional methods or designing completely novel methods to study biofilm mechanics.

References

Aggarwal S, Hozalski RM (2010) Determination of biofilm mechanical properties from tensile tests performed using a micro-cantilever method, 1029–2454. Biofouling: J Bioadhesion Biofilm Res 26(4):479–486

Aggarwal S, Poppele EH, Hozalski RM (2009) Development and testing of a novel microcantilever technique for measuring the cohesive strength of intact biofilms. Biotechnol Bioeng 105(5):924–934

Ahimou F, Semmens MJ, Novak PJ, Haugstad G (2007) Biofilm cohesiveness measurement using a novel atomic force microscopy methodology. Appl Environ Microbiol 73(9):2897–2904

Aravas N, Laspidou CS (2008) On the calculation of the elastic modulus of a biofilm streamer. Biotechnol Bioeng 101(1):196–200

Barraud N, Hassett DJ, Hwang SH, Rice SA, Kjelleberg S, Webb JS (2006) Involvement of nitric oxide in biofilm dispersal of *Pseudomonas aeruginosa*. J Bacteriol 188(21):7344–7353

Bartowsky EJ, Henschke PA (2008) Acetic acid bacteria spoilage of bottled red wine. Int J Food Microbiol 125(1):60–70

Beech IB, Sunner JA, Hiraoka K (2005) Microbe-surface interactions in biofouling and biocorrosion processes. Int Microbiol 8(3):157–168

Bradshaw DJ, Marsh PD (1999) Use of continuous flow techniques in modeling dental plaque biofilms. In: Doyle RJ (ed) Methods in enzymology, vol 310, Biofilms., pp 279–296

Cense AW, Van Dongen MEH, Gottenbos B, Nuijs AM, Shulepov SY (2006a) Removal of biofilms by impinging water droplets. J Appl Phys 100(12):124701–124708

Cense AW, Peeters EAG, Gottenbos B, Baaijens FPT, Nuijs AM, van Dongen MEH (2006b) Mechanical properties and failure of Streptococcus mutans biofilms, studied using a microindentation device. J Microbiol Meth 67(3):463–472

Chaboche JL (2008) A review of some plasticity and viscoplasticity constitutive theories. Int J Plast 24(10):1642–1693

Chen MJ, Zhang Z, Bott TR (1998) Direct measurement of the adhesive strength of biofilms in pipes by micromanipulation. Biotechnol Tech 12(12):875–880

Chen MJ, Zhang Z, Bott TR (2005) Effects of operating conditions on the adhesive strength of Pseudomonas fluorescens biofilms in tubes. Colloids Surf B 43(2):61–71

Chew JYM, Paterson WR, Wilson DI (2004a) Fluid dynamic gauging for measuring the strength of soft deposits. J Food Eng 65(2):175–187

Chew JYM, Cardoso SSS, Paterson WR, Wilson DI (2004b) CFD studies of dynamic gauging. Chem Eng Sci 59(16):3381–3398

Coetser SE, Cloete TE (2005) Biofouling and biocorrosion in industrial water systems. Crit Rev Microbiol 31(4):213–232

Costerton JW, Lewandowski Z, Caldwell DE, Korber DR, Lappin-Scott HM (1995) Microbial biofilms. Annu Rev Microbiol 49:711–745

Couarraze G, Grossiord J (1991) Initiation à la rhéologie, 2nd ed, Lavoisier – Tec & Doc, Paris

Coufort C, Derlon N, Ochoa-Chaves J, Liné A, Paul E (2007) Cohesion and detachment in biofilm systems for different electron acceptor and donors. Water Sci Technol 55(8–9):421–428

Davies DG, Marques CN (2009) A fatty acid messenger is responsible for inducing dispersion in microbial biofilms. J Bacteriol 191(5):1393–1403

de Beer D, Stoodley P, Roe F, Lewandowski Z (1994) Effects of biofilm structures on oxygen distribution and mass transport. Biotech Bioeng 43:1131–1138

Derlon N, Masse´ A, Escudie´ R, Bernet N, Paul E (2008) Stratification in the cohesion of biofilms grown under various environmental conditions. Water Res 42(8–9):2102–2110

Dunsmore BC, Jacobsen A, Hall-Stoodley L, Bass CJ, Lappin-Scott HM, Stoodley P (2002) The influence of fluid shear on the structure and material properties of sulphate-reducing bacterial biofilms. J Ind Microbiol Biotechnol 29(6):347–353

Godon J-J, Zumstein E, Dabert P, Habouzit F, Moletta R (1997) Molecular microbial diversity of an anaerobic digestor as determined by small-subunit rDNA sequence analysis. Appl Environ Microbiol 63(7):2802–2813

Grédiac M, Pierron F, Avril S, Toussaint E (2006) The virtual fields method for extracting constitutive parameters from full-field measurements: a review. Strain Int J Exp Mech 42:233–253, DOI:dx.doi.org

Hohne DN, Younger JG, Solomon MJ (2009) Flexible microfluidic device for mechanical property characterization of soft viscoelastic solids such as bacterial biofilms. Langmuir 25 (13):7743–7751

Klapper I, Rupp CJ, Cargo R, Purvedorj B, Stoodley P (2002) Viscoelastic fluid description of bacterial biofilm material properties. Biotechnol Bioeng 80(3):289–296

Körstgens V, Flemming H-C, Wingender J, Borchard W (2001a) Uniaxial compression measurement device for investigation of the mechanical stability of biofilms. J Microbiol Meth 46 (1):9–17

Körstgens V, Flemming H-C, Wingender J, Borchard W (2001b) Influence of calcium ions on the mechanical properties of a model biofilm of mucoid *Pseudomonas aeruginosa*. Water Sci Technol 43(6):49–57

Lau PCY, Dutcher JR, Beveridge TJ, Lam JS (2009) Absolute quantitation of bacterial biofilm adhesion and viscoelasticity by microbead force spectroscopy. Biophys J 96(7):2935–2948

Lemaitre J, Chaboche JL (1988) Mécanique des matériaux solides, Dunod, 2ème édition

Liao Q, Wang Y-J, Wang Y-Z, Zhu X, Tian X, Li J (2010) Formation and hydrogen production of photosynthetic bacterial biofilm under various illumination conditions. Bioresour Technol 101 (14):5315–5324

Mandel J (1958) Théorie générale de la viscoélasticité linéaire. Cahier du Groupe Français de Rhéologie 3(4):21–35

Mathias JD, Stoodley P (2009) Applying the digital image correlation method to estimate the mechanical properties of bacterial biofilms subjected to a wall shear stress. Biofouling 25(8):695–703

Mignot F, Puel JP, Suquet PM (1980) Bifurcation and homogenization. Int J Eng Sci 18 (2):409–414

Möhle RB, Langemann T, Haesner M, Augustin W, Scholl S, Neu TR, Hempel DC, Horn H (2007) Structure and shear strength of microbial biofilms as determined with confocal laser scanning microscopy and fluid dynamic gauging using a novel rotating disc biofilm reactor. Biotechnol Bioeng 98(4):747–755

Nowacki W (1965) Théorie du fluage. Eyrolles Ed, Paris, 219p

Ohashi A, Harada H (1994) Adhesion strength of biofilm developed in an attached growth reactor. Water Sci Technol 29(10–11):281–288

Ohashi A, Harada H (1996) A novel concept for evaluation of biofilm adhesion strength by applying tensile force and shear force. Water Sci Technol 34(5–6):201–211

Ohashi A, Koyama T, Syutsubo K, Harada H (1999) A novel method for evaluation of biofilm tensile strength resisting erosion. Water Sci Technol 39(7):261–268

Ortiz M, Pandolfi A (1999) Finite-deformation irreversible cohesive elements for three-dimensional crack-propagation analysis. Int J Numer Methods Eng 44:1267–1282

Picologlou BF, Zelver N, Characklis WG (1980) Biofilm growth and hydraulic performance. J Hydraulics Division Am Soc Civ Eng 106(HY5):733–746

Poppele EH, Hozalski RM (2003) Micro-cantilever method for measuring the tensile strength of biofilms and microbial flocs. J Microbiol Meth 55(3):607–615

Rickard AH, McBain AJ, Stead AT, Gilbert P (2004) Shear rate moderates community diversity in freshwater biofilms. Appl Environ Microbiol 70(12):7426–7435

Rochex A, Godon J-J, Bernet N, Escudié R (2008) Role of shear stress on composition, diversity and dynamics of biofilm bacterial communities. Water Res 42(20):4915–4922

Socransky SS, Haffajee AD (2002) Dental biofilms: difficult therapeutic targets. Periodontol 2000 28(1):12–55

Stoodley P, Lewandowski Z, Boyle JD, Lappin-Scott HM (1999a) Structural deformation of bacterial biofilms caused by short-term fluctuations in fluid shear: an in situ investigation of biofilm rheology. Biotechnol Bioeng 65(1):83–92

Stoodley P, Lewandowski Z, Boyle JD, Lappin-Scott HM (1999b) The formation of migratory ripples in a mixed species bacterial biofilm growing in turbulent flow. Environ Microbiol 1(5):447–455

Stoodley P, Wilson S, Hall-Stoodley L, Boyle JD, Lappin-Scott HM, Costerton JW (2001a) Growth and detachment of cell clusters from mature mixed-species biofilms. Appl Environ Microbiol 67(12):5608–5613

Stoodley P, Cargo R, Rupp CJ, Wilson S, Klapper I (2002) Biofilm material properties as related to shear-induced deformation and detachment phenomena. J Ind Microbiol Biotechnol 29(6):361–367

Sutton M, Wolters W, Perters W, Ranson W, McNeill S (1983) Determination of displacements using an improved digital correlation method. Image Vis Comput 1(3):133–139

Taherzadeh D, Picioreanu C, Küttler U, Simone A, Wall WA, Horn H (2010) Computational study of the drag and oscillatory movement of biofilm streamers in fast flows. Biotechnol Bioeng 105(3):600–610

Towler BW, Rupp CJ, Cunningham ALB, Stoodley P (2003) Viscoelastic properties of a mixed culture biofilm from rheometer creep analysis. Biofouling 19(5):279–285

Vinogradov AM, Winston M, Rupp CJ, Stoodley P (2004) Rheology of biofilms formed from the dental plaque pathogen *Streptococcus mutans*. Biofilms 1:49–56

Wagner M, Loy A (2002) Bacterial community composition and function in sewage treatment systems. Curr Opin Biotechnol 13(3):218–227

Willcock L, Gilbert P, Holah J, Wirtanen G, Allison DG (2000) A new technique for the performance evaluation of clean-inplace disinfection of biofilms. J Ind Microbiol Biotechnol 25(5):235–241

Woolard CR, Irvine RL (1994) Biological treatment of hypersaline wastewater by a biofilm of halophilic bacteria. Water Environ Res 66(3):230–235

Xu X-P, Needleman A (1994) Numerical simulations of fast crack growth in brittle solids. J Mech Phys Solids 42:1397–1434

Yeung AKC, Pelton R (1996) Micromechanics: a new approach to studying the strength and breakup of flocs. J Colloid Interface Sci 184(2):579–585

Wound Healing by an Anti-Staphylococcal Biofilm Approach

Randall D. Wolcott, Florencia Lopez-Leban, Madanahally Divakar Kiran, and Naomi Balaban

Abstract In this chapter, we discuss the use of anti-biofilm agents as critical players for achieving healing of chronic wounds. These types of wounds are typically persistent due to the presence of biofilms that become a major barrier toward healing.

RNAIII inhibiting peptide (RIP) is the main anti-biofilm agent described in this chapter. By functional genomics (microarray analysis), RIP was shown in *Staphylococcus aureus* to downregulate the expression of genes involved in biofilm formation. Such genes participate both in neutralizing acidic pH and enabling survival in an anaerobic environment. RIP was also shown to downregulate the production of multiple toxins, yielding nonpathogenic bacteria.

The implementation of the general use of RIP in managing complex wounds, along with other anti-biofilm strategies, dramatically changed the outcomes for many of the most desperate wounds. The amazing improvement in wound healing trajectory is best explained by the suppression of staphylococcal biofilms.

Hopefully, the evidence gathered will help establish changes in the way clinicians tackle chronic persistent wounds. Instead of amputation, the concept of healing might become a better alternative with the addition of anti-biofilm agents to the existing protocols.

R.D. Wolcott (✉)
Southwest Regional Wound Care Center, Medical Biofilm Research Institute, 2002 Oxford Ave, Lubbock, TX 79410, USA
e-mail: randy@randallwolcott.com

F. Lopez-Leban • M.D. Kiran
Tufts University, Cummings School of Veterinary Medicine, Department of Biomedical Sciences, North Grafton, MA USA

N. Balaban (✉)
Tufts University, Cummings School of Veterinary Medicine, Department of Biomedical Sciences, North Grafton, MA USA
and
Yale University, Department of Chemical Engineering, 9 Hillhouse Avenue, New Haven, CT 06520-8286, USA
e-mail: naomi.balaban@yale.edu; naomibalaban@yahoo.com

1 Introduction

1.1 The Problem: Chronic Biofilm-Based, Nonhealing Wounds

Complex nonhealing wounds are often associated with the presence of a biofilm containing multiple bacterial species and cultured with staphylococci on at least one culture (Wolcott et al. 2009). In patients with diabetic neuropathy, for example, a small cut can often become a life-threatening wound due to the common underlying conditions of an immunocompromised host who also has bad perfusion. These situations usually lead to severely infected diabetic foot ulcers that generally become chronic and are very difficult to treat. These chronic wounds often do not respond to commercially available agents such as appropriate antibiotics, selective biocides, and advanced dressings. Eradicating bacterial biofilms from these types of chronically infected wounds through the use of anti-biofilm agents is key for their healing (Wolcott and Rhoads 2008). Otherwise, the limb is typically amputated, or the infection might aggressively spread, creating a high risk of death. Such infections, using conventional treatment modalities, now result in more than 100,000 limb amputations and killing tens of thousands of people a year in the USA alone (Ziegler-Graham et al. 2008). An APIC Guide 2010. Therefore, a new treating agent, one that would specifically target staphylococcal biofilm, is urgently needed.

The National Institutes of Health estimates that up to 80% of human infectious diseases are biofilm based (Costerton et al. 1999; James et al. 2008). However, biofilm-based infection management is very new to the medical community, and often comes with scrutiny (Amy 2008). Nevertheless, series of treatments that allow successful eradication of biofilm bacteria is already saving many limbs and lives. Under the novel standard of care that includes anti-biofilm strategies (see below), up to 91% of the wounds that would have been deemed "unhealable" and would have resulted in a major limb amputation are now healed. In addition, with the use of anti-biofilm strategies, the use of conventional antibiotics has dropped by 85%, indicating that the key to enhanced recoveries was the implementation of anti-biofilm treatments (see case histories below) (Wolcott and Rhoads 2008). The use of an anti-biofilm strategy is described below.

1.2 Staphylococci and Biofilms

Staphylococcus aureus and *S. epidermidis* are Gram-positive bacteria that are part of the normal flora of the skin but can become pathogenic and cause devastating diseases, which have led to 90,000 deaths per year in just the United States (Chang et al. 2006). Diseases caused by these bacteria vary from food poisoning to cellulitis, mastitis, arthritis, osteomyelitis, keratitis, pneumonia, endocarditis, device-associated and wound-associated infections, toxic shock syndrome, and sepsis. Staphylococci survive in the host and cause disease through the formation

Table 1 Some of the virulence factors produced by *Staphylococcus aureus*

Virulence factor	Activity in host tissue
Cell wall components	
Capsular polysaccharide	Anti-phagocytic
Cell surface proteins	
Protein A	Interacts with the Fc region of IgG
Fibrinogen, fibronectin, collagen-binding protein	Facilitate cell adhesion
Extraceullular toxins and enzymes	
Hemolysins ($\alpha,\beta,\delta,\gamma$)	Cytotoxic to tissue cells and leukocytes
P-V leukocidin	Destroys leukocytes
Toxic Shock syndrome toxin	Superantigen
Enterotoxins	Superantigen
Coagulase	Catalyzes conversion of fibrinogen to fibrin
Proteases	Degrade proteins
Nucleases	Cleave DNA and RNA
Staphylokinase	Converts plasminogen to fibrinolytic plasmin
Hyaluronidase	Degrades hylauronic acid
Siderophores	Capture iron from the host
Factors important in a biofilm	
Urease	Maintenance of pH
Arginine deiminase (ADI)	Use of arginine in anaerobic conditions as a source of energy

of biofilms and especially *S. aureus* by the production of numerous toxins, such as hemolysins, proteases, and lipases (Table 1) (Lowy 1998).

Many of the virulence factors are regulated by quorum-sensing mechanisms, where bacteria coordinate their activities. This coordinated behavior is acquired by cell- to-cell communication, or quorum-sensing, where bacteria secrete molecules that activate signal transduction pathways, leading to the expression of genes to enhance their ability to survive in a specific environment (Novick and Geisinger 2008). In staphylococci, quorum-sensing influences the formation of biofilms and production of multiple toxins that allow the bacteria to survive within the host and cause disease. Some of the molecules involved are the *agr* gene products, which include the autoinducing cyclical octapeptide (AIP), which activates the phosphorylation of the two component system AgrC/AgrA, which in turn activates the production of a regulatory mRNA molecule (RNAIII) that induces in *S. aureus* the production of numerous toxins while upregulating the expression of surface molecules such as Protein A (Dunman et al. 2001).

1.3 RNAIII Inhibiting Peptide: An Inhibitor of Quorum-Sensing and Biofilm Formation

Our approach is to suppress infections using RIP. RIP was initially discovered in the supernatants of a nonhemolytic staphylococcus spp (Balaban and Novick 1995) and has since been used in its synthetic amide form YSPWTNF-$_{NH2}$.

RIP inhibits staphylococcal colonization and toxin production. It has been shown to be extremely effective in preventing and treating drug-resistant staphylococcal infections in multiple experimental infection models. Those include the murine cellulitis, arthritis, wound and sepsis models, the rabbit osteomyelitis and keratitis models, the rat wound and device associated infection models, and the cow mastitis model. RIP has been shown to suppress infection caused by any staphylococcal strain tested, including various drug-resistant strains such as methicillin- or vancomycin-resistant *S. aureus* (MRSA, VISA) and *S. epidermidis* (MRSE, VISE) (Balaban et al. 1998, 2007, 2000, 2003a; Anguita-Alonso et al. 2007; Cirioni et al. 2003, 2006; Balaban and Stoodley 2005; Kiran et al. 2008a; Ghiselli et al. 2006; Giacometti et al. 2005, 2003; Dell'Acqua et al. 2004; Schierle et al. 2009; Lopez-Leban et al. 2010).

While the molecular mechanisms of RIP are not completely known, RIP was shown to repress *agr* expression and RNAIII production (Kiran et al. 2008b) and to repress *agr*-regulated toxin production in vitro (Vieira-da-Motta and Ribeiro 2001). RIP was also shown to suppress biofilm formation in an *agr*-independent manner, since RIP inhibited biofilm formation in *agr*-deficient strains (Novick and Geisinger 2008; Balaban et al. 2003b). Additionally, RIP was shown to repress TRAP phosphorylation. TRAP (Target of RNAIII Activating Protein) is a 167 amino acid residue protein whose phosphorylation is naturally activated by the autoinducer RNAIII activating protein (RAP) (Balaban et al. 2001). The sequence of TRAP is highly conserved among staphylococci (Gov et al. 2004) and has structural homologues in other Gram-positive bacteria, such as YhgC in bacilli. Both TRAP and YhgC have recently been shown to bind DNA and to protect it from stress. TRAP was also shown to protect DNA from spontaneous and adaptive (*agr*) mutations, and like *S. aureus* in the presence of RIP, YhgC– mutant *B. anthracis* were defective in their ability to form a biofilm (Kiran and Balaban 2009; Kiran et al. 2010). The observation that RIP suppresses both *agr* and TRAP function suggests that RIP may regulate *S. aureus* virulence using multiple independent molecular mechanisms, some through *agr* and some through TRAP. Although RIP does not have the properties of a conventional antibiotic, being neither bacteriocidal nor bacteriostatic (Kiran et al. 2008b), it does reduce the ability of the bacteria to survive within the host, allowing even the most meager immune responses to overcome an infection.

2 Results

2.1 *Clinical Cases*

The implementation of the general use of RIP in managing complex wounds, along with other anti-biofilm strategies, dramatically changed the outcomes for many of the most desperate wounds. The following cases were chosen to illustrate that wounds treated with RIP healed much better than would be expected by conventional techniques or other anti-biofilm approaches alone. The amazing improvement in wound healing trajectory is best explained by suppression of staphylococcal biofilms.

2.1.1 Case 1 (Fig. 1)

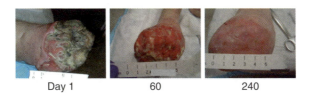

Day 1   60   240

Fig. 1 Case report

An 83-year-old male had a nonhealing wound, which included MRSA biofilm on the forefoot that produced osteomyelitis and a progressive dying back of the forefoot. The patient had failed to respond to previous commercially available treatments including Iodosorb, Acticoat, and Hydroferra Blue. He had also failed to respond to initial anti-biofilm efforts using lactoferrin and xylitol. On a compassionate use basis, this was one of our first cases to use RIP. The results were remarkable. Once RIP was added to the treatment protocol, the dying back stopped almost immediately, and several weeks later the patient had granulation tissue covering all the exposed bone that had previously been present. The patient went on to heal his wound in a total of 9 months. This was accomplished in a limb with critical limb ischemia, osteomyelitis, and severe peripheral neuropathy in an older man with longstanding diabetes. Our conclusion was that since lactoferrin and xylitol gel alone did not produce such dramatic positive outcomes, RIP is the most likely explanation for the dramatic improvement. Thereafter, subsequent cases were treated with lactoferrin and xylitol gel in conjunction with RIP (see cases below).

2.1.2 Case 2 (Fig. 2)

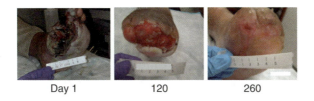

Day 1   120   260

Fig. 2 Case report

A 63-year-old patient presented with a severe diabetic foot ulcer. He started treatment with RIP, lactoferrin, and xylitol gel with frequent debridement, selective biocides, and antibiotics. Very quickly the necrotic tissue resolved, and the dying back ceased. Within weeks, the exposed bone was covered, and the wound showed contraction and slow but persistent healing. By 9 months, the patient's foot was completely healed, and he was able to return to his job.

2.1.3 Case 3 (Fig. 3)

Day 1   20   90

Fig. 3 Case report

A 41-year-old male was a juvenile onset diabetic with a renal transplant on immunosuppressives. The patient had peripheral arterial occlusive disease and peripheral neuropathy of his right foot. He had his left leg amputated approximately two years prior to this presentation of a surgical site infection. The patient was not a candidate for revascularization and had to remain on his immunosuppressants. Yet with the inclusion of RIP with concurrent use of frequent debridement, selective biocides, lactoferrin/xylitol, and antibiotics, the patient was able to heal his wound in 12 weeks.

2.1.4 Case 4 (Fig. 4)

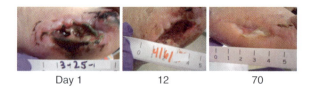

Day 1   12   70

Fig. 4 Case report

A 56-year-old female with a 20-year history of diabetes mellitus presented to her podiatrist with necrotic fourth and fifth toes. The patient had adequate perfusion and underwent a fourth and fifth toe resection. This surgical site dehisced, and there was significant necrosis; the patient was referred for comprehensive wound care. She had MRSA, and RIP was instituted on her first visit together with lactoferrin, xylitol gel, and Acticoat. The patient went on to heal in 8–10 weeks.

2.1.5 Case 5 (Fig. 5)

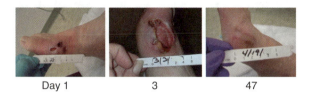

Day 1　　　　　　3　　　　　　47

Fig. 5 Case report

A 45-year-old male presented with a longstanding diabetic foot ulcer on the first metatarsal head region of his right foot. The patient had gone golfing and developed two new necrotic lesions. There was significant swelling. The patient was a diabetic but without significant peripheral arterial occlusive disease or peripheral neuropathy. The wound was painful. Early cultures showed Coagulase-Negative Staphylococci, resistant to methicillin (MRSE). The patient was started on RIP therapy with lactoferrin and xylitol gel, and Acticoat along with aggressive debridement, including opening of the undermining in the area. The patient went on to heal his wound in 7 weeks. It is interesting to note that his pain dramatically decreased within 2–3 days of starting the topical use of RIP gel.

2.1.6 Case 6 (Fig. 6)

Day 1　　　　　　14　　　　　　21

Fig. 6 Case report

A 91-year-old female was admitted to a long-term acute care facility after nearly dying from a volvulus. The patient had a colostomy and a nonhealing surgical wound for more than 60 days that had cultured MRSA, even though the patient had been on Vancomycin for several weeks. Anti-biofilm agents (Lactoferrin and xylitol gel) along with RIP were applied. Two weeks later the patient showed dramatic closing of the wound and was healed within 21 days. In the presence of RIP gel, the healing trajectory was much more consistent with an acute wound than the chronic wound (James et al. 2008) she possessed.

2.1.7 Case 7 (Fig. 7)

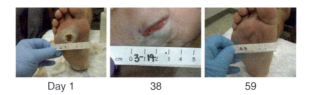

Day 1   38   59

Fig. 7 Case report

This 46-year-old diabetic male first noticed his plantar diabetic foot ulcer approximately 3 weeks before being referred to the wound care center. The wound was found to track 2–3 cm toward the great toe. There was copious drainage, quite a bit of slough, and the wound was deep, but no bone involvement was noted. The patient had a culture result showing methicillin-resistant *S. epidermidis* (MRSE). He was started on daptomycin along with RIP and other anti-biofilm strategies (Lactoferrin and xylitol gel). The undermining was opened. Within 5 weeks, the patient demonstrated almost complete healing.

2.1.8 Case 8 (Fig. 8)

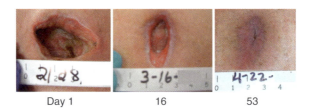

Day 1   16   53

Fig. 8 Case report

Wounds other than diabetic foot ulcers also showed improved healing with suppression of staphylococcal biofilm agent RIP. For example, a 34-year-old graduate student with an infected nonhealing surgical wound in the neck had been managed for more than 3 months without success with intermittent IV antibiotics (vancomycin for 14 days followed by linezolid for 10 days) and wet-to-dry dressings. He had MRSA in the wound and also fragments of bone. Two spinous processes were exposed. He was started on Linezolid and topical RIP with lactoferrin xylitol gel along with frequent debridement. In 8 weeks, the patient was healed.

2.1.9 Case 9 (Fig. 9)

A 33-year-old male developed a furuncle on his right chest. He was treated with outpatient antibiotics and then spent a week in the hospital on IV antibiotics

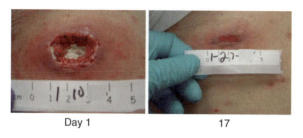

Day 1   17

Fig. 9 Case report

(daptomycin). His culture showed MRSA, and he was referred for chronic wound management. The wound was intensely painful to the patient. The application of topical RIP along with IV daptomycin brought the pain under control within 24–48 h. The patient demonstrated near complete healing of a severely infected, very deep wound in 20 days.

2.1.10 Case 10 (Fig. 10)

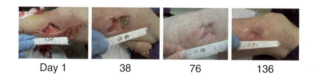

Day 1   38   76   136

Fig. 10 Case report

Hamamelitannin is a nonpeptide analog of RIP. Hamamelitannin was identified based on the sequence homology of RIP to the N terminal part of RAP and by virtual screening of a RIP-based pharmacophore against a database of commercially available small-molecule compounds. Hamamelitannin, a natural product found in the bark of witch hazel, has no effect on staphylococcal growth in vitro, but like RIP, it inhibits the expression of *agr* (RNAIII) and formation of staphylococcal biofilms in vitro (Kiran et al. 2008b). Below is a case history of a person treated with hamamelitannin to salvage limb and save life.

A 50-year-old male was working his cattle when a steer kicked him in his left lower leg. This shattered the tibia and fibula. The patient was hospitalized and underwent open reduction and internal fixation of his fracture. The site where the plate was placed dehisced, and the patient was left with a large nonhealing surgical wound over the metal plate about 5–7 days after his surgery. The surgeons felt like the metal plate needed to be removed, and if it was, they would have to amputate his leg above the knee.

The patient decided to leave the metal plate and try to heal the wound that was cultured with *S. epidermidis*. The patient was treated with a topical preparation, which included hamamelitannin and Triclosan. The patient was treated orally with Doxycycline for 8 weeks. The patient showed rapid closure of the wound over the implanted medical device. It took a total of 16 weeks to heal the wound over the plate. The patient has been doing his farm and ranch activities, and x-rays show the fracture is healed. Clearly, hamamelitannin was important in suppressing the *S. epidermidis* biofilm to the point where the patient could heal his wound.

In conclusion, many of the cases described above are some of the most severe diabetic foot ulcers because these occurred in older individuals with longstanding diabetes, peripheral neuropathy, and severe peripheral arterial occlusive disease with very low transcutaneous oximetry measurements (TCOMs), neither of which could be revascularized. They had a number of other comorbidities, yet by simply suppressing the staphylococcus biofilm, even the meager host defenses and host healing mechanisms that remained in these individuals were enough that over time the patients could heal their wounds. These cases suggest that wound biofilm is a very important barrier to wound healing, and its suppression may be sufficient to obtain wound healing even in the most desperate of wounds.

An important point to make about the efficacy of RIP is the speed of healing. When wound biofilm is suppressed, chronic wounds tend to act as their "acute" counterparts, meaning that the healing trajectory most closely correlates with that of acute wounds, which heal faster (Wolcott and Rhoads 2008).

3 Results

3.1 Molecular Mechanisms of RIP

Although RIP has been demonstrated as a very effective molecule that can prevent and treat persistent infections, the lack of detailed information on its molecular mechanisms has created much controversy and has greatly hindered the clinical development of the molecule. To gain knowledge on the genes regulated by RIP and thus to better understand its activity as an anti-biofilm agent, microarray analysis of RNA isolated from *S. aureus* grown in the presence or absence of RIP was performed.

Specifically, *S. aureus* lab strain RN6390B was grown in LB overnight at 37°C while shaking. Culture was then diluted 1:100 in LB, grown for about 2 h to OD_{600} of 0.3 (early exponential phase of growth). Cells (300 μl ~ 3 × 10^8 CFU) were grown with 10 μl water or RIP (340 μg, i.e., ~3 pmoles/10^3 bacteria) for an additional 60 min at 37°C while shaking (to OD_{600} of 1.2). Cells were placed on ice, collected by centrifugation at 4°C, and supernatant removed. Cell pellet was resuspended in 40 μl TE (10 mM Tris, 1 mM EDTA pH 7.2), frozen, and thawed several times with vigorous mixing. Lysostaphin was added (final concentration 100 μg/ml) and cells incubated at 37°C for 30 min. SDS and proteinase K were added to a final concentration of 1% SDS/500 μg/ml proteinase K, and sample vigorously mixed and incubated at 22°C for 10 min. RNA was isolated using Qiagen RNeasy™ Protect (Qiagen) according to the manufacturer's instructions. Purified RNA was Dnase I-treated and RNAsin® (Ribonuclease Inhibitor,

Promega). Total RNA was used both to test for RNAIII production by northern blotting as described (Korem et al. 2003), as well as for microarray analysis. Microarray analysis was carried out by Genome Explorations using GeneChip® *S. aureus* Genome Array (Affymetrix) that contains sequences of *S. aureus* N315, Mu50, NCTC 8325, and COL. The array contains probesets to over 3,300 *S. aureus* open reading frames and probes to study over 4,800 intergenic regions sequences. Analyses of microarray results were carried out by ArrayStar v2.0.

3.1.1 RIP Downregulates Toxin Genes

Microarray analysis of cells treated or not treated with RIP indicates that in the presence of RIP, important virulence factors such as alpha and beta-hemolysins, V8 proteases, and phospholipase C are significantly downregulated. Several genes involved in capsular polysaccharide production showed a tendency to be downregulated as well (Table 2). At the same time, cell surface molecule Protein A was upregulated. No significant change in the expression of cell surface proteins such as fibrinogen-fibronectin or collagen binding proteins was observed. The pattern of virulence gene expression is very similar to that of *agr* mutant strains (Dunman et al. 2001).

As shown in Fig. 11, RNAIII production is reduced in the presence of RIP. RNAIII is the effector molecule of the *agr*, and when their expression is disrupted, toxin genes are downregulated while protein A is upregulated. We thus conclude that RIP downregulates toxin production by suppressing the *agr*.

3.1.2 RIP Regulates Genes Important for Biofilm Formation and Stress Response

Microarray analysis also indicates that in the presence of RIP, multiple genes important in formation of biofilms and in survival under stress conditions were also regulated (Table 3). For example, genes that were upregulated include: clpP, clpX, hrcA/dnaK/grpE, groES/groEL, ctsR, sodM. These genes are known to be

Table 2 Virulence genes regulated by RIP as well as by *agr* (Dunman et al. 2001)

Gene	Description	RIP+	Agr–
set	Exotoxin 2	Down	Down
hlb	Beta-hemolysin	Down	Down
hlg	Gamma-hemolysin	Down	Down
aur	Zinc metalloproteinase aureolysin	Down	Down
spl	Serine protease V8	Down	Down
ssp	Cysteine proteinase	Down	Down
plc	Phospholipase C	Down	?
vwb	von Willebrand factor-binding protein VWbp (coagulase)	Down	?
cap	Capsular polysaccharide synthesis enzymes	Down	Down
spa	Protein A	Up	Up

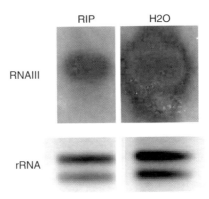

Fig. 11 RIP suppresses RNAIII production: *S. aureus* were grown with or without RIP, RNA isolated, and northern blotted. The presence of RNAIII was detected using RNAIII-specific radiolabeled DNA and membrane autoradiographed. Loading control is demonstrated as ethidium bromide-staining of ribosomal RNA (rRNA)

Table 3 Stress response genes regulated in *S. aureus* by RIP or TRAP and in *B. anthracis* by YhgC

Class	Gene	Description	RIP±	TRAP–	YhgC–
Stress					
	clpP	ATP-dependent Clp protease proteolytic subunit	Up	Up	Up
	clpX	ATP-dependent Clp protease ATP-binding subunit	Up	Up	Up
	ctsR	Transcription repressor of class III stress genes	Up	Up	Up
	hrcA	Heat-inducible transcription repressor	Up	Up	Up
	grpE	Heat shock protein (HSP24)	Up	Up	Up
	dnaK	Chaperone protein (HSP70)	Up	Up	Up
	groES	Chaperonin (HSP10)	Up	Up	Up
	groEL	Heat shock 61 kDa protein	Up	Up	Up
	sodM	Manganese superoxide dismutase	Up	Up	Up
Biofilm					
	ureABCEFGD	Urease gene cluster	Down	Down	Down
	arcABDC	Arginine deiminase (ADI) pathway	Down	Down	?
Anaerobiosis					
	nrdDG	Ribonucleoside-triphosphate reductase system	Down	Down	Up
	nreABC	Oxygen-responsive nitrogen regulation system	Down	Down	?
	narGHJI	Nitrate reductase	Down	Down	Up

involved in stress response (Frees et al. 2004, 2005; Chatterjee et al. 2009, 2010; Derre et al. 1999; Chastanet et al. 2003).

Genes that were downregulated include *ureABCEFGD*, *arcABDC*, *nrdDG*, *narGHJI*, *nreABC*. Most of these genes are known to be important for biofilm

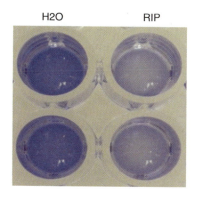

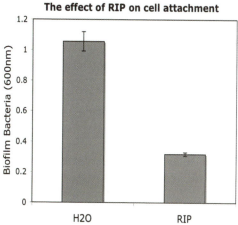

Fig. 12 RIP inhibits cell attachment in vitro: *S. aureus* were placed in polystyrene plates and incubated with RIP or H2O for 3 h at 37° C without shaking. Attached cells were stained (*Top*) and OD$_{600nm}$ determined (*Bottom*) as described (Kiran et al. 2008b)

survival since they aid in bacterial growth under anaerobic conditions and low pH, as occurs in the biofilm environment (Beenken et al. 2004).

Most of these genes are known to be important for biofilm survival (Beenken et al. 2004). Interestingly, a similar pattern of regulation was observed in *traP*– *S. aureus* mutants as well as in *yhgC*– *B. anthracis* mutants (Table 3). This suggests that RIP may regulate biofilm formation and stress response via TRAP. The downregulation observed in the genes involved in biofilm formation possibly explains why in the presence of RIP biofilm production is reduced (Fig. 12).

4 Discussion

Because bacterial biofilm is the major contributor to nonhealing chronic wounds, there is a specific demand for biofilm-targeted strategies to tackle this problem. Although the standard care of wound treatment regime is critical, it is still not enough to allow proper healing. It is in this context that the addition of anti-biofilm agents such as RIP can be the final decisive factor needed to reach a positive outcome.

Although RIP has been used successfully to treat chronic wounds and has been shown to enhance the sensitivity of biofilms to antibiotics (Cirioni et al. 2006; Balaban et al. 2003b), its clinical development has been hindered by the lack of the basic understanding of its effect on single or polymicrobial biofilms (Novick and Geisinger 2008). To try to better understand the effect of RIP on gene expression in *S. aureus*, functional transcriptomics experiments were carried out and showed that RIP downregulates toxin genes and genes important for biofilm formation. At the same time, RIP was shown to upregulate stress response genes.

4.1 *RIP Reduces Toxin Production*

RIP was shown here by transcriptional profiling and by other experimental methods (Vieira-da-Motta and Ribeiro 2001) to inhibit the production of various toxins important for pathogenesis (Lowy 1998). At the same time, protein A was upregulated but no significant upregulation of adhesion molecules was observed (such as fibrinogen, fibronectin, or collagen-binding protein). This pattern of gene expression suggests that in the presence of RIP, the bacteria reduce toxin production without increasing cell adhesion, while maintaining their ability to protect themselves from host immunoglobulin by expressing protein A. We assume that RIP affects virulence by suppressing the *agr* or its effector molecule RNAIII. This is supported both by northern blotting (Fig. 11) and by the transcriptional profiling that is strikingly similar to that of the *agr* (Dunman et al. 2001).

4.1.1 RIP Reduces Genes Important for Biofilm Survival

RIP was shown to downregulate *ureABCEFGD*, *arcABDC*, *nrdDG*, *narGHJI*, *nreABC*, collectively important for biofilm mode of growth, which requires adaptation both to anaerobic conditions and to an acidic pH (Beenken et al. 2004). The maintenance of pH homeostasis is achieved by producing alkaline compounds such as ammonia that buffer the surrounding microenvironment by neutralizing the acids. Ammonia is generated through two pathways: the urease (urea amidohydrolase encoded by *ureABCEFGD*) and arginine deiminase (ADI) pathways (encoded by *arcABDC*). The ADI pathway also enables bacterial growth under anaerobic conditions by using arginine as an energy source(Makhlin et al. 2007). The expression of the *nrdDG* genes, which encode for Class III ribonucleotide reductases that were shown to be required for anaerobic growth, was also shown to be important as a virulence determinant in the pathogenesis of staphylococcal arthritis in a mouse model. (Kirdis et al. 2007). The *nreABC* genes that encode for members of the two-component regulatory system are involved in the control of dissimilatory nitrate/nitrite reduction in response to oxygen. NreB/nreC activate the expression of the nitrate (narGHJI) and nitrite (nir) reductase operons, important for growth of *S. aureus* in anaerobic environments (Schlag et al. 2008).

All these genes, *ureABCEFGD, arcABDC, nrdDG, narGHJI, nreABC*, are downregulated in the presence of RIP, indicating that the ability of these bacteria to form and survive in a biofilm is diminished. The cells are unable to maintain a stable pH or reducing conditions compatible with an anaerobic environment, and thus biofilms are not produced. Of note is that the same genes were downregulated in *traP– S. aureus* and in *yhgc– B. anthracis* mutants, and like with RIP, *yhgC– B. anthracis* were also defective in their ability to produce a biofilm (tested for 3 h in vitro). Put together this suggests that RIP may regulate the expression of genes important for biofilm formation via the TRAP/YhgC family of proteins.

4.1.2 RIP Increases Bacterial Stress Response

RIP upregulates genes important for survival under stress conditions, such as *clpP, clpX, hrcA/dnaK/grpE, groES/groEL, ctsR*, and *sodM*. In *S. aureus*, stress response was shown to be controlled by six different classes of genes distinguished by their regulatory mechanisms (Derre et al. 1999; Schumann 2003; Clements and Foster 1999). Our results indicate that in the presence of RIP members of Class I, Class III, and class VI stress protein genes are upregulated Class I genes encode classical chaperones such as DnaK, DnaJ, GrpE, GroES and GroEL. Their expression involves a σ^A-dependent promoter and the highly conserved CIRCE (controlling inverted repeat of chaperone expression) operator sequence, which is the binding site for the HrcA repressor (Fleury et al. 2009). Class III genes are defined as those devoid of the CIRCE operator sequence and whose induction by heat shock and general stress conditions is σ^B-independent. They are controlled by the Class III stress gene repressor, CtsR, which recognizes a repeated heptad operator sequence (Derre et al. 1999). Class III genes encode the ATP-dependent protease ATP-binding subunit ClpC (Msadek et al. 1994; Kruger et al. 1994), ATP-dependent protease proteolytic subunit ClpP (Gerth et al. 1998; Msadek et al. 1998) and ATP-dependent protease ATP-binding subunit ClpE. Class VI contain the ClpX ATPases, FtsH, LonA, HtpG, AhpC, and TrxA heat shock proteins (Derre et al. 1999). Collectively, such responses to a stress allow the refolding or elimination of damaged proteins.

An additional protection mechanism is the production of a nonspecific DNA binding protein, Dps (DNA binding protein from stationary phase cells), which protects DNA from free radicals, radiation and heat shock (Gupta and Chatterji 2003; Nair and Finkel 2004; Liu et al. 2006a; Liu et al. 2006b) by the formation of a biocrystalline complex with DNA (Gupta and Chatterji 2003; Almiron et al. 1992; Wolf et al. 1999; Liu et al. 2006a; Imlay 2003; Levin-Zaidman et al. 2000; Wolf et al. 1999; Frenkiel-Krispin et al. 2004). DNA is thus protected through structural transitions that lead from an open, dispersed DNA conformation (which is vulnerable to damaging agents) into ordered structures. DNA molecules in these ordered structures are tightly packed, and this limits the accessibility of the DNA to detrimental factors such as oxidizing agents and nucleases. MrgA is a Dps homolog in *S. aureus* (Morikawa et al. 2006). Bacteria that are deficient in DNA protection or repair mechanisms have increased susceptibility to reactive oxygen species (ROS), have increased susceptibility to phagocytes and reduced virulence (Fang 2004). Both TRAP in *S. aureus* and YhgC in *B. anthracis* have been shown to bind DNA and protect it from oxidative stress response Kim, personal communication, (Kiran and Balaban 2009; Kiran et al. 2010), essentially qualifying these molecules as belonging to the Dps family of proteins. RIP inhibits the phosphorylation of TRAP. Interestingly, the same pattern of gene expression was found by microarray analysis in *S. aureus* treated with RIP, in *S. aureus* TRAP- mutant, and in *B. anthracis* YhgC– mutant. This supports the notion that TRAP and YhgC are functional homologues and that RIP probably regulates stress response via TRAP.

In an attempt to try to better understand stress response in bacteria, let us take a look at one of the more primitive eukaryotic cells, the yeast *Saccharomyces cerevisiae*, in which stress has been extensively studied. It is known that in the adaptation of the cell to environments that have poor conditions for growth and survival, the environmental stress response program (ESR) is triggered (Gasch et al. 2000). This is a survival skill of the yeast cell in which a broad spectrum of genes are turned on to respond in a positive way, adapting the cell's metabolism to the new physiological requirements. Mechanisms of DNA repair, protein folding, defense against reactive oxygen species, energy generation, and storage are increased functions, while cell growth and protein synthesis are arrested (Gasch et al. 2000). Although ESR is a typical response in which the cell will automatically turn on and off particular sets of genes, other environmental stimuli will specifically target other genes, rendering a unique gene expression pattern customized for each specific condition (DeRisi et al. 1997). A similar situation might be encountered by different bacterial pathogens during the process of infection. The coordinated regulation of apparently unrelated gene networks is key for the microorganism's survival, achieving the appropriate and rapid response to changing adverse environmental conditions in the host (Boor 2006).

4.1.3 Functional Equivalent Pathogroups and Synergistic Treatment Protocols

Chronic wounds often contain polymicrobial biofilms composed of Gram-positive and Gram-negative bacteria (Kiran et al. 2008a; Wolcott and Ehrlich 2008; Narisawa et al. 2008). The persistence and tenacity of these wounds are thought to be functions of the populations' range of phenotypes and genotypes (Ehrlich et al. 2005). Observed patterns of population diversity in chronic wounds have led to the concept of a functional equivalent pathogroups (FEPs) (Dowd et al. 2008). FEPs are a consortium of genotypically distinct bacteria that in monoculture do not cause disease but as an aggregate, collectively possess the necessary factors to maintain chronic infections (O'Connell et al. 2006). Based on the knowledge of phenotypic diversification existing in the microbial biofilm ecosystem that ensures its survival, it is then not surprising the need of diverse and potentiating techniques to achieve healing. The successful treatment of nonhealing chronic wounds therefore relies on the combinatorial and cooperative effect of antibiotics, cleaning procedures, better perfusion, and immune response in conjunction with a variety of anti-biofilm agents such as RIP, lactoferrin, xylitol, silver, copper, zinc, nitric oxide, quaternary ammonium compounds (QACs), etc. (Hetrick et al. 2009; Harrison et al. 2008; Bjarnsholt et al. 2007; Baker et al. 2010). The cooperative activity between anti-biofilm agents is well known. In the cases shown in this chapter, most of the treatments described included RIP, lactoferrin, and xylitol as anti-biofilm agents acting synergistically. Another example is copper and QACs that can act alone or in concert, interfering with normal biofilm growth by reducing the activity of nitrate reductases, thus interfering with its metabolism (Harrison et al. 2008). RIP showed a tendency of reduced activity of the nitrate reductases as well, implying that perhaps one of the reasons RIP inhibits biofilms is by the same molecular mechanisms as seen for these compounds. Similar to RIP, excess copper has been shown to upregulate oxidative stress response genes and the misfolded protein response genes, while downregulating genes involved in the production of virulence factors as well as biofilm formation (Baker et al. 2010). This might make sense from the bacterial point of view, as a desperate attempt of the bacteria to try to thrive for survival.

In summary, the similarities between gene expression in bacteria treated with RIP or mutated in *agr* or *traP*, we hypothesize that RIP directly affects staphylococci by interfering with stress responses (TRAP-mediated) and quorum-sensing (*agr*-mediated)-based gene regulation (Fig. 13). This interferes with a key component of the FEP and collapses the collective ability of the consortia to persist. When the biofilm collapses, the once protected bacterial community can be targeted by the immune system and by antibiotics, allowing complete recovery of the otherwise nonhealing wound.

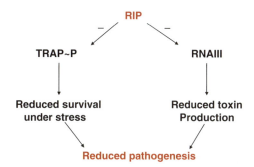

Fig. 13 Studies carried out in vitro indicate that RIP may regulate pathogenesis by downregulating TRAP phosphorylation (which leads to the reduced ability of the bacteria to survive under stress conditions) and by down regulating RNAIII production (which leads to reduced toxin production)

References

Almiron M, Link AJ, Furlong D, Kolter R (1992) A novel DNA-binding protein with regulatory and protective roles in starved *Escherichia coli*. Genes Dev 6:2646–54

Amy P (2008) Interview with Dr. Randall Wolcott, bacterial biofilm wound specialist. Bacteriality 2008, April 13

An APIC Guide, 2010, Guide to the Elimination of Methicillin-Resistant Staphylococcus aureus (MRSA) Transmission in Hospital Settings (http://www.apic.org/downloads/MRSA_elimination_guide_27030.pdf

Anguita-Alonso P, Giacometti A, Cirioni O et al (2007) RNAIII-inhibiting-peptide-loaded polymethylmethacrylate prevents in vivo *Staphylococcus aureus* biofilm formation. Antimicrob Agents Chemother 51:2594–6

Baker J, Sitthisak S, Sengupta M, Johnson M, Jayaswal RK, Morrissey JA (2010) Copper stress induces a global stress response in Staphylococcus aureus and represses sae and agr expression and biofilm formation. Appl Environ Microbiol 76:150–160

Balaban N, Novick RP (1995) Autocrine regulation of toxin synthesis by *Staphylococcus aureus*. Proc Natl Acad Sci U S A 92:1619–23

Balaban N, Goldkorn T, Nhan RT et al (1998) Autoinducer of virulence as a target for vaccine and therapy against *Staphylococcus aureus*. Science 280:438–40

Balaban N, Collins LV, Cullor JS et al (2000) Prevention of diseases caused by Staphylococcus aureus using the peptide RIP. Peptides 21:1301–11

Balaban N, Goldkorn T, Gov Y et al (2001) Regulation of *Staphylococcus aureus* pathogenesis via target of RNAIII-activating protein (TRAP). J Biol Chem 276:2658–67

Balaban N, Giacometti A, Cirioni O et al (2003a) Use of the quorum-sensing inhibitor RNAIII-inhibiting peptide to prevent biofilm formation in vivo by drug-resistant *Staphylococcus epidermidis*. J Infect Dis 187:625–30

Balaban N, Gov Y, Bitler A, Boelaert JR (2003b) Prevention of *Staphylococcus aureus* biofilm on dialysis catheters and adherence to human cells. Kidney Int 63:340–5

Balaban N, Stoodley P, Fux CA, Wilson S, Costerton JW, Dell'Acqua G (2005) Prevention of staphylococcal biofilm-associated infections by the quorum sensing inhibitor RIP. Clin Orthop Relat Res 437:48–54

Balaban N, Cirioni O, Giacometti A et al (2007) Treatment of *Staphylococcus aureus* biofilm infection by the quorum-sensing inhibitor RIP. Antimicrob Agents Chemother 51:2226–9

Beenken KE, Dunman PM, McAleese F et al (2004) Global gene expression in *Staphylococcus aureus* biofilms. J Bacteriol 186:4665–84

Bjarnsholt T, Kirketerp-Moller K, Kristiansen S et al (2007) Silver against *Pseudomonas aeruginosa* biofilms. APMIS 115:921–8

Boor KJ (2006) Bacterial stress responses: what doesn't kill them can make then stronger. PLoS Biol 4:e23

Chang W, Toghrol F, Bentley WE (2006) Toxicogenomic response of *Staphylococcus aureus* to peracetic acid. Environ Sci Technol 40:5124–31

Chastanet A, Fert J, Msadek T (2003) Comparative genomics reveal novel heat shock regulatory mechanisms in Staphylococcus aureus and other gram-positive bacteria. Mol Microbiol 47:1061–73

Chatterjee I, Schmitt S, Batzilla CF et al (2009) Staphylococcus aureus ClpC ATPase is a late growth phase effector of metabolism and persistence. Proteomics 9:1152–76

Chatterjee I, Neumayer D, Herrmann M (2010) Senescence of staphylococci: using functional genomics to unravel the roles of ClpC ATPase during late stationary phase. Int J Med Microbiol 300:130–136

Cirioni O, Giacometti A, Ghiselli R et al (2003) Prophylactic efficacy of topical temporin A and RNAIII-inhibiting peptide in a subcutaneous rat Pouch model of graft infection attributable to staphylococci with intermediate resistance to glycopeptides. Circulation 108:767–71

Cirioni O, Giacometti A, Ghiselli R et al (2006) RNAIII-inhibiting peptide significantly reduces bacterial load and enhances the effect of antibiotics in the treatment of central venous catheter-associated Staphylococcus aureus infections. J Infect Dis 193:180–6

Clements MO, Foster SJ (1999) Stress resistance in *Staphylococcus aureus*. Trends Microbiol 7:458–62

Costerton JW, Stewart PS, Greenberg EP (1999) Bacterial biofilms: a common cause of persistent infections. Science 284:1318–22

Dell'Acqua G, Giacometti A, Cirioni O et al (2004) Suppression of drug-resistant Staphylococcal infections by the quorum-sensing inhibitor RNAIII-inhibiting peptide. J Infect Dis 190: 318–20

DeRisi JL, Iyer VR, Brown PO (1997) Exploring the metabolic and genetic control of gene expression on a genomic scale. Science 278:680–6

Derre I, Rapoport G, Msadek T (1999) CtsR, a novel regulator of stress and heat shock response, controls clp and molecular chaperone gene expression in gram-positive bacteria. Mol Microbiol 31:117–31

Dowd SE, Wolcott RD, Sun Y, McKeehan T, Smith E, Rhoads D (2008) Polymicrobial nature of chronic diabetic foot ulcer biofilm infections determined using bacterial tag encoded FLX amplicon pyrosequencing (bTEFAP). PLoS ONE 3:e3326

Dunman PM, Murphy E, Haney S et al (2001) Transcription profiling-based identification of *Staphylococcus aureus* genes regulated by the agr and/or sarA loci. J Bacteriol 183:7341–53

Ehrlich GD, Hu FZ, Shen K, Stoodley P, Post JC (2005) Bacterial plurality as a general mechanism driving persistence in chronic infections. Clin Orthop Relat Res 20–24

Fang FC (2004) Antimicrobial reactive oxygen and nitrogen species: concepts and controversies. Nat Rev Microbiol 2:820–32

Fleury B, Kelley WL, Lew D, Gotz F, Proctor RA, Vaudaux P (2009) Transcriptomic and metabolic responses of *Staphylococcus aureus* exposed to supra-physiological temperatures. BMC Microbiol 9:76

Frees D, Chastanet A, Qazi S et al (2004) Clp ATPases are required for stress tolerance, intracellular replication and biofilm formation in Staphylococcus aureus. Mol Microbiol 54: 1445–62

Frees D, Sorensen K, Ingmer H (2005) Global virulence regulation in Staphylococcus aureus: pinpointing the roles of ClpP and ClpX in the sar/agr regulatory network. Infect Immun 73: 8100–8

Frenkiel-Krispin D, Ben-Avraham I, Englander J, Shimoni E, Wolf SG, Minsky A (2004) Nucleoid restructuring in stationary-state bacteria. Mol Microbiol 51:395–405

Gasch AP, Spellman PT, Kao CM et al (2000) Genomic expression programs in the response of yeast cells to environmental changes. Mol Biol Cell 11:4241–57

Gerth U, Kruger E, Derre I, Msadek T, Hecker M (1998) Stress induction of the *Bacillus subtilis* clpP gene encoding a homologue of the proteolytic component of the Clp protease and the involvement of ClpP and ClpX in stress tolerance. Mol Microbiol 28:787–802

Ghiselli R, Giacometti A, Cirioni O et al (2006) RNAIII-inhibiting peptide in combination with the cathelicidin BMAP-28 reduces lethality in mouse models of staphylococcal sepsis. Shock 26: 296–301

Giacometti A, Cirioni O, Gov Y et al (2003) RNA III inhibiting peptide inhibits in vivo biofilm formation by drug-resistant *Staphylococcus aureus*. Antimicrob Agents Chemother 47:1979–83

Giacometti A, Cirioni O, Ghiselli R et al (2005) RNAIII-inhibiting peptide improves efficacy of clinically used antibiotics in a murine model of staphylococcal sepsis. Peptides 26:169–75

Gov Y, Borovok I, Korem M et al (2004) Quorum sensing in Staphylococci is regulated via phosphorylation of three conserved histidine residues. J Biol Chem 279:14665–72

Gupta S, Chatterji D (2003) Bimodal protection of DNA by Mycobacterium smegmatis DNA-binding protein from stationary phase cells. J Biol Chem 278:5235–41

Harrison JJ, Turner RJ, Joo DA et al (2008) Copper and quaternary ammonium cations exert synergistic bactericidal and antibiofilm activity against Pseudomonas aeruginosa. Antimicrob Agents Chemother 52:2870–81

Hetrick EM, Shin JH, Paul HS, Schoenfisch MH (2009) Anti-biofilm efficacy of nitric oxide-releasing silica nanoparticles. Biomaterials 30:2782–9

Imlay JA (2003) Pathways of oxidative damage. Annu Rev Microbiol 57:395–418

James GA, Swogger E, Wolcott R et al (2008) Biofilms in chronic wounds. Wound Repair Regen 16:37–44

Kiran MD, Bala S, Hirshberg M, Balaban N (2010) YhgC protects Bacillus anthracis from oxidative stress. Int J Artif Organs. 33:590–607

Kiran MD, Balaban N (2009) TRAP plays a role in stress response in *Staphylococcus aureus*. Int J Artif Organs 32:592–9

Kiran MD, Giacometti A, Cirioni O, Balaban N (2008a) Suppression of biofilm related, device-associated infections by staphylococcal quorum sensing inhibitors. Int J Artif Organs 31: 761–70

Kiran MD, Adikesavan NV, Cirioni O et al (2008b) Discovery of a quorum-sensing inhibitor of drug-resistant staphylococcal infections by structure-based virtual screening. Mol Pharmacol 73:1578–86

Kirdis E, Jonsson IM, Kubica M et al (2007) Ribonucleotide reductase class III, an essential enzyme for the anaerobic growth of Staphylococcus aureus, is a virulence determinant in septic arthritis. Microb Pathog 43:179–88

Korem M, Sheoran AS, Gov Y, Tzipori S, Borovok I, Balaban N (2003) Characterization of RAP, a quorum sensing activator of *Staphylococcus aureus*. FEMS Microbiol Lett 223:167–75

Kruger E, Volker U, Hecker M (1994) Stress induction of clpC in *Bacillus subtilis* and its involvement in stress tolerance. J Bacteriol 176:3360–7

Levin-Zaidman S, Frenkiel-Krispin D, Shimoni E, Sabanay I, Wolf SG, Minsky A (2000) Ordered intracellular RecA-DNA assemblies: a potential site of in vivo RecA-mediated activities. Proc Natl Acad Sci U S A 97:6791–6

Lopez-Leban F, Kiran MD, Wolcott R, Balaban N (2010) Molecular mechanisms of RIP, an effective inhibitor of chronic infections. Int J Artif Organs 33:582–9

Liu X, Hintze K, Lonnerdal B, Theil EC (2006a) Iron at the center of ferritin, metal/oxygen homeostasis and novel dietary strategies. Biol Res 39:167–71

Liu X, Kim K, Leighton T, Theil EC (2006b) Paired *Bacillus anthracis* Dps (mini-ferritin) have different reactivities with peroxide. J Biol Chem 281:27827–35

Lowy FD (1998) *Staphylococcus aureus* infections. N Engl J Med 339:520–32

Makhlin J, Kofman T, Borovok I et al (2007) *Staphylococcus aureus* ArcR controls expression of the arginine deiminase operon. J Bacteriol 189:5976–86

Morikawa K, Ohniwa RL, Kim J, Maruyama A, Ohta T, Takeyasu K (2006) Bacterial nucleoid dynamics: oxidative stress response in *Staphylococcus aureus*. Genes Cells 11:409–23

Msadek T, Kunst F, Rapoport G (1994) MecB of *Bacillus subtilis*, a member of the ClpC ATPase family, is a pleiotropic regulator controlling competence gene expression and growth at high temperature. Proc Natl Acad Sci U S A 91:5788–92

Msadek T, Dartois V, Kunst F, Herbaud ML, Denizot F, Rapoport G (1998) ClpP of *Bacillus subtilis* is required for competence development, motility, degradative enzyme synthesis, growth at high temperature and sporulation. Mol Microbiol 27:899–914

Nair S, Finkel SE (2004) Dps protects cells against multiple stresses during stationary phase. J Bacteriol 186:4192–8

Narisawa N, Haruta S, Arai H, Ishii M, Igarashi Y (2008) Coexistence of antibiotic-producing and antibiotic-sensitive bacteria in biofilms is mediated by resistant bacteria. Appl Environ Microbiol 74:3887–94

Novick RP, Geisinger E (2008) Quorum sensing in staphylococci. Annu Rev Genet 42:541–64

O'Connell HA, Kottkamp GS, Eppelbaum JL, Stubblefield BA, Gilbert SE, Gilbert ES (2006) Influences of biofilm structure and antibiotic resistance mechanisms on indirect pathogenicity in a model polymicrobial biofilm. Appl Environ Microbiol 72:5013–9

Schlag S, Fuchs S, Nerz C et al (2008) Characterization of the oxygen-responsive NreABC regulon of *Staphylococcus aureus*. J Bacteriol 190:7847–58

Schierle CF, De la Garza M, Mustoe TA, Galiano RD (2009) Staphylococcal biofilms impair wound healing by delaying reepithelialization in a murine cutaneous wound model. Wound Rep Reg 17:354–59

Schumann W (2003) The *Bacillus subtilis* heat shock stimulon. Cell Stress Chaperones 8:207–17

Vieira-da-Motta O, Ribeiro PD, Dias da Silva W, Medina-Acosta E (2001) RNAIII inhibiting peptide (RIP) inhibits agr-regulated toxin production. Peptides 22:1621–7

Wolcott RD, Ehrlich GD (2008) Biofilms and chronic infections. JAMA 299:2682–4

Wolcott RD, Rhoads DD (2008) A study of biofilm-based wound management in subjects with critical limb ischaemia. J Wound Care 17:145–8, 150–152, 154–155

Wolcott RD, Gontcharova V, Sun Y, Zischakau A, Dowd SE (2009) Bacterial diversity in surgical site infections: not just aerobic cocci any more. J Wound Care 18:317–23

Wolf SG, Frenkiel D, Arad T, Finkel SE, Kolter R, Minsky A (1999) DNA protection by stress-induced biocrystallization. Nature 400:83–5

Ziegler-Graham K, MacKenzie EJ, Ephraim PL, Travison TG, Brookmeyer R (2008) Estimating the prevalence of limb loss in the United States: 2005 to 2050. Arch Phys Med Rehabil 89:422–429

Interfering with "Bacterial Gossip"

Thomas Bjarnsholt, Tim Tolker-Nielsen, and Michael Givskov

Abstract Biofilm resilience poses major challenges to the development of novel antimicrobial agents. Biofilm bacteria can be considered small groups of "Special Forces" capable of infiltrating the host and destroying important components of the cellular defense system with the aim of crippling the host defense. Antibiotics exhibit a rather limited effect on biofilms. Furthermore, antibiotics have an "inherent obsolescence" because they select for development of resistance. Bacterial infections with origin in bacterial biofilms have become a serious threat in developed countries. *Pseudomonas aeruginosa* biofilms are thought to be the dominant agent in many chronic infections including those in cystic fibrosis lungs and chronic wounds. With the present day's awareness of biofilms, the future task is to exploit this knowledge for development and application of antimicrobial intervention strategies that appropriately target bacteria in their relevant habitat with the aim of mitigating their destructive impact on patients. In this review, we describe molecular mechanisms involved in "bacterial gossip" (more scientifically referred to as quorum sensing (QS) and c-di-GMP signaling), virulence, biofilm formation, resistance and QS inhibition as future antimicrobial targets, in particular those that would work to minimize selection pressures for the development of resistant bacteria.

T. Bjarnsholt
Department of International Health, Immunology and Microbiology, University of Copenhagen, Blegdamsvej 3, DK 2200, Copenhagen, Denmark
and
Department of Clinical Microbiology, University Hospital, Rigshospitalet, Denmark

T. Tolker-Nielsen • M. Givskov (✉)
Department of International Health, Immunology and Microbiology, University of Copenhagen, Blegdamsvej 3, DK 2200, Copenhagen, Denmark
e-mail: ttn@sund.ku.dk; mgivskov@sund.ku.dk

1 Introduction

Bacteria appear in nature as single, free floating (planktonic) cells, or organized as sessile, matrix embedded aggregates. The latter form is commonly referred to as the biofilm mode of growth. These two life forms have serious implications for infections in humans. To generalize, acute infections involve planktonic bacteria and they are often treatable with either one or the other antibiotic. However, in the cases where bacteria succeed in forming biofilms within the host, the infection may, despite massive antibiotic therapy, develop into a chronic state. The important hallmarks of chronic, biofilm infections are development of local inflammations, extreme tolerance to the action of conventional antimicrobial agents, and an almost infinite capacity to evade the host defense systems in particular innate immunity. To device novel intervention principles for biofilm control, it is important to understand the molecular mechanisms underlying biofilm formation, in particular under in vivo conditions.

Key to successfully establish a chronic infection is the ability of bacteria to interact with each other and make collective decisions as well as the capability to perceive evocative signals and stimuli from the host organism and vice versa. Where conventional antibiotics have been designed to hinder bacterial growth by blocking basal life processes, new ways of attenuating bacteria may take into account important interactive control elements such as those involved in signaling known as Quorum Sensing (QS) and c-di-GMP. QS constitutes a cell–cell signaling-based regulatory mechanism, which becomes operational when cells have amassed at high densities (Allison and Matthews 1992) (which is the case in biofilm aggregates) (see Fig. 1). In particular, *N*-acyl-L-homoserine lactone (AHL)-based systems are the most well-studied examples of QS. QS, or autoinduction as it was initially called, was first reported in the late 1960s and early 1970s by Tomasz (1965) and Nealson et al. (1970). Nealson et al. (1970) studied the biology of light-producing organelles in deep-sea fish, the light was produced by bacteria of the species *Vibrio fischeri,* by a cell density dependent reaction. Bioluminescence originates from the expression of two luciferase genes, the *lux* genes, which rapidly increases when the growth of the bacteria enters the late exponential phase and early lag phase of growth. The signaling dependency was elucidated by the luminescence of bacteria in early log phase was inducible by the addition of cell-free medium derived from stationary phase bacteria. However, QS involvement in more complex, corporative behavior such as swarming motility was first described for *Serratia liquefaciens* by Eberl et al. (1999). Today, more than 70 Gram-negative bacterial species have been reported to harbor AHL-based systems (Fuqua et al. 2001; Taga and Bassler 2003). Except from basal life processes, QS controls a wide range of functions in Gram-negative bacteria, such as Ti-plasmid conjugation in *Agrobacterium tumefaciens*, antibiotic production in *Erwinia carotovora* (Taga and Bassler 2003), virulence gene expression in *Vibrio cholerae* (Zhu et al. 2002), *Burkholderia cepacia* (Gilligan 1991), and *P. aeruginosa* (Whiteley et al. 1999), and surface motility by means of swarming in *S. liquefaciens* (Eberl et al. 1996, 1999), *P. aeruginosa* (Kohler et al. 2000), and *B. cepacia* (Huber et al. 2001).

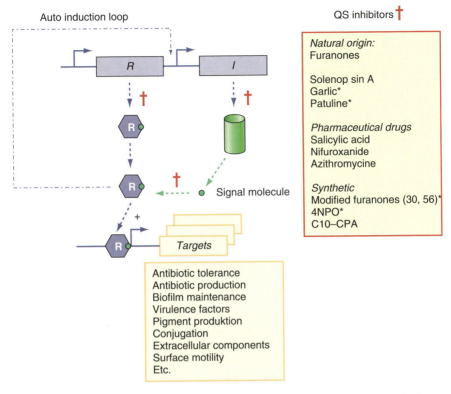

Fig. 1 The basic QS regulator, with the *R*-gene encoding the regulator protein and the *I*-gene encoding the signal molecule synthetase. The red crosses symbolizes the putative sites for QS inhibition, with known QS inhibitors listed in the red table (*marks QS inhibitors verified in animal models). The active complex of the R-protein and signal molecule activates numerous targets, as listed in the orange table

Instead of AHL molecules, the Gram-positives employ small peptides as signaling molecules, usually 5–17 amino acids in length (Federle and Bassler 2003). Examples of Gram-positive QS controlled behavior are development of genetic competence and sporulation in *Bacillus subtilis* (Magnuson et al. 1994), virulence expression in *Enterococcus faecalis* (Qin et al. 2000) and *Staphylococcus aureus* (Kong et al. 2006) and biofilm development in *S. aureus* (Boles and Horswill 2008) as described in another chapter of this book.

2 *Pseudomonas aeruginosa*

P. aeruginosa is a well-characterized bacterium capable of causing chronic infections, and it is considered the archetypical lab bacterium for in vitro biofilm studies and cell–cell communication, in particular QS. The genome of the sequenced

P. aeruginosa PA01 consists of 5770 open reading frames (ORF) (Stover et al. 2000c), which make it one of the largest known bacterial genomes, around 50% larger than *Escherichia coli*. Based on its genetic complexity, *P. aeruginosa* resembles simple eukaryotes such as *Saccharomyces cerevisia*. Numerous experimental tools are available for *P. aeruginosa* in particular those required for genetic engineering, controlled gene expression, reporters, and a variety of in vivo models of infection including plants (O'Sullivan and O'Gara 1992), nematodes (Rasmussen et al. 2005a), and animals (Bjarnsholt et al. 2005a, b; Christensen et al. 2007; Hentzer et al. 2003; Jensen et al. 2007a; Rasmussen et al. 2005c; van Gennip et al. 2009; Wu et al. 2001, 2004b) as well as DNA-microarray chips for transcriptomics.

Pseudomonas can be isolated from most environments including soil, marshes, costal marine habitats, plants, and mammal tissue (Stover et al. 2000b). In hospital environments, it grows in moist reservoirs such as food, cut flowers, sinks, toilets, floor mops, dialysis equipment, and even in disinfectant solutions (Murray et al. 2002). *P. aeruginosa* is an opportunistic human pathogen, which is not part of the normal human flora, but it may give rise to acute infections in hospitalized patients in particular immunocompromised individuals, such as AIDS and neutropenic patients undergoing chemotherapy. *P. aeruginosa* is a significant cause of infections in burn victims and patients having foreign bodies inserted. It causes pneumonia in patients receiving artificial ventilation in intensive care units as well as keratitis in users of extended-wear soft contact lenses (Lyczak et al. 2000; Stover et al. 2000a; van Delden and Iglewski 1998b). It is also the predominant cause of chronic lung infection and mortality in CF patients (Frederiksen et al. 1997; Hoiby 1974; Koch and Hoiby 1993). Other susceptible patients include those suffering from venous leg ulcers and diabetic foot ulcers. Its ability to cause disease relays on biofilm formation and production of a cocktail of extracellular virulence factors, many of which expression is coordinated by the QS systems (Rumbaugh et al. 2000; van Delden and Iglewski 1998a; Williams et al. 2000).

3 QS in *P. aeruginosa*

The first reports of QS to be a controller of virulence surfaced in the mid-1990 (Davies et al. 1998b; Latifi et al. 1995). Classical genetics and later, transcriptomic analysis (Hentzer et al. 2003; Schuster et al. 2003; Wagner et al. 2003) as well as in vivo models of infections in plants (O'Sullivan and O'Gara 1992), nematodes (Rasmussen et al. 2005b) and animals (Bjarnsholt et al. 2005a, b; Christensen et al. 2007; Hentzer et al. 2003; Jensen et al. 2007a; Rasmussen et al. 2005c; van Gennip et al. 2009; Wu et al. 2001, 2004b) have substantiated this (see Fig. 1). *P. aerugonosa* QS builds on two quorum sensors composed of a signal generators (I) and receptors (R) encoded by the gene clusters *lasI,R* and *rhlI,R* (Pesci et al. 1997a) (see Fig. 1). The systems employ a 3-oxo-C12-HSL specific signal for the LasR receptor and a C4-HSL specific signal for the RhlR receptor. A second receptor for 3-oxo-C12-HSL is the QscR receptor (Lequette et al. 2006). The *las*- and *rhl*-encoded quorum sensors were identified as being hierarchically ordered with the *rhl* quorum sensor underlain control by the *las* system (Pesci et al. 1997b). The current model is as follows: at low cell densities, the *las* quorum sensor is expressed constitutively at an idle level, producing the LasI synthetase (which in turn produces the 3-oxo-C12-HSL signal molecule), and the LasR transcriptional regulator. The majority of produced 3-oxo-C12-HSL signal molecules are pumped out of the cell by efflux pumps (Pearson et al. 1999a), which are likely to produce an equilibrium of signal molecules between the inside and the outside of the cell. When the cell density increases, the concentration of signal molecules increases inside and outside of the cells. Consequently, the transcriptional inducing complex of LasR with bound 3-oxo-C12-HSL increases because of the self-inducing effect (Seed et al. 1995). At the threshold level, the 3-oxo-C12-HSL signal molecule triggers LasR proteins to form the active multimeric transcriptional induction complexes. These complexes acts to induce the transcription of target genes among them the expression of *lasI* and *lasR* (see Fig. 1). In batch cultures, genes regulated by QS are not induced all at once, they exhibit a temporal expression pattern over the remaining cell cycle (Hentzer et al. 2003). The *las* system also acts as a positive regulator of the *rhl* quorum sensor (de Kievit et al. 2002). 3-oxo-C12-HSL associated with the QscR receptor controls a regulon comprising a number of specific genes as well as it overlaps with the two other LasR and RhlR controlled regulons (Lequette et al. 2006). Since a QscR null-mutant appears hypervirulent, this indicates that one role of QscR is to carefully adjust maximum expression of LasR- and RhlR-controlled virulence.

It has been shown that the *rhl* QS system induces the expression of many genes involved in virulence including the signal generator *rhlI*, and despite previous genetic analysis stated complete dependency of a functional *las* quorum sensor, transcriptomic analysis has demonstrated that *rhl* QS can operate independently (Medina et al. 2003; Skindersoe et al. 2008) of the *las* system. Interestingly, a recent article describes the activity of the *rhl* system to only be delayed in the absence of a functional *las* system and even overcoming the absence of the *las* activation by activating *las*-specific functions such as generation of signal molecules both 3-oxo-C12-HSL and PQS (Dekimpe and Deziel 2009). Apart from this, the expression of genes regulated by the *rhl* system follows the same pattern as described for the *las* system. In contrast to the *las* signal molecule the *rhl* signal molecule C4-HSL diffuses freely in and out of the cells due to its small sized acyl chain (Pearson et al. 1999b). The active complex of RhlR-BHL acts as a positive transcriptional regulator, self-inducing its own synthesis and a number of target genes (Skindersoe et al. 2008).

A third signal molecule, the *Pseudomonas* Quinolone signal (PQS) is a 4-quinolone-based entity encoded by the biosynthetic *pqsABCDE* operon and *pqsH*, that through condensation of anthranilate and β-keto dodecanoate generate more than 50 2-alkyl-4-quinolones out of which HHQ (2-heptyl-4-quinolone) is converted to PQS by PqsH. HHQ is thought to diffuse between cells and to be converted to PQS in the target cells. Of all the synthesized 2-alkyl-4-quinolones,

HHQ and PQS are the dominant molecules, and especially the PQS molecule is bioactive. PQS is produced during logarithmic growth (Dubern and Diggle 2008) through early stationary phase peaking in late stationary phase of growth (McKnight et al. 2000b). The PQS system has been shown to be an integrated part of the *P. aeruginosa* QS hierarchy. Expression of the *pqsABCDE* regulon is regulated via PqsR and depends on the ratio between the two AHLs; 3-oxo-C12-HSL induces and BHL-HSL reduces expression of the operon (McGrath et al. 2004). Pesci et al. (1999) reported that synthesis of PQS required a functional LasR protein. However, later it was shown that a *lasR* mutant is capable of producing PQS in the "late" stationary phase, indicating that only "early" (or transition phase) PQS synthesis requires LasR (Diggle et al. 2003a). McKnight et al. (2000a) showed that PQS imposes a significant transcription initiation effect on *rhlI*. Exogenous addition of PQS at the time of inoculation initiates the production of several virulence factors such as elastase, lectin A, and pyocyanin, independent of the density or growth phase of the bacteria (Diggle et al. 2003b). It has been reported that a functional PQS system is required for *P. aeruginosa* pathogenicity in a nematode model of infection (Gallagher and Manoil 2001), and the PQS signal of *P. aeruginosa* has been detected in the lungs of CF patients (Collier et al. 2002).

4 *P. aeruginosa* QS and Biofilm Differentiation

P. aeruginosa forms biofilms which during infection enable them to cope with hostile environments in particular those of relevant host organisms. In the host, biofilm formation may therefore be considered as an important pathogenicity trait. *P. aeriginosa* biofilms are bacteria encapsulated in extrapolymeric substances (EPS) produced by the bacteria and additionally adapted from the host. The EPS consists of polysaccharides, extracellular DNA and other macromolecular components such as proteins, lipids and biosurfactants (Pamp et al. 2007). The EPS seems to constitute the scaffolding component of the aggregating bacteria in the biofilm (Allesen-Holm et al. 2005; Bjarnsholt et al. 2005a; Nivens et al. 2001), acts as scavenger of free oxygen radicals (Simpson et al. 1989), and annihilates the effect of antibiotics including aminoglycosides (Allison and Matthews 1992; Hentzer 2001).

Many believe that biofilm formation is a developmental progression that involves a number of complex and highly regulated processes (O'Toole et al. 2000) including cell–cell signaling. Sauer (2003) have generated proteomic data suggesting that in vitro *P. aeruginosa* biofilm formation proceeds through an ordered, temporal series of events which inevitably relies on the ordered expression of a collection of biofilm-specific genes (the biofilm program) (O'Toole et al. 2000; Sauer 2003). What at the present speaks against a specific genetic biofilm program is that specific genes, expressed independently of environmental or the laboratory setups under which biofilms have been formed, still have not been identified in *P. aeruginosa* (Hentzer et al. 2003). In our view, the available data support a model by which *P. aeruginosa* biofilm formation, and the successive growth and development may proceed through different but converging paths each of which may contain chaotic elements. So instead of one particular, predestined biofilm program, much of the available data suggest that the *P. aeruginosa* biofilm transcriptome reflects adaptation to nutritional and environmental conditions (Hentzer et al. 2005a, 2005; Purevdorj et al. 2002; Reimmann et al. 2002). However, what we need to get the full picture is increased sensitivity of the present transcriptomic protocols and omics analysis performed under relevant in vivo conditions.

Throughout the years, great emphasis has been put on mushroom-shaped multicellular structures as a phenomenon that underscores the presence of confined cell differentiation and diverge activities of subpopulations in development of biofilm architecture. Davies et al. (1998a) concluded that *P. aeruginosa* required QS signaling to enable development of differentiated biofilms with mushroom-shaped multicellular structures. There has been a lot of controversy about this dogma. Heydorn et al. (2002) showed that with citrate as the carbon source the wild-type and QS mutant form similar flat biofilms. Apparently, the involvement of signaling in the biofilm forming process depends on the specific conditions of each in vitro laboratory setup and is therefore conditional (Kjelleberg and Molin 2002). Another contribution to the apparent inconsistency of the involvement of QS in biofilm development is the application of QS mutants with uncertain genetic backgrounds. It was reported by Beatson et al. (2002) that QS mutants accumulate secondary mutations, which leads to misinterpretation of the QS effect on biofilm formation. This is further supported by Schaber et al. (2007) who observed that *P. aeruginosa* forms a biofilm in a thermal induced infection of mice independently of QS. Apparently, nutrient availability is the major factor in determining development of mushroom structures in *P. aeruginosa* biofilm in vitro. Klausen et al. (2003b) showed that with glucose as carbon source *P. aeruginosa* forms a highly differentiated biofilm composed of mushroom structures, but with citrate as carbon source the same strain forms a flat and undifferentiated biofilm. With glucose as carbon source the cells differentiate into a nonmotile subpopulation engaged in forming the stalks of the mushrooms, and a motile subpopulation subsequently engaged in forming the caps of the mushrooms (Klausen et al. 2003a). With citrate as carbon source, the entire population is motile in the initial phase of biofilm formation and microcolonies that could serve as mushroom stalks are not formed (Klausen et al. 2003b). When citrate is used as carbon source, it is present in much higher concentration than iron in the medium. Because citrate chelates iron, it is possible that the use of citrate as carbon source in flow-through systems reduces the level of iron available to the bacteria. Iron limitation has been shown to induce surface motility in *P. aeruginosa* (Singh et al. 2002), which may explain why the entire population is motile in *P. aeruginosa* biofilms irrigated with citrate medium. We would like to emphasize, therefore, that the available evidence suggests that formation of the mushroom-shaped structures in *P. aeruginosa* biofilms is first and foremost subject to the physical and chemical conditions present in the laboratory flow-cells, which argues against the hypothesis that biofilm formation in essence follows a predestined developmental program which as indispensable components include surface sensing and concomitant QS signaling. As we will discuss later, the role of QS is to offer protection and enable the biofilm bacteria evade host immunity and detrimental action of antimicrobial compounds. Furthermore, the presence of mushroom structures has to our knowledge never been reported in ex vivo samples. However, although they may be limited to flow-cells, the studies involving these structures have provided important insights regarding the fundamental processes of biofilm formation and tolerance development that may be of relevance outside the laboratory model systems.

5 *P. aeruginosa*, Intracellular Signaling and Biofilm Differentiation

The role of intracellular signaling in the biofilm lifecycle has gained momentum. Biofilms are thought to maintain "steady state" by means of growth and dispersal. The dispersal occurs either as single cells or as small microcolonies ripped-off from the biofilm (Costerton et al. 1999; Stapper et al. 2004; Stoodley et al. 2001; Webb et al. 2003). Evidence is accruing that the shift between the biofilm lifestyle (with production of matrix components such as Pel/Psl polysaccharides and Cup fimbriae) and the planktonic lifestyle (without production of these matrix components) is regulated via proteins that contain diguanylate cyclase or phosphodiesterase activities. Such proteins influence the intracellular level of the second messenger molecule cyclic diguanosine-$5'$-monophosphate (c-di-GMP) (Hickman et al. 2005; Meissner et al. 2007). It appears that high intracellular c-di-GMP levels upregulate matrix production and biofilm formation, whereas low intracellular c-di-GMP levels downregulate matrix production and induce a planktonic lifestyle. Several factors may contribute to dispersal of biofilms including phage induction (Webb et al. 2003), carbon starvation Gjermansen et al. (2005, 2009); Schleheck et al. 2009), and nitric oxide signaling (Barraud et al. 2006, 2009). Carbon starvation and nitric oxide signaling were shown to mediate phosphodiesterase activity in *P. aeruginosa* causing decreased intracellular c-di-GMP levels. In addition, Davies and colleagues have recently shown that *P. aeruginosa* produces a small organic compound, *cis*-2-decanoic acid which, when added exogenously at native concentrations (2.5 nM), was able to induce dispersion in an in vitro *P. aeruginosa* biofilm (Davies and Marques 2009). Taken together, this suggests that dispersal is both a programmed process and subject to physical conditions such as shear forces and sloughing.

These conclusions are based solely on in vitro observations, but if massive dispersal can be chemically promoted and occurs in medical biofilms, it may have severe implications since it is a mechanism by which biofilm bacteria might disseminate through the infected organ to the entire body and thereby cause rigorous systemic infections. For this "worst case scenario" to develop, it would require flow conditions as those present in blood vessels or in the synovial fluid surrounding a moving artificial limp. If the biofilm, however, is located within the tissue or organs such as in the CF lung, a chronic wound or permanent tissue fillers, the absence of direct flow limits the dissemination of bacteria, but localized increase in inflammation may occur as a consequence of biofilm spreading to vacant areas.

6 Signaling and the EPS Matrix

P. aeruginosa QS mutants evidently are able to form mushroom-shaped structures in vitro (Allesen-Holm et al. 2005). This can be explained by the fact that bacteria in the upper part of *P. aeruginosa* biofilm have a growth advantage (Pamp et al. 2008), it may be expected that there is a strong selection in *P. aeruginosa* QS mutant flow-chamber biofilms for secondary mutants that are able to form extended multicellular structures. Recent work has suggested that *P. aeruginosa* QS mutants can form microcolonies in biofilms, but that subsequent formation of mushroom caps requires QS-regulated components such as rhamnolipid and extracellular DNA (eDNA) (Pamp and Tolker-Nielsen 2007; Yang et al. 2009a). In addition, evidence has been shown that chemical inhibition of the PQS system causes *P. aeruginosa* to form microcolonies on which mushroom caps are not subsequently formed (Yang et al. 2007, 2009a). Transcriptomic analysis unquestionably reveals that QS controls gene expression in biofilms. Hentzer et al. (2005a) showed that the majority of QS-regulated genes are expressed 16 h after inoculation in flow chambers and are inducible in a *lasI*, *rhlI* background by the addition of exogenous OdDHL and BHL. With background in comparison of the transcriptome at various stages of the growth cycle of planktonic cells (in our laboratory settings), there appears to be a subclass of QS regulated biofilm genes that are not expressed in the planktonic mode of growth according to Hentzer et al. (2005a). A number of genes involved in iron scavenging or metabolism appear to be dependent on a functional QS system for expression in the biofilm mode of growth, which supports the findings that iron availability plays a role in the development of biofilm architecture (Banin et al. 2005). A recent study has provided evidence that, after formation of the initial microcolonies in *P. aeruginosa* biofilm, PQS chelation of iron helps inducing genes involved in iron scavenging, which promotes subsequent structural development (Yang et al. 2009a). Hentzer et al. (2004a) have shown that older (10 days or more) QS-deficient biofilms differ in volume compared to the wild type. In line with that, Bjarnsholt et al. (2005a) showed that QS deficiency impairs the structural stability of the biofilm. Microscopic inspections revealed that the towers and mushrooms of QS-deficient biofilms appeared wobbly and were less robust than their wild-type counterparts. An important link between biofilm stability and QS is the production of the eDNA, which is QS regulated (Allesen-Holm et al. 2005). Whitchurch et al. (2002) reported that eDNA plays a role as matrix component in *P. aeruginosa* biofilm formation. The eDNA has been shown to offer rigidity to the biofilm architechture (Jensen et al. 2007b; Bjarnsholt et al. 2005a), which might explain why QS-deficient mutants are often reported "biofilm negative" in microtiter assays as the wash and staining procedures mechanically remove the less robust QS-deficient biofilm.

In conclusion, it appears that although QS contributes actively to the regulation of gene expression in the biofilms under in vitro conditions, a functional QS system is not required for *P. aeruginosa* biofilm formation per se. QS gets activated as a consequence of increased cell density, which in turn leads to increased concentration of QS signal molecules (MacLehose et al. 2004). The spatial build-up of signal molecules is likely to be more pronounced in biofilms compared with planktonic cultures, due to the confined space of the biofilm matrix (Charlton et al. 2000).

7 How *P. aeruginosa* Resists Antibiotics and Immune Defenses

Numerous investigations have fueled the dogma that biofilm bacteria are superior to their planktonic counterparts when it comes to coping with detrimental effects of antimicrobials (Anwar et al. 1989; Bjarnsholt et al. 2005a, 2007; Costerton et al. 1995; Donlan and Costerton 2002; Drenkard 2003). In the recent years, a number of molecular mechanisms have started to surface which seem to enforce and combine the "mid nineties" hypothesis of the biofilm matrix providing an extensive barrier for penetration in combination with classical resistance mechanisms and physiological adaptations. Chemostat experiments performed during the 1970s and 1980s demonstrated that tolerance can at least in part be attributed to the physiology of slow growth, a condition observed in large areas of in vitro biofilms. Slow growth is likely to be caused by nutrient-limiting gradients present in the biofilm (Roberts and Stewart 2004). DNA array-based analysis performed by us has shown that the transcriptome of a biofilm (showing increased tolerance to antibiotic treatments) resembles the transcriptome of a transition phase planktonic culture including expression of genes involved in anaerobiosis (Hentzer et al. 2005a). This can explain the reduced antimicrobial properties of most antibiotics, since they target biological processes, which are most predominant in actively growing cells (Donlan and Costerton 2002; Hoyle and Costerton 1991). This is in accordance with observations by Pamp et al. (2008) and Haagensen et al. (2006), showing that in vitro *P. aeruginosa* consists of at least two sub-populations one growing more actively than the other. Accordingly, the outer layer being most similar to exponential growing cells are killed by conventional antibiotics whereas cells in the more central parts due to conditions of slow growth tolerate these antibiotics. These cells, however, are killed by membrane targeting antibiotics such as Colistin and probably also sensitive to heavy metals which in contrast are actively pumped out by efflux pumps in the growing outer subpopulation. The microenvironment of the biofilm such as a difference in the pH, pCO_2, pO_2, etc. may affect the efficacy of antimicrobial compounds, for example, macrolides and tetracyclines are compromised at low pH (Dunne 2002; Walters et al. 2003).

QS is involved in the development of antibiotic and biocide tolerance, but the full mechanistic pathway is not yet known (Bjarnsholt et al. 2005a; Hassett et al. 1999;

Hentzer and Givskov 2003; Hentzer et al. 2002, 2003; Shih and Huang 2002). However, eDNA release is QS regulated and recently Mulcahy et al. (2008) demonstrated that eDNA by means of its cationic chelating properties can neutralize the activity of many antimicrobials including tobramycin. This may explain our previously published observations that in vitro tobramycin tolerance is in part QS dependent and may explain why treatment with QS inhibitors (QSI) show synergistic action with tobramycin (Bjarnsholt et al. 2005a). The protective role of the biofilm matrix has also been shown to provide the biofilm embedded cells resistance against heavy metals such as zinc, copper, lead (Teitzel and Parsek 2003), and silver (Bjarnsholt et al. 2007). Teitzel and Parsek (2003) observed that exponentially growing cells were more tolerant to heavy metals than stationary phase planktonic cells.

P. aeruginosa biofilms are now believed to play a key role in persistent infections, in particular in nonhealing leg ulcers (chronic wounds) (Bjarnsholt et al. 2008; Kirketerp-Moller et al. 2008b) and the chronic lung infection of CF patients (Bjarnsholt et al. 2009a) (see Fig. 2). Chronic wounds may arise from pressure wounds or cuts in patients with reduced or impaired blood circulation in the lower extremities. The Chronic wound remain at an inflammatory stage, which is characterized by a continuing influx of polymorphonuclear neutrophils (PMNs) (Banin et al. 2005) that release cytotoxic enzymes, free oxygen radicals, and

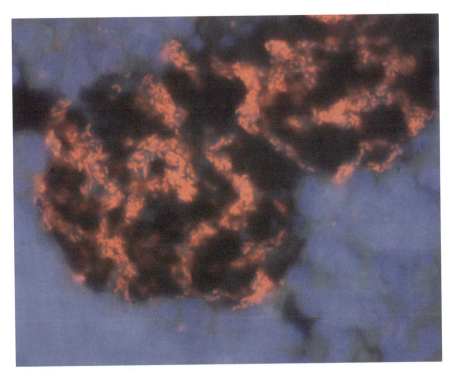

Fig. 2 *P. aeruginosa* biofilm in a chronic infected CF lung. The bacteria are detected by specific PNA FISH (*red*) and the leucocytes are detected by DAPI (*blue*)

inflammatory mediators causing extensive collateral damage to the host tissue. We have recently put forward that, within a chronic wound, the healing and destructive processes are out of balance (Bjarnsholt et al. 2008). Consequently, manipulating and counterbalancing these processes may promote healing of the chronic wound. Furthermore, the *P. aeruginosa* infected wounds have a larger surface area than wounds without *P. aeruginosa* (Amara et al. 2009) infection, and it is our view that the presence of *P. aeruginosa* plays a significant role in delaying or even preventing the healing process (Anwar et al. 1989).

CF is a monogenic autosomal recessive, multiorganic disease though the main cause of morbidity and mortality has its root in the patients increased susceptibility to lung infections. The genetic cause of CF was identified in 1989 as a defect in the Cystic Fibrosis Transmembrane Conductance Regulator (CFTR) gene, located on chromosome 7 (Kerem et al. 1989; Riordan et al. 1989; Rommens et al. 1989). This causes a dysfunction of a chloride ion channel, depleting the otherwise normal mucus of water which impairs the mucociliary clearing mechanism of the bronchia. Here, inhaled microorganisms are able to multiply which in turn recruits an inflammatory response. The lower respiratory tract in patients with CF is recurrently colonized with a variety of bacteria, such as *Haemophilus influenzae*, *S. aureus*, and *P. aeruginosa* before it is chronically infected with *P. aeruginosa* (80% of the adult patients). However, aggressive antibiotic treatments, which delay the onset of chronic infection, remarkably prolong the life expectancy of patients with CF (Barraud et al. 2006).

Microscopic inspections of ex vivo samples from wounds and lungs of CF patients reveal that bacteria are assembled in structures consisting of aggregated cells, which are surrounded but not penetrated by inflammatory cells, in particular phagocytic cells such as PMNs (see Fig. 2). We propose that bacterial persistence, including the extreme tolerance to antibiotics and evasion by the immune system, is caused by the bacteria's ability to form these aggregations which represent bacterial biofilms in vivo. In these settings, the presence of the EPS component alginate can be demonstrated by fluorescent antibodies (Bjarnsholt et al. 2009a; Kirketerp-Moller et al. 2008a). Second, both kinds of patients are immunocompromised suffering from defects in the primary line of defenses. Chronic wounds consist primarily of granulation tissue composed of a network of collagen fibers, new capillaries, and extracellular matrix, together with PMNs, macrophages, and fibroblasts. Embedded in this, we have found the aggregated microcolonies of bacteria. This is in accordance with our observations from the chronic infected CF lung; here within the lumen of the bronchial airways, *P. aeruginosa* is also detected in aggregated microcolonies. Since the patients in both diseases do not suffer from defects in the cellular defense, the PMNs would be expected to eradicate the bacteria.

The key to this persistence is caused by particular components of the EPS matrix. *P. aeruginosa* produces rhamnolipids (Jensen et al. 2007b) powerful detergents, which cause cellular necrosis and concomitant elimination of the PMNs (Alhede et al. 2009; Jensen et al. 2007b; van Gennip et al. 2009). Accordingly, the rhamnolipids function as a PMN shield, and we put forward that the capability of mounting this shield significantly contributes to the ability of *P. aeruginosa* to persist in the lung as well as in the wound (see Fig. 3). The implications of sustained PMN lysis are that antimicrobials as well as tissue-devastating compounds spill out. Examples of such compounds are myeloperoxidase, elastase, and matrix metallopeptidase 9 (MMP-9) in which chronic wound fluids are particularly rich, in contrast to fluid from acute wounds in humans (Wysocki et al. 1993). This also applies for CF patients, in contrast to patients suffering from acute respiratory failure (Gaggar et al. 2007). We also suggest that the elimination of PMNs may facilitate further colonization.

Regulation of rhamnolipid synthesis is governed by QS, in particular the PQS and *rhl* systems (Alhede et al. 2009). Rhamnolipid production constitutes a "biofilm shield" functions to promote necrosis of PMN-leukocytes (upon biofilm contact), which subsequently causes collateral tissue damage and promotes development of inflammation. This suggests that quora (which are likely to be represented by the observed bacterial biofilm aggregates) have amassed at certain locations in chronic wounds (Fazli et al. 2009; Kirketerp-Moller et al. 2008b) and in the CF lung (Bjarnsholt et al. 2009a). Such quora are then capable of eliminating PMNs by the production of rhamnolipid, which in turn would reduce the number of functional PMNs at their location (Alhede et al. 2009; Jensen et al. 2007b). Interestingly, AlgR was recently shown to reduce expression of RhlR-controlled gene expression in the biofilm mode (Morici et al. 2007). Concurrently, we found that the content of rhamnolipids was below the detection limit in our in vitro biofilms (Alhede et al. 2009). However, exposure to freshly isolated PMNs superseded AlgR repression and significantly increased expression of rhamnolipids up to a level of 50 µg/g biofilm, which in pure form would causes lysis of PMN within 30 min of exposure (Alhede et al. 2009) (Jensen et al. 2007b). The inducing effect was found to require a functional QS system to induce transcription initiation of the RhlR and PQS regulated genes, and it indicates that *P. aerginosa* to upregulate its PMN shield receives and responds to signal molecules originating from the host defense cells. Specific effects of cytokines such as interferon gamma (IFN-g) have recently been shown to be transmitted through IFN-g binding to OprF, resulting in the expression of the QS-dependent virulence determinants PA-I lectin, BHL and Pyocyanin (Wu et al. 2005). The observation that lungs of mice infected with wild-type *P. aeruginosa* contain significantly less intact PMN compared with lungs of mice infected with a *rhlA* mutant supports our current "shield model" (Alhede et al. 2009); in the host similar to the in vitro (Jensen et al. 2007a), *P. aeruginosa* biofilms resist phagocytosis by eradicating the PMNs on contact.

Cross-kingdom signaling plays a significant role in bacteria–host interactions. Recent evidence from us also supports the view that QS signal molecules have distinct targets and alter gene expression in mammalian cells (Kristiansen et al. 2008). These alterations include modulation of pro-inflammatory cytokines and induction of apoptosis. In particular, we found that *P. aeruginosa* QS signals decrease the production of IL-12 by murine bone marrow-derived dendritic cells (BM-DCs) without altering the IL-10 release. BM-DCs exposed to QS signals during antigen stimulation exhibit a decreased ability to induce specific T-cell

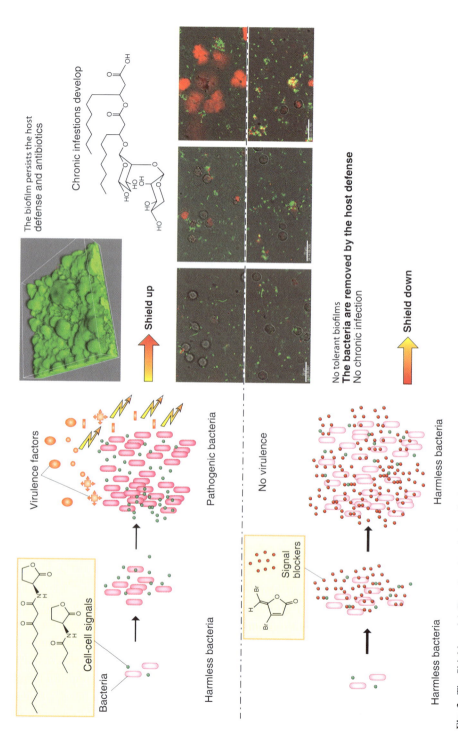

Fig. 3 The Shield model. Top situation depicts the buildup of a pathogenic biofilm with the rhamnolipid shield. The lower situation depicts the same buildup of bacteria but the shield is disrupted by QS inhibitors, preventing the tolerant and pathogenic biofilm

proliferation. These in vitro results suggest that the *P. aeruginosa* QS signal molecules impede DCs in exerting their T-cell-stimulatory effects and function as immunomodulators changing the milieu from the host-protecting proinflammatory TH1, thereby possibly enabling the establishment of bacterial infection within the host (Skindersoe et al. 2009).

8 Interference with Bacterial Gossip and Why It Works as and Antimicrobial Strategy

Knowledge of the above described bacteria–host interactions and shielding makes the development of new treatment scenarios possible. Lately, several studies have proven the concept of inhibiting QS as a functional and relevant antimicrobial strategy (Bjarnsholt et al. 2005b; Hentzer et al. 2003; Rasch et al. 2004; Wu et al. 2004b) (see Figs. 1 and 3). Despite the fact that QS systems do not control expression of essential genes per se, QS systems have been shown to constitute new promising antimicrobial drug targets (Bjarnsholt et al. 2005a, b; Hentzer and Givskov 2003; Rasch et al. 2004; Wu et al. 2004b). We also recently demonstrated that knockout of the *rhlA* (leaving an otherwise functional QS system) leads to significant attenuation in two infectious animal models (van Gennip et al. 2009). We propose that the main reason why QSI conduct functional antimicrobial effects in vivo is by preventing the establishment of the PMN shield, and consequently reinstate proper antimicrobial functions of the PMNs (see Fig. 3). Since PMN lysis is then likely to be reduced, local tissue destruction and inflammation will be kept to a minimum. This in turn reduces release of PMN DNA, which will be expected to further contribute to reduction of the viscosity of the bronchial secretion in CF. This may help increasing and maintain lung function (Hodson et al. 2003). Both effects will be expected to tip the balance in favor of bacterial eradication and subsequent healing. Disruption of the QS signaling cascade may be a particularly valuable strategy in the treatment of chronic, biofilm infections. Likewise, it has to be emphasized that the rationale behind this approach is to specifically interfere with disease-causing traits and development of inflammation rather than restricting the growth of the bacteria. As production of virulence factors is inhibited, the pathogen can no longer maintain its position in an increasingly more hostile host environment and consequently become cleared by the innate host defenses. Given that the mere attenuation of the pathogen does not impose a harsh and direct selective pressure on the cells, it can be anticipated that resistant mutants will not arise as readily as in the case of conventional antibiotic treatment. We have denoted this as the anti-pathogenic drug principle (Hentzer and Givskov 2003).

The first example that nature has already applied QSIs as an antibacterial strategy was the halogenated furanones produced by the Australian macroalga *Delisea pulchra* (de Nys et al. 1993; Givskov et al. 1996; Kjelleberg et al. 1997). The authors showed that the halo-furanones, administered within a certain concentration window, specifically inhibit bacterial QS, thereby blocking surface colonization and virulence factor expression in *S. liquefaciens*. Western blot analysis indicated that the halo-furanones are able to promote rapid turnover of LuxR homologues proteins, thereby reducing the amount of protein available to interact with AHL and act as transcriptional regulator (Manefield et al. 2002). These observations taken together with a mutational and in silico docking analysis of compounds into the LuxR crystal structure indicate a direct interaction with the receptor protein (Koch et al. 2005). The efficacy against *P. aeruginosa* QS and virulence was greatly enhanced after subsequent development of a series of synthetic furanone analogues. Among these, compound C-30 (Hentzer et al. 2003) and C-56 (Hentzer et al. 2002) showed QSI drug potential in *P. aeruginosa*. Hentzer et al. (2003) showed that biofilms grown with and without furanone C-30 were not significantly different with respect to formation and biomass. However, when treating 3-day-old biofilms with 100 μg/ml tobramycin, grown either in the presence or in the absence of furanone C-30, a significant number of dead cells in the furanone C-30 treated biofilm were detected. The antimicrobial properties of both furanone C-30 and furanone C-56 were also demonstrated in a chronic pulmonary mouse model (Hentzer et al. 2003; Wu et al. 2004a), an implant model (Christensen et al. 2007; Rasch et al. 2004), and a fish model for Vibriosis 141. The bacteria were cleared significantly faster in mice treated with 1 μg/g body weight furanone compared to placebo-treated mice. Other furanone derivatives have been synthesized by Kim et al. (2008); however, these furanones also block in vitro biofilm formation of *P. aeruginosa*. This indicated that these furanone derivatives are highly toxic to the bacteria since genetic analysis shows that a functional QS system is not required for biofilm formation in vitro. Since the discovery of the halo-furanones, a number of papers have reported on synthesis of QSI compounds by means of organic chemistry approach (Borlee et al. 2008; Geske et al. 2007, 2008a, b, c; Mattmann et al. 2008; Muh et al. 2006; Olsen et al. 2002; Persson et al. 2005a, b; Riedel et al. 2006). However, a detailed overview of this is not the scope of this chapter.

Many of these compounds have been developed based on agonist-based designs and probed for their action by means of specific QS screens (see below). The Greenberg group has employed in silico analysis for drug development (Muh et al. 2006) were the 12 akyl chain substituted tetrazol PD12 showed an ID_{50} of staggering 30 nM. Except for that and the fact that the target specificity and functionality of only C-30 has been validated by transcriptomic analysis and animal experiments, this compound is so far the most active, functional QSI drug ever designed with an ID_{50} of 2 μM and full QS inhibition (without any growth inhibition) exerted even at 10 μM. In connection with that, we emphasize that physiological relevant concentrations of *P. aeruginosa* QS signals are likely to be in the range of 1–20 μM. Recently, a crystal structure of a LasR-peptide with OdDHL bound became available (Bottomley et al. 2007). For the first time it has become possible to apply in silico analysis of receptor–ligand interactions such as those performed by Amara et al. (2009). The authors demonstrated direct covalent interaction of a reactive isothiocyanate AHL derivative with an amino acid present in the binding site. Based on the 3-dimensional structure of the LasR protein (without ligand), we have used computer-based virtual screening in a search for putative quorum-sensing inhibitors from a database comprising approved drugs and natural compounds (Yang et al. 2009b). Three potential QSIs, which showed effects on *P. aeruginosa* virulence factor expression, were identified but much less active compared with C-30.

QSI compounds are present in food, herbal, and fungal sources. Several years ago, we developed a new tool for high-throughput screening of fresh samples as well as crude extracts and pure compounds of QSI (Bjarnsholt et al. 2009b; Rasmussen et al. 2005a). The presence of QSI compounds in natural foods is extremely interesting because, in most cases, vegetables and herbs are non toxic to humans and are easily available. One of the most potent QSIs we have detected is contained in extracts of garlic (Bjarnsholt et al. 2005b; Rasmussen et al. 2005a). Also fungi produce QSIs (Rasmussen et al. 2005c) and supports the view that organisms devoid of functional immune systems relay on chemical defense systems against competing or detrimental bacteria.

In 2001, Teteda et al. (2001), published results showing the antibiotic azithromycin inhibits QS of *P. aeruginosa*. This led us to investigate whether QS inhibition is a common feature of antibiotics. We screened 12 antibiotics for their QS-inhibitory activities using the QSI selector system (Rasmussen et al. 2005a). Three of the antibiotics tested (azithromycin, ceftazidime, and ciprofloxacin) were active in the assay and were further examined for their effects on QS-regulated virulence factor production in *P. aeruginosa*. Consistent results from the virulence factor assays, reverse transcription-PCR, and the DNA microarrays support the finding that azithromycin, ceftazidime, and ciprofloxacin decrease the expression of a range of QS-regulated virulence factors. The data suggest that the underlying mechanism may be mediated by changes in membrane permeability, thereby influencing the flux of 3-oxo-C12-HSL (Skindersoe et al. 2008). Azithromycin inhibits or reduce the production of several of the virulence factors of *P. aeruginosa*, such as elastase, rhamnolipids (Skindersoe et al. 2008) and interfere with alginate polymerization (Hoffmann et al. 2007). Recently, another group reported that azithromycin inhibits the transcription AHL synthesis by blocking unknown *lasI* and *rhlI* regulation (Kai et al. 2009). It is proposed that the azithromycin interferes with one of the precursors of the AHL molecules indicating that the QSI activity is a secondary effect (Kai et al. 2009). Azithromycin does improve the respiratory function in CF patients (Saiman 2004; Southern and Barker 2004). The principal role of QSIs is to attenuate the virulence of the bacteria, and in our opinion if the excreted rhamnolipids are reduced *P. aeruginosa* biofilms are dramatically weakened. Due to this, it is of great interest that Park et al. (2008) published Solenopsin A extracted from the fire ant *Solenopsis invicta* to target the *rhl* QS system specifically.

With the present-day knowledge the central role of "bacterial gossip" in pathogenesis is evident. In particular QS seems to be an obvious new drug target to the control of unwanted microbial activity. This implies both the production of virulence factors, antibiotic resistance, and modulation of the host immune system. We believe that QSI and also c-di-GMP modulating drugs could be beneficial in the fight against biofilm infections, though much effort and enthusiasm from the industry is needed to develop pharmaceutical relevant molecules. This endeavor is of global, societal, and economic significance. Despite the selective force imposed by signaling drugs may be less stringent compared with conventional antibiotics, it will however be unwise to assume that resistance at some point will not occur, and this probably is why we will never end the battle against infections.

References

Alhede M, Bjarnsholt T, Jensen PO, Phipps RK, Moser C, Christophersen L, Christensen LD, van GM, Parsek M, Hoiby N, Rasmussen TB, Givskov M (2009) Pseudomonas aeruginosa recognizes and responds aggressively to the presence of polymorphonuclear leukocytes. Microbiology 155:3500–3508

Allesen-Holm M, Barken Bundvig K, Yang L, Klausen M, Webb JS, Kjelleberg S, Molin S, Givskov M, Tolker-Nielsen T (2005) A characterization of DNA release in *Pseudomonas aeruginosa* cultures and biofilms. Mol Microbiol 59(4):1114–1128, Online early

Allison DG, Matthews MJ (1992) Effect of polysaccharide interactions on antibiotic susceptibility of *Pseudomonas aeruginosa*. J Appl Bacteriol 73:484–488

Amara N, Mashiach R, Amar D, Krief P, Spieser SA, Bottomley MJ, Aharoni A, Meijler MM (2009) Covalent inhibition of bacterial quorum sensing. J Am Chem Soc 131:10610–10619

Anwar H, van Biesen T, Dasgupta M, Lam K, Costerton JW (1989) Interaction of biofilm bacteria with antibiotics in a novel in vitro chemostat system. Antimicrob Agents Chemother 33: 1824–1826

Banin E, Vasil ML, Greenberg EP (2005) Iron and *Pseudomonas aeruginosa* biofilm formation. Proc Natl Acad Sci USA 102:11076–11081

Barraud N, Hassett DJ, Hwang SH, Rice SA, Kjelleberg S, Webb JS (2006) Involvement of nitric oxide in biofilm dispersal of *Pseudomonas aeruginosa*. J Bacteriol 188:7344–7353

Barraud N, Schleheck D, Klebensberger J, Webb JS, Hassett DJ, Rice SA, Kjelleberg S (2009) Nitric oxide signaling in *Pseudomonas aeruginosa* biofilms mediates phosphodiesterase activity, decreased cyclic diguanosine-5′-monophosphate levels and enhanced dispersal. J Bacteriol 191(23):7333–7342

Beatson SA, Whitchurch CB, Semmler AB, Mattick JS (2002) Quorum sensing is not required for twitching motility in *Pseudomonas aeruginosa*. J Bacteriol 184:3598–3604

Bjarnsholt T, Jensen PO, Burmolle M, Hentzer M, Haagensen JA, Hougen HP, Calum H, Madsen KG, Moser C, Molin S, Høiby N, Givskov M (2005a) *Pseudomonas aeruginosa* tolerance to tobramycin, hydrogen peroxide and polymorphonuclear leukocytes is quorum-sensing dependent. Microbiology 151:373–383

Bjarnsholt T, Jensen PO, Rasmussen TB, Christophersen L, Calum H, Hentzer M, Hougen HP, Rygaard J, Moser C, Eberl L, Høiby N, Givskov M (2005b) Garlic blocks quorum sensing and promotes rapid clearing of pulmonary *Pseudomonas aeruginosa* infections. Microbiology 151: 3873–3880

Bjarnsholt T, Kirketerp-Moller K, Kristiansen S, Phipps R, Nielsen AK, Jensen PO, Høiby N, Givskov M (2007) Silver against *Pseudomonas aeruginosa* biofilms. APMIS 115:921–928

Bjarnsholt T, Kirketerp-Moller K, Jensen PO, Madsen KG, Phipps R, Krogfelt K, Høiby N, Givskov M (2008) Why chronic wounds will not heal: a novel hypothesis. Wound Repair Regen 16:2–10

Bjarnsholt T, Jensen PO, Fiandaca MJ, Pedersen J, Hansen CR, Andersen CB, Pressler T, Givskov M, Hoiby N (2009a) *Pseudomonas aeruginosa* biofilms in the respiratory tract of cystic fibrosis patients. Pediatr Pulmonol 44:547–558

Bjarnsholt T, van Gennip M, Jakobsen TH, Christensen LD, Jensen PO, Givskov M (2009b) In vitro screens for quorum sensing inhibitors and in vivo confirmation of their effect. Nat Protoc 5(2):282–293

Boles BR, Horswill AR (2008) Agr-mediated dispersal of *Staphylococcus aureus* biofilms. PLoS Pathog 4:e1000052

Borlee BR, Geske GD, Robinson CJ, Blackwell HE, Handelsman J (2008) Quorum-sensing signals in the microbial community of the cabbage white butterfly larval midgut. ISME J 2:1101–1111

Bottomley MJ, Muraglia E, Bazzo R, Carfi A (2007) Molecular insights into quorum sensing in the human pathogen Pseudomonas aeruginosa from the structure of the virulence regulator LasR bound to its autoinducer. J Biol Chem 282:13592–13600

Charlton TS, de Nys R, Netting A, Kumar N, Hentzer M, Givskov M, Kjelleberg S (2000) A novel and sensitive method for the quantification of N-3-oxoacyl homoserine lactones using gas chromatography-mass spectrometry: application to a model bacterial biofilm. Environ Microbiol 2:530–541

Christensen LD, Moser C, Jensen PO, Rasmussen TB, Christophersen L, Kjelleberg S, Kumar N, Hoiby N, Givskov M, Bjarnsholt T (2007) Impact of *Pseudomonas aeruginosa* quorum sensing on biofilm persistence in an in vivo intraperitoneal foreign-body infection model. Microbiology 153:2312–2320

Collier DN, Anderson L, McKnight SL, Noah TL, Knowles M, Boucher R, Schwab U, Gilligan P, Pesci EC (2002) A bacterial cell to cell signal in the lungs of cystic fibrosis patients. FEMS Microbiol Lett 215:41

Costerton JW, Lewandowski Z, Caldwell DE, Korber DR, Lappin-Scott HM (1995) Microbial biofilms. Annu Rev Microbiol 49:711–745

Costerton JW, Stewart PS, Greenberg EP (1999) Bacterial biofilms: a common cause of persistent infections. Science 284:1318–1322

Davies DG, Marques CN (2009) A fatty acid messenger is responsible for inducing dispersion in microbial biofilms. J Bacteriol 191:1393–1403

Davies DG, Parsek MR, Pearson JP, Iglewski BH, Costerton JW, Greenberg EP (1998) The involvement of cell-to-cell signals in the development of a bacterial biofilm. Science 280: 295–298

de Kievit TR, Kakai Y, Register JK, Pesci EC, Iglewski BH (2002) Role of the *Pseudomonas aeruginosa* las and rhl quorum-sensing systems in rhlI regulation. FEMS Microbiol Lett 212: 101–106

de Nys R, Wright AD, König GM, Sticher O (1993) New halogenated furanones from the Marine alga *Delisea pulchra*. Tetrahedron 49:11213–11220

Dekimpe V, Deziel E (2009) Revisiting the quorum-sensing hierarchy in *Pseudomonas aeruginosa*: the transcriptional regulator RhlR regulates LasR-specific factors. Microbiology 155:712–723

Diggle SP, Winzer K, Chhabra SR, Worrall KE, Camara M, Williams P (2003) The *Pseudomonas aeruginosa* quinolone signal molecule overcomes the cell density-dependency of the quorum sensing hierarchy, regulates rhl-dependent genes at the onset of stationary phase and can be produced in the absence of LasR. Mol Microbiol 50:29–43

Donlan RM, Costerton JW (2002) Biofilms: survival mechanisms of clinically relevant microorganisms. Clin Microbiol Rev 15:167–193

Drenkard E (2003) Antimicrobial resistance of *Pseudomonas aeruginosa* biofilms. Microbes Infect 5:1213–1219

Dubern JF, Diggle SP (2008) Quorum sensing by 2-alkyl-4-quinolones in *Pseudomonas aeruginosa* and other bacterial species. Mol Biosyst 4:882–888

Dunne WM Jr (2002) Bacterial adhesion: seen any good biofilms lately? Clin Microbiol Rev 15: 155–166

Eberl L, Winson MK, Sternberg C, Stewart GS, Christiansen G, Chhabra SR, Bycroft B, Williams P, Molin S, Givskov M (1996) Involvement of N-acyl-L-hormoserine lactone autoinducers in controlling the multicellular behaviour of *Serratia liquefaciens*. Mol Microbiol 20:127–136

Eberl L, Molin S, Givskov M (1999) Surface motility of *Serratia liquefaciens* MG1. J Bacteriol 181:1703–1712

Fazli M, Bjarnsholt T, Kirketerp-Moller K, Jorgensen B, Andersen AS, Krogfelt K, Givskov M, Tolker-Nielsen T (2009) Non-random distribution of *Pseudomonas aeruginosa* and *Staphylococcus aureus* in chronic wounds. J Clin Microbiol 47(12):4084–4089

Federle MJ, Bassler BL (2003) Interspecies communication in bacteria. J Clin Invest 112: 1291–1299

Frederiksen B, Koch C, Hoiby N (1997) Antibiotic treatment of initial colonization with *Pseudomonas aeruginosa* postpones chronic infection and prevents deterioration of pulmonary function in cystic fibrosis. Pediatr Pulmonol 23:330–335

Fuqua C, Parsek MR, Greenberg EP (2001) Regulation of gene expression by cell-to-cell communication: acyl- homoserine lactone quorum sensing. Annu Rev Genet 35:439–468

Gaggar A, Li Y, Weathington N, Winkler M, Kong M, Jackson P, Blalock JE, Clancy JP (2007) Matrix metalloprotease-9 dysregulation in lower airway secretions of cystic fibrosis patients. Am J Physiol Lung Cell Mol Physiol 293:L96–L104

Gallagher LA, Manoil C (2001) *Pseudomonas aeruginosa* PAO1 kills *Caenorhabditis elegans* by cyanide poisoning. J Bacteriol 183:6207–6214

Geske GD, O'Neill JC, Miller DM, Mattmann ME, Blackwell HE (2007) Modulation of bacterial quorum sensing with synthetic ligands: systematic evaluation of N-acylated homoserine lactones in multiple species and new insights into their mechanisms of action. J Am Chem Soc 129:13613–13625

Geske GD, Mattmann ME, Blackwell HE (2008a) Evaluation of a focused library of N-aryl L-homoserine lactones reveals a new set of potent quorum sensing modulators. Bioorg Med Chem Lett 18:5978–5981

Geske GD, O'Neill JC, Blackwell HE (2008b) Expanding dialogues: from natural autoinducers to non-natural analogues that modulate quorum sensing in Gram-negative bacteria. Chem Soc Rev 37:1432–1447

Geske GD, O'Neill JC, Miller DM, Wezeman RJ, Mattmann ME, Lin Q, Blackwell HE (2008c) Comparative analyses of N-acylated homoserine lactones reveal unique structural features that dictate their ability to activate or inhibit quorum sensing. Chembiochem 9:389–400

Gilligan PH (1991) Microbiology of airway disease in patients with cystic fibrosis. Clin Microbiol Rev 4:35–51

Givskov M, de Nys R, Manefield M, Gram L, Maximilien R, Eberl L, Molin S, Steinberg PD, Kjelleberg S (1996) Eukaryotic interference with homoserine lactone-mediated prokaryotic signalling. J Bacteriol 178:6618–6622

Gjermansen M, Ragas P, Sternberg C, Molin S, Tolker-Nielsen T (2005) Characterization of starvation-induced dispersion in *Pseudomonas putida* biofilms. Environ Microbiol 7:894–906

Gjermansen M, Nilsson M, Yang L, Tolker-Nielsen T (2009) Characterization of starvation-induced dispersion in *Pseudomonas putida* biofilms: genetic elements and molecular mechanisms. Mol Microbiol 75(4):815–826

Haagensen JA, Klausen M, Tolker-Nielsen T, Ernst RK, Miller SI, Molin S (2006) Differentiation and distribution of colistin/SDS tolerant cells in *Pseudomonas aeruginosa* flow-cell biofilms. J Bacteriol 189(1):28–37

Hassett DJ, Ma JF, Elkins JG, McDermott TR, Ochsner UA, West SE, Huang CT, Fredericks J, Burnett S, Stewart PS, McFeters G, Passador L, Iglewski BH (1999) Quorum sensing in *Pseudomonas aeruginosa* controls expression of catalase and superoxide dismutase genes and mediates biofilm susceptibility to hydrogen peroxide. Mol Microbiol 34:1082–1093

Hentzer M (2001) PhD, Technical University of Denamrk

Hentzer M, Givskov M (2003) Pharmacological inhibition of quorum sensing for the treatment of chronic bacterial infections. J Clin Invest 112:1300–1307

Hentzer M, Riedel K, Rasmussen TB, Heydorn A, Andersen JB, Parsek MR, Rice SA, Eberl L, Molin S, Høiby N, Kjelleberg S, Givskov M (2002) Inhibition of quorum sensing in *Pseudomonas aeruginosa* biofilm bacteria by a halogenated furanone compound. Microbiology 148:87–102

Hentzer M, Wu H, Andersen JB, Riedel K, Rasmussen TB, Bagge N, Kumar N, Schembri MA, Song Z, Kristoffersen P, Manefield M, Costerton JW, Molin S, Eberl L, Steinberg P,

Kjelleberg S, Høiby N, Givskov M (2003) Attenuation of *Pseudomonas aeruginosa* virulence by quorum sensing inhibitors. EMBO J 22:3803–3815

Hentzer M, Eberl L, Givskov M (2004a) Qourum sensing in biofilms: gossip in the slime city. In: Mahmoud G (ed) Microbial biofilms. ASM, Washington, DC

Hentzer M, Eberl L, Givskov M (2005) Transcriptome analysis of *Pseudomonas aeruginosa* biofilm development: anaerobic respiration and iron limitation. Biofilms 2:37–61

Hentzer M, Eberl L, Givskov M (2005a) Transcriptome analysis of *Pseudomonas aeruginosa* biofilm development: anaerobic respiration and iron limitation. Biofilms 2(01):37–61

Heydorn A, Ersboll B, Kato J, Hentzer M, Parsek MR, Tolker-Nielsen T, Givskov M, Molin S (2002) Statistical analysis of *Pseudomonas aeruginosa* biofilm development: impact of mutations in genes involved in twitching motility, cell-to-cell signaling, and stationary-phase sigma factor expression. Appl Environ Microbiol 68:2008–2017

Hickman JW, Tifrea DF, Harwood CS (2005) A chemosensory system that regulates biofilm formation through modulation of cyclic diguanylate levels. Proc Natl Acad Sci USA 102: 14422–14427

Hodson ME, McKenzie S, Harms HK, Koch C, Mastella G, Navarro J, Strandvik B (2003) *Dornase alfa* in the treatment of cystic fibrosis in Europe: a report from the epidemiologic registry of cystic fibrosis. Pediatr Pulmonol 36:427–432

Hoffmann N, Lee B, Hentzer M, Rasmussen TB, Song Z, Johansen HK, Givskov M, Høiby N (2007) Azithromycin blocks quorum sensing and alginate polymer formation and increases the sensitivity to serum and stationary growth phase killing of *P. aeruginosa* and attenuates chronic *P. aeruginosa* lung infection in Cftr -/-mic. Antimicrob Agents Chemother 51(10):3677–3687

Hoiby N (1974) Epidemiological investigations of the respiratory tract bacteriology in patients with cystic fibrosis. Acta Pathol Microbiol Scand B Microbiol Immunol 82:541–550

Hoyle BD, Costerton JW (1991) Bacterial resistance to antibiotics: the role of biofilms. Prog Drug Res 37:91–105

Huber B, Riedel K, Hentzer M, Heydorn A, Gotschlich A, Givskov M, Molin S, Eberl L (2001) The cep quorum-sensing system of *Burkholderia cepacia* H111 controls biofilm formation and swarming motility. Microbiology 147:2517–2528

Jensen PO, Bjarnsholt T, Phipps R, Rasmussen TB, Calum H, Christoffersen L, Moser C, Williams P, Pressler T, Givskov M, Hoiby N (2007a) Rapid necrotic killing of polymorphonuclear leukocytes is caused by quorum-sensing-controlled production of rhamnolipid by *Pseudomonas aeruginosa*. Microbiology 153:1329–1338

Jensen PO, Bjarnsholt T, Phipps R, Rasmussen TB, Calum H, Christoffersen L, Moser C, Williams P, Pressler T, Givskov M, Høiby N (2007b) Rapid necrotic killing of polymorphonuclear leukocytes is caused by quorum-sensing-controlled production of rhamnolipid by *Pseudomonas aeruginosa*. Microbiology 153:1329–1338

Kai T, Tateda K, Kimura S, Ishii Y, Ito H, Yoshida H, Kimura T, Yamaguchi K (2009) A low concentration of azithromycin inhibits the mRNA expression of N-acyl homoserine lactone synthesis enzymes, upstream of lasI or rhlI, in *Pseudomonas aeruginosa*. Pulm Pharmacol Ther 22(6):483–486

Kerem B, Rommens JM, Buchanan JA, Markiewicz D, Cox TK, Chakravarti A, Buchwald M, Tsui LC (1989) Identification of the cystic fibrosis gene: genetic analysis. Science 245:1073–1080

Kim C, Kim J, Park HY, Park HJ, Lee JH, Kim CK, Yoon J (2008) Furanone derivatives as quorum-sensing antagonists of *Pseudomonas aeruginosa*. Appl Microbiol Biotechnol 80: 37–47

Kirketerp-Moller K, Jensen PO, Fazli M, Madsen KG, Pedersen J, Moser C, Tolker-Nielsen T, Hoiby N, Givskov M, Bjarnsholt T (2008a) Distribution, organization, and ecology of bacteria in chronic wounds. J Clin Microbiol 46:2717–2722

Kirketerp-Moller K, Jensen PO, Fazli M, Madsen KG, Pedersen J, Moser C, Tolker-Nielsen T, Høiby N, Givskov M, Bjarnsholt T (2008b) Distribution, organization, and ecology of bacteria in chronic wounds. J Clin Microbiol 46:2717–2722

Kjelleberg S, Molin S (2002) Is there a role for quorum sensing signals in bacterial biofilms? Curr Opin Microbiol 5:254–258

Kjelleberg S, Steinberg P, Givskov M, Gram L, Manefield M, de Nys R (1997) Do marine natural products interfere with prokeryotic AHL regulatory systems? Aquat Microb Ecol 13:85–93

Klausen M, Aes-Jorgensen A, Molin S, Tolker-Nielsen T (2003a) Involvement of bacterial migration in the development of complex multicellular structures in *Pseudomonas aeruginosa* biofilms. Mol Microbiol 50:61–68

Klausen M, Heydorn A, Ragas P, Lambertsen L, Aes-Jorgensen A, Molin S, Tolker-Nielsen T (2003b) Biofilm formation by *Pseudomonas aeruginosa* wild type, flagella and type IV pili mutants. Mol Microbiol 48:1511–1524

Koch C, Hoiby N (1993) Pathogenesis of cystic fibrosis. Lancet 341:1065–1069

Koch B, Liljefors T, Persson T, Nielsen J, Kjelleberg S, Givskov M (2005) The LuxR receptor: the sites of interaction with quorum-sensing signals and inhibitors. Microbiology 151:3589–3602

Kohler T, Curty LK, Barja F, van Delden C, Pechere JC (2000) Swarming of *Pseudomonas aeruginosa* is dependent on cell-to-cell signaling and requires flagella and pili. J Bacteriol 182:5990–5996

Kong KF, Vuong C, Otto M (2006) Staphylococcus quorum sensing in biofilm formation and infection. Int J Med Microbiol 296:133–139

Kristiansen S, Bjarnsholt T, Adeltoft D, Ifversen P, Givskov M (2008) The *Pseudomonas aeruginosa* autoinducer dodecanoyl-homoserine lactone inhibits the putrescine synthesis in human cells. APMIS 116:361–371

Latifi A, Winson MK, Foglino M, Bycroft BW, Stewart GS, Lazdunski A, Williams P (1995) Multiple homologues of LuxR and LuxI control expression of virulence determinants and secondary metabolites through quorum sensing in *Pseudomonas aeruginosa* PAO1. Mol Microbiol 17:333–343

Lequette Y, Lee JH, Ledgham F, Lazdunski A, Greenberg EP (2006) A distinct QscR regulon in the Pseudomonas aeruginosa quorum-sensing circuit. J Bacteriol 188:3365–3370

Lyczak JB, Cannon CL, Pier GB (2000) Establishment of Pseudomonas aeruginosa infection: lessons from a versatile opportunist. Microbes Infect 2:1051–1060

MacLehose HG, Gilbert P, Allison DG (2004) Biofilms, homoserine lactones and biocide susceptibility. J Antimicrob Chemother 53:180–184

Magnuson R, Solomon J, Grossman AD (1994) Biochemical and genetic characterization of a competence pheromone from B. subtilis. Cell 77:207–216

Manefield M, Rasmussen TB, Henzter M, Andersen JB, Steinberg P, Kjelleberg S, Givskov M (2002) Halogenated furanones inhibit quorum sensing through accelerated LuxR turnover. Microbiology 148:1119–1127

Mattmann ME, Geske GD, Worzalla GA, Chandler JR, Sappington KJ, Greenberg EP, Blackwell HE (2008) Synthetic ligands that activate and inhibit a quorum-sensing regulator in Pseudomonas aeruginosa. Bioorg Med Chem Lett 18:3072–3075

McGrath S, Wade DS, Pesci EC (2004) Dueling quorum sensing systems in Pseudomonas aeruginosa control the production of the Pseudomonas quinolone signal (PQS). FEMS Microbiol Lett 230:27–34

McKnight SL, Iglewski BH, Pesci EC (2000) The Pseudomonas quinolone signal regulates rhl quorum sensing in Pseudomonas aeruginosa. J Bacteriol 182:2702–2708

Medina G, Juarez K, Diaz R, Soberon-Chavez G (2003) Transcriptional regulation of Pseudomonas aeruginosa rhlR, encoding a quorum-sensing regulatory protein. Microbiology 149:3073–3081

Meissner A, Wild V, Simm R, Rohde M, Erck C, Bredenbruch F, Morr M, Romling U, Haussler S (2007) Pseudomonas aeruginosa cupA-encoded fimbriae expression is regulated by a GGDEF and EAL domain-dependent modulation of the intracellular level of cyclic diguanylate. Environ Microbiol 9:2475–2485

Morici LA, Carterson AJ, Wagner VE, Frisk A, Schurr JR, Honer zu Bentrup K, Hassett DJ, Iglewski BH, Sauer K, Schurr MJ (2007) Pseudomonas aeruginosa AlgR represses the Rhl quorum-sensing system in a biofilm-specific manner. J Bacteriol 189:7752–7764

Muh U, Schuster M, Heim R, Singh A, Olson ER, Greenberg EP (2006) Novel Pseudomonas aeruginosa quorum-sensing inhibitors identified in an ultra-high-throughput screen. Antimicrob Agents Chemother 50:3674–3679

Mulcahy H, Charron-Mazenod L, Lewenza S (2008) Extracellular DNA chelates cations and induces antibiotic resistance in Pseudomonas aeruginosa biofilms. PLoS Pathog 4:e1000213

Murray PR, Rosenthal KS, Kobayashi GS, Pfaller MA (2002) Medical microbiology. Mosby, St. Louis

Nealson KH, Platt T, Hastings JW (1970) Cellular control of the synthesis and activity of the bacterial luminescent system. J Bacteriol 104:313–322

Nivens DE, Ohman DE, Williams J, Franklin MJ (2001) Role of alginate and its O acetylation in formation of Pseudomonas aeruginosa microcolonies and biofilms. J Bacteriol 183:1047–1057

O'Sullivan DJ, O'Gara F (1992) Traits of fluorescent Pseudomonas spp. involved in suppression of plant root pathogens. Microbiol Rev 56:662–676

O'Toole G, Kaplan HB, Kolter R (2000) Biofilm formation as microbial development. Annu Rev Microbiol 54:49–79

Olsen JA, Severinsen R, Rasmussen TB, Hentzer M, Givskov M, Nielsen J (2002) Synthesis of new 3- and 4-substituted analogues of acyl homoserine lactone quorum sensing autoinducers. Bioorg Med Chem Lett 12:325–328

Pamp SJ, Tolker-Nielsen T (2007) Multiple roles of biosurfactants in structural biofilm development by Pseudomonas aeruginosa. J Bacteriol 189:2531–2539

Pamp S, Gjermansen M, Tolker-Nielsen T (2007) The biofilm matrix: a sticky framework. In: Kjelleberg S, Givskov M (eds) Bacterial biofilm formation and adaptation. Horizon BioScience, Wymondham, UK, pp 37–69

Pamp SJ, Gjermansen M, Johansen HK, Tolker-Nielsen T (2008) Tolerance to the antimicrobial peptide colistin in Pseudomonas aeruginosa biofilms is linked to metabolically active cells, and depends on the pmr and mexAB-oprM genes. Mol Microbiol 68:223–240

Park J, Kaufmann GF, Bowen JP, Arbiser JL, Janda KD (2008) Solenopsin A, a venom alkaloid from the fire ant Solenopsis invicta, inhibits quorum-sensing signaling in Pseudomonas aeruginosa. J Infect Dis 198:1198–1201

Pearson JP, van Delden C, Iglewski BH (1999) Active efflux and diffusion are involved in transport of Pseudomonas aeruginosa cell-to-cell signals. J Bacteriol 181:1203–1210

Persson T, Givskov M, Nielsen J (2005a) Quorum sensing inhibition: targeting chemical communication in gram-negative bacteria. Curr Med Chem 12:3103–3115

Persson T, Hansen TH, Rasmussen TB, Skinderso ME, Givskov M, Nielsen J (2005b) Rational design and synthesis of new quorum-sensing inhibitors derived from acylated homoserine lactones and natural products from garlic. Org Biomol Chem 3:253–262

Pesci EC, Pearson JP, Seed PC, Iglewski BH (1997) Regulation of las and rhl quorum sensing in Pseudomonas aeruginosa. J Bacteriol 179:3127–3132

Pesci EC, Milbank JB, Pearson JP, McKnight S, Kende AS, Greenberg EP, Iglewski BH (1999) Quinolone signaling in the cell-to-cell communication system of Pseudomonas aeruginosa. Proc Natl Acad Sci USA 96:11229–11234

Purevdorj B, Costerton JW, Stoodley P (2002) Influence of hydrodynamics and cell signaling on the structure and behavior of Pseudomonas aeruginosa biofilms. Appl Environ Microbiol 68:4457–4464

Qin X, Singh KV, Weinstock GM, Murray BE (2000) Effects of Enterococcus faecalis fsr genes on production of gelatinase and a serine protease and virulence. Infect Immun 68:2579–2586

Rasch M, Buch C, Austin B, Slierendrecht WJ, Ekmann KS, Larsen JL, Johansen C, Riedel K, Eberl L, Givskov M, Gram L (2004) An inhibitor of bacterial quorum sensing reduces mortalities caused by Vibriosis in rainbow trout (Oncorhynchus mykiss, Walbaum). Syst Appl Microbiol 27:350–359

Rasmussen TB, Bjarnsholt T, Skindersoe ME, Hentzer M, Kristoffersen P, Kote M, Eberl L, Nielsen J, Givskov M (2005a) Screening for quorum sensing inhibitors using a novel genetic system: the QSI selector. J Bacteriol 187:1799–1814

Rasmussen TB, Bjarnsholt T, Skindersoe ME, Hentzer M, Kristoffersen P, Kote M, Nielsen J, Eberl L, Givskov M (2005b) Screening for quorum-sensing inhibitors (QSI) by use of a novel genetic system, the QSI selector. J Bacteriol 187:1799–1814

Rasmussen TB, Skindersoe ME, Bjarnsholt T, Christensen KB, Andersen JB, Ostenfeld-Larsen T, Hentzer M, Givskov M (2005c) Idendity and effects of quorum sensing inhibitors produced by *Penicillum* species. Microbiology 151:1325–1340

Reimmann C, Ginet N, Michel L, Keel C, Michaux P, Krishnapillai V, Zala M, Heurlier K, Triandafillu K, Harms H, Defago G, Haas D (2002) Genetically programmed autoinducer destruction reduces virulence gene expression and swarming motility in Pseudomonas aeruginosa PAO1. Microbiology 148:923–932

Riedel K, Kothe M, Kramer B, Saeb W, Gotschlich A, Ammendola A, Eberl L (2006) Computer-aided design of agents that inhibit the cep quorum-sensing system of Burkholderia cenocepacia. Antimicrob Agents Chemother 50:318–323

Riordan JR, Rommens JM, Kerem B, Alon N, Rozmahel R, Grzelczak Z, Zielenski J, Lok S, Plavsic N, Chou JL (1989) Identification of the cystic fibrosis gene: cloning and characterization of complementary DNA. Science 245:1066–1073

Roberts ME, Stewart PS (2004) Modeling antibiotic tolerance in biofilms by accounting for nutrient limitation. Antimicrob Agents Chemother 48:48–52

Rommens JM, Iannuzzi MC, Kerem B, Drumm ML, Melmer G, Dean M, Rozmahel R, Cole JL, Kennedy D, Hidaka N (1989) Identification of the cystic fibrosis gene: chromosome walking and jumping. Science 245:1059–1065

Rumbaugh KP, Griswold JA, Hamood AN (2000) The role of quorum sensing in the in vivo virulence of Pseudomonas aeruginosa. Microbes Infect 2:1721–1731

Saiman L (2004) The use of macrolide antibiotics in patients with cystic fibrosis. Curr Opin Pulm Med 10:515–523

Sauer K (2003) The genomics and proteomics of biofilm formation. Genome Biol 4:219

Schaber JA, Triffo WJ, Suh SJ, Oliver JW, Hastert MC, Griswold JA, Auer M, Hamood AN, Rumbaugh KP (2007) Pseudomonas aeruginosa forms biofilms in acute infection independent of cell-to-cell signaling. Infect Immun 75:3715–3721

Schleheck D, Barraud N, Klebensberger J, Webb JS, McDougald D, Rice SA, Kjelleberg S (2009) Pseudomonas aeruginosa PAO1 preferentially grows as aggregates in liquid batch cultures and disperses upon starvation. PLoS ONE 4:e5513

Schuster M, Lostroh CP, Ogi T, Greenberg EP (2003) Identification, timing, and signal specificity of Pseudomonas aeruginosa quorum-controlled genes: a transcriptome analysis. J Bacteriol 185:2066–2079

Seed PC, Passador L, Iglewski BH (1995) Activation of the Pseudomonas aeruginosa lasI gene by LasR and the Pseudomonas autoinducer PAI: an autoinduction regulatory hierarchy. J Bacteriol 177:654–659

Shih PC, Huang CT (2002) Effects of quorum-sensing deficiency on Pseudomonas aeruginosa biofilm formation and antibiotic resistance. J Antimicrob Chemother 49:309–314

Simpson JA, Smith SE, Dean RT (1989) Scavenging by alginate of free radicals released by macrophages. Free Radic Biol Med 6:347–353

Singh PK, Parsek MR, Greenberg EP, Welsh MJ (2002) A component of innate immunity prevents bacterial biofilm development. Nature 417:552–555

Skindersoe ME, Alhede M, Phipps R, Yang L, Jensen PO, Rasmussen TB, Bjarnsholt T, Tolker-Nielsen T, Hoiby N, Givskov M (2008) Effects of antibiotics on quorum sensing in Pseudomonas aeruginosa. Antimicrob Agents Chemother 52:3648–3663

Skindersoe ME, Zeuthen LH, Brix S, Fink LN, Lazenby J, Whittall C, Williams P, Diggle SP, Froekiaer H, Cooley M, Givskov M (2009) Pseudomonas aeruginosa quorum-sensing signal molecules interfere with dendritic cell-induced T-cell proliferation. FEMS Immunol Med Microbiol 55:335–345

Southern KW, Barker PM (2004) Azithromycin for cystic fibrosis. Eur Respir J 24:834–838

Stapper AP, Narasimhan G, Ohman DE, Barakat J, Hentzer M, Molin S, Kharazmi A, Hoiby N, Mathee K (2004) Alginate production affects Pseudomonas aeruginosa biofilm development and architecture, but is not essential for biofilm formation. J Med Microbiol 53:679–690

Stoodley P, Wilson S, Hall-Stoodley L, Boyle JD, Lappin-Scott HM, Costerton JW (2001) Growth and detachment of cell clusters from mature mixed-species biofilms. Appl Environ Microbiol 67:5608–5613

Stover CK, Pham XQ, Erwin AL, Mizoguchi SD, Warrener P, Hickey MJ, Brinkman FS, Hufnagle WO, Kowalik DJ, Lagrou M, Garber RL, Goltry L, Tolentino E, Westbrock-Wadman S, Yuan Y, Brody LL, Coulter SN, Folger KR, Kas A, Larbig K, Lim R, Smith K, Spencer D, Wong GK, Wu Z, Paulsen IT, Reizer J, Saier MH, Hancock RE, Lory S, Olson MV (2000) Complete genome sequence of Pseudomonas aeruginosa PA01, an opportunistic pathogen. Nature 406: 959–964

Taga ME, Bassler BL (2003) Chemical communication among bacteria. Proc. Natl. Acad. Sci. USA

Tateda K, Comte R, Pechere JC, Kohler T, Yamaguchi K, van Delden C (2001) Azithromycin inhibits quorum sensing in Pseudomonas aeruginosa. Antimicrob Agents Chemother 45: 1930–1933

Teitzel GM, Parsek MR (2003) Heavy metal resistance of biofilm and planktonic Pseudomonas aeruginosa. Appl Environ Microbiol 69:2313–2320

Tomasz A (1965) Control of the competent state in Pneumococcus by a hormone-like cell product: an example for a new type of regulatory mechanism in bacteria. Nature 208:155–159

van Delden C, Iglewski BH (1998) Cell-to-cell signaling and Pseudomonas aeruginosa infections. Emerg Infect Dis 4:551–560

van Gennip M, Christensen LD, Alhede M, Phipps R, Jensen PO, Christophersen L, Pamp SJ, Moser C, Mikkelsen PJ, Koh AY, Tolker-Nielsen T, Pier GB, Hoiby N, Givskov M, Bjarnsholt T (2009) Inactivation of the rhlA gene in Pseudomonas aeruginosa prevents rhamnolipid production, disabling the protection against polymorphonuclear leukocytes. APMIS 117: 537–546

Wagner VE, Bushnell D, Passador L, Brooks AI, Iglewski BH (2003) Microarray analysis of Pseudomonas aeruginosa quorum-sensing regulons: effects of growth phase and environment. J Bacteriol 185:2080–2095

Walters MC III, Roe F, Bugnicourt A, Franklin MJ, Stewart PS (2003) Contributions of antibiotic penetration, oxygen limitation, and low metabolic activity to tolerance of Pseudomonas aeruginosa biofilms to ciprofloxacin and tobramycin. Antimicrob Agents Chemother 47:317–323

Webb JS, Thompson LS, James S, Charlton T, Tolker-Nielsen T, Koch B, Givskov M, Kjelleberg S (2003) Cell death in Pseudomonas aeruginosa biofilm development. J Bacteriol 185:4585–4592

Whitchurch CB, Tolker-Nielsen T, Ragas PC, Mattick JS (2002) Extracellular DNA required for bacterial biofilm formation. Science 295:1487

Whiteley M, Lee KM, Greenberg EP (1999) Identification of genes controlled by quorum sensing in Pseudomonas aeruginosa. Proc Natl Acad Sci USA 96:13904–13909

Williams P, Camara M, Hardman A, Swift S, Milton D, Hope VJ, Winzer K, Middleton B, Pritchard DI, Bycroft BW (2000) Quorum sensing and the population-dependent control of virulence. Philos Trans R Soc Lond B Biol Sci 355:667–680

Wu H, Song Z, Givskov M, Doring G, Worlitzsch D, Mathee K, Rygaard J, Høiby N (2001) Pseudomonas aeruginosa mutations in lasI and rhlI quorum sensing systems result in milder chronic lung infection. Microbiology 147:1105–1113

Wu H, Song Z, Hentzer M, Andersen JB, Molin S, Givskov M, Hoiby N (2004a) Synthetic furanones inhibit quorum-sensing and enhance bacterial clearance in Pseudomonas aeruginosa lung infection in mice. J Antimicrob Chemother 53:1054–1061

Wu H, Song Z, Hentzer M, Andersen JB, Molin S, Givskov M, Høiby N (2004b) Synthetic furanones inhibit quorum-sensing and enhance bacterial clearance in Pseudomonas aeruginosa lung infection in mice. J Antimicrob Chemother 53:1054–1061

Wu L, Estrada O, Zaborina O, Bains M, Shen L, Kohler JE, Patel N, Musch MW, Chang EB, Fu YX, Jacobs MA, Nishimura MI, Hancock RE, Turner JR, Alverdy JC (2005) Recognition of host immune activation by Pseudomonas aeruginosa. Science 309:774–777

Wysocki AB, Staiano-Coico L, Grinnell F (1993) Wound fluid from chronic leg ulcers contains elevated levels of metalloproteinases MMP-2 and MMP-9. J Invest Dermatol 101:64–68

Yang L, Barken KB, Skindersoe ME, Christensen AB, Givskov M, Tolker-Nielsen T (2007) Effects of iron on DNA release and biofilm development by Pseudomonas aeruginosa. Microbiology 153:1318–1328

Yang L, Nilsson M, Gjermansen M, Givskov M, Tolker-Nielsen T (2009a) Pyoverdine and PQS mediated subpopulation interactions involved in *Pseudomonas aeruginosa* biofilm formation. Mol Microbiol 74(6):1380–1392

Yang L, Rybtke MT, Jakobsen TH, Hentzer M, Bjarnsholt T, Givskov M, Tolker-Nielsen T (2009b) Computer-aided identification of recognized drugs as Pseudomonas aeruginosa quorum-sensing inhibitors. Antimicrob Agents Chemother 53:2432–2443

Zhu J, Miller MB, Vance RE, Dziejman M, Bassler BL, Mekalanos JJ (2002) Quorum-sensing regulators control virulence gene expression in Vibrio cholerae. Proc Natl Acad Sci USA 99:3129–3134

Hygienically Relevant Microorganisms in Biofilms of Man-Made Water Systems

Jost Wingender

Abstract In recent years, it has become evident that biofilms in drinking water distribution networks and other man-made water systems can become transient or long-term habitats for hygienically relevant microorganisms. Important categories of these organisms include faecal indicator bacteria (e.g. *Escherichia coli*), obligate bacterial pathogens of faecal origin (e.g. *Campylobacter* spp.), opportunistic bacteria of environmental origin (e.g. *Legionella* spp., *Pseudomonas aeruginosa*), enteric viruses and parasitic protozoa (e.g. *Cryptosporidium parvum*). These organisms can attach to preexisting biofilms, where they become integrated and survive for days to weeks or even longer, depending on the biology and ecology of the organism and the environmental conditions. There are indications that at least part of the biofilm populations of pathogenic bacteria persist in a viable but non-culturable state. Thus, biofilms in man-made water systems can function as an environmental reservoir for pathogenic microorganisms and present a potential source of water contamination, resulting in a health risk for humans. This review outlines the current knowledge of the integration and fate of hygienically relevant microorganisms in biofilms of man-made water systems, with consideration of the physicochemical and biological factors that govern these processes.

1 Introduction

Biofilms are microbial populations that adhere to each other and/or to interfaces, and are typically surrounded by a matrix of extracellular polymeric substances (EPS) (Hall-Stoodley et al. 2004). In man-made water systems, interfaces for

J. Wingender (✉)
Department of Aquatic Microbiology, Faculty of Chemistry, Biofilm Centre, University of Duisburg-Essen, Universitätsstraße 5, 45141 Essen, Germany
e-mail: jost.wingender@uni-due.de

biofilm growth are usually solid–liquid as in piped water distribution systems, but biofilms can also develop at air–liquid interfaces (pellicles or floating biofilms) as in water storage tanks or water basins of cooling towers (Declerck et al. 2007a). EPS mainly consist of polysaccharides, proteins, DNA and lipids; they determine the structural and functional integrity of microbial biofilms, and contribute significantly to the organization of the biofilm community (Flemming and Wingender 2010). Although the details of biofilm development processes vary depending on the microbial species and the prevailing environmental conditions, general distinct developmental steps have been recognized in biofilm formation (Stoodley et al. 2002; Hall-Stoodley et al. 2004). These include the initial attachment of cells to a surface, followed by the aggregation of cells into microcolonies, subsequent growth and maturation of microcolonies into an established biofilm, and the passive detachment or active release of single cells or aggregates of cells into the surrounding environment. Biofilms develop on virtually any surface in natural soil and aquatic environments, on tissues of plants, animals and humans as well as in all types of man-made water systems (Costerton et al. 1987). Biofilms provide a protected habitat, allowing survival of microorganisms under adverse environmental conditions and offering tolerance to antimicrobial compounds (Hall-Stoodley et al. 2004).

Biofouling, which is defined as the undesirable development of biofilms on surfaces, can be of substantial hygienic, operational and economic relevance in technical water systems (Mittelman 1995). In recent years, it has become obvious that microorganisms with pathogenic properties can persist and multiply in biofilms of man-made water systems. In drinking water distribution systems (DWDS) and domestic plumbing, but also in other technical water systems, biofilms can function as an environmental reservoir for pathogenic microorganisms and then are potential sources of contamination. Biofilms present a health risk when pathogens are released from the biofilms and are transmitted to susceptible human hosts upon exposure to contaminated water (Fig. 1).

Important modes of transmission are ingestion of contaminated water, inhalation of pathogen-containing aerosols or infection by contact of skin, mucous membranes, eyes and ears with contaminated water (WHO 2008). Other hygienic aspects of biofilms in water systems are the production of biofilm-associated metabolites of health significance such as H_2S and nitrite, the release of endotoxins, the deterioration of the aesthetic water quality through discolouration, turbidity and malodours, and the build-up of trophic food webs leading to the multiplication of free-living protozoa and invertebrate animals. The general causes and consequences of biofouling in man-made water systems as well as biofouling control strategies have been reviewed previously (Mittelman 1995; Flemming 2002; Batté et al. 2003; Bachmann and Edyvean 2005). The focus of this review is on the aspect of hygienically relevant microorganisms in biofilms of DWDS and plumbing systems, but also other man-made water systems that supply water for diverse human uses will be included.

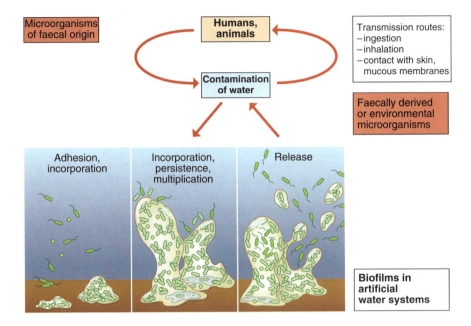

Fig. 1 Role of biofilms as environmental reservoirs of hygienically relevant microorganisms and as sources of contamination and infection in man-made water systems

2 General Characteristics of Biofilms in Man-Made Water Systems

DWDS belong to the most extensively investigated, regulated and monitored man-made water systems in terms of the presence of microorganisms relevant to public health. Since drinking water should meet the necessary quality requirements at the point of consumption, it is important to understand the development of biofilms and their role as reservoirs for pathogenic microorganisms both in water distribution mains and in consumers' plumbing systems. However, most studies have focused on biofilms on materials relevant to public DWDS, while biofilms on materials used to construct domestic plumbing systems have less often been considered (Eboigbodin et al. 2008). Other technical water systems with a potential risk of human exposure to water-related pathogens include recreational and therapeutic water systems (e.g. swimming pools, spas, hydrotherapy pools), reclaimed water systems, cooling water systems in industry and as components of air-conditioning systems, water systems in the food, dairy and beverage industry or water systems of paper manufacturing in paper mills.

Independent of the type of water system, biofilms are present on all surfaces in water treatment, distribution and storage. Biofilm formation is especially critical in water systems of hospitals and other health-care facilities, where biofilm-derived pathogens can contribute to water-associated nosocomial infections (Anaissie et al. 2002). At the point-of-use, biofilms can serve as a source for the continual presentation of microorganisms to patients, care givers, and environmental surfaces with which water may come into contact (Ortolano et al. 2005). Man-made water systems include secondary circulatory systems, which display conditions conducive to biofilm formation. Swimming pools, spas and pools used for therapeutic purposes are usually equipped with circulatory systems including piping and filters; elevated temperatures and sufficient nutrients from bathers promote biofilm formation in these systems, especially on filter media such as sand, which provide large surface areas for attachment and growth of microorganisms (Unhoch and Vore 2005). Cooling water systems provide favourable conditions for the accumulation and significant growth of microbial biofilms (Harris 1999).

In man-made water systems, surfaces are usually colonized at higher levels compared with the bulk water phase. It has been estimated that 95% of the overall bacterial biomass in DWDS can be located at drinking water pipe walls, while only 5% occur in the water phase (Flemming et al. 2002); in a domestic hot water system, most of the culturable bacteria (72%) were found to be surface associated (Bagh et al. 2004). Another important type of microbial accumulation in DWDS and possibly other water systems occurs on loose deposits, which can be colonized by microorganisms in high concentration (Zacheus et al. 2001; Batté et al. 2003). Thus, in man-made water systems, most of the numbers and activity of microorganisms reside in biofilms on material surfaces and deposits; as a consequence, biofilms represent an important potential for microbiological contamination and deterioration of the hygienic water quality.

New materials exposed to water systems are rapidly colonized reaching maximal cell counts within weeks, but a real equilibrium is never reached since species composition and density can vary depending on biofilm age (Martiny et al. 2003), material which serves as a substratum and sometimes as a transient nutrient source (Schwartz et al. 1998a; Norton and LeChevallier 2000), nutrient availability (Norton and LeChevallier 2000), hydraulic conditions, water temperature, type and level of disinfectant residuals (Norton and LeChevallier 2000) as well as protozoan grazing, which can control biofilm growth (Pedersen 1990).

In DWDS, total cell counts of several months to years old biofilms on pipe surfaces or coupons exposed to drinking water range in the order of 10^4–10^8 cells per cm^2 (Block et al. 1993; Servais et al. 1995; Kalmbach et al. 1997; Wingender and Flemming 2004; Långmark et al. 2005a), while the numbers of culturable heterotrophic plate count (HPC) bacteria in established biofilms can vary between approximately 10^1 and 10^6 colony-forming units (cfu) per cm^2 (LeChevallier et al. 1987; Block et al. 1993; Wingender and Flemming 2004; Långmark et al. 2005a). Culturable bacteria usually represent a small fraction of the total cell numbers, frequently varying between approximately 0.01% and a few percent of the total cell counts (Kalmbach et al. 1997; Wingender and Flemming 2004). Low culturability is a typical property of bacteria in oligotrophic water environments, and seems to be characteristic of bacteria in months to years old drinking water biofilms.

Occasionally, enhanced levels of biofilm bacteria with high proportions of culturable bacteria can be found under conditions of sufficient nutrient levels in the form of biodegradable organic substances entering the water system or leaching from materials within the distribution system. Certain types of plastic and elastomeric materials can promote biofilm formation due to the release of biodegradable compounds (Keevil 2002; Rogers et al. 1994b; Kilb et al. 2003).

In general, biofilms in man-made water systems predominantly consist of environmental microorganisms without any relevance for human health (autochthonous microflora). These natural populations usually develop and constitute the biofilms, and are mostly non-pathogenic. However, aquatic biofilms can occasionally become habitats for pathogenic microorganisms, which may be either faecally derived (allochthonous organisms) or originate from natural soil and water environments (autochthonous organisms).

3 Categories of Hygienically Relevant Biofilm Microorganisms

Diseases associated with exposure to contaminated drinking water, swimming pool water or other types of water from man-made water systems are caused by pathogenic heterotrophic bacteria, enteric viruses as well as parasitic protozoa and some free-living amoebae (Percival et al. 2004); fungi may be involved in water-related infections in certain settings such as in hospitals or in swimming pool environments.

Multiple routes exist through which pathogens can enter man-made water systems, before they form biofilms or become incorporated into established biofilms. They may originate from the source water, and pass water treatment barriers and disinfection, resulting in a breakthrough of the bacteria and an inoculation of the distribution system. Alternatively, they may infiltrate water distribution networks through point source contaminations by leaking or broken pipes, installation and repair works, cross connections or backflow.

Two categories of hygienically relevant microorganisms can be distinguished; on the one hand, microorganisms with pathogenic properties, which have been shown to be associated with water-related illness and outbreaks, and on the other hand, bacteria which are primarily used as index and indicator organisms in water analysis, pointing to the presence of pathogenic organisms of faecal origin (index organisms) or indicating the effectiveness of water treatment processes as well as integrity of water distribution systems (indicator organisms) (WHO 2008). Among water-related pathogenic microorganisms are primary (obligate) pathogens, which cause disease in humans independent of their health status and as enteric pathogens are usually faecally derived, while others are potentially pathogenic organisms (opportunistic pathogens), which cause disease in sensitive human subgroups including the elderly, children, immunocompromised individuals, patients with pre-existing disease or other predisposing conditions, which facilitate infection by these organisms. Opportunistic pathogens are frequently natural aquatic organisms, and thus adapted to oligotrophic environmental conditions, which are typical of many man-made water systems. A number of pathogenic microorganisms have been recognized as emerging pathogens, either as newly discovered pathogens (e.g. *Campylobacter* spp., *H. pylori*, *Legionella* spp., *Cryptosporidium* spp.) or as new variants of already known species (e.g., enterohaemorrhagic *Escherichia coli* O157:H7) (Szewzyk et al. 2000).

Depending on the biological properties and ecology of the microbial pathogens, the organisms may attach to surfaces as primary colonizers and actively establish biofilms alone or in combination with other microorganisms, and/or the organisms may enter pre-existing biofilms and become integrated as secondary colonizers in these biofilms (Fig. 1). Multiplication in biofilms can be expected for heterotrophic bacteria, free-living protozoa and fungi which adapt to the oligotrophic conditions characteristic of many artificial water systems. Under suitable laboratory conditions, all relevant water-related pathogenic bacterial species have actually been shown to be able to adhere to solid surfaces and/or to form monospecies biofilms, indicating their potential as biofilm organisms. However, enteric viruses and parasitic protozoa are obligate parasites and dependent on multiplication in animal or human hosts. Thus, these organisms can only be expected to attach to and persist in biofilms without being able to proliferate.

Traditionally, pathogenic bacteria in water are analyzed by cultural methods. However, pathogenic bacteria may enter a viable but non-culturable (VBNC) state. Bacteria in the VBNC state cannot grow on conventional microbiological media on which they would normally grow and develop into colonies, but are still alive and are characterized by low levels of metabolic activity (Oliver 2010). The conversion to the VBNC state is supposed to be a response to adverse environmental conditions such as lack of nutrients, unfavourable water temperature, the presence of disinfectants or toxic metal ions such as copper. VBNC bacteria can become culturable again upon resuscitation under favourable conditions. Oliver (2010) provided a list of pathogens known to enter the VBNC state, in which all relevant water-associated bacterial pathogens are included. Investigations of the VBNC state are usually performed using planktonic cells, so it is largely unknown whether the VBNC state can also be induced in biofilm environments. To circumvent the shortcomings of non-culturability of biofilm organisms, culture-independent methods are increasingly used to characterize the composition and diversity of microbial biofilm communities, and to identify pathogens in biofilms of man-made water systems. Of relevance are immunological (antibody-based) techniques and nucleic acid-based methods, which include fluorescence in situ hybridization (FISH) or peptide nucleic acid FISH (PNA-FISH) with ribosomal RNA as a target for group- or species-specific fluorescent oligonucleotide probes, and polymerase chain reaction (PCR) targeted at specific DNA sequences, alone or in combination with denaturant gradient gel electrophoresis, cloning and sequencing of 16S rRNA genes. The use of fluorescently labelled antibodies or 16S rRNA probes in combination with microscopic techniques (scanning confocal laser microscopy, episcopic differential interference contrast microscopy) allow the in situ visualization of pathogens in biofilms (Keevil 2003). In Table 1, hygienically relevant bacterial

Table 1 Hygienically relevant bacteria that were detected in drinking water biofilms by culture-independent methods

Organism	Type of drinking water biofilm	Culture-independent method	References
Campylobacter jejuni	Experimental flow-through system	FISH	Buswell et al. (1998), Lehtola et al. (2006a)
Campylobacter coli	Experimental flow-through system	PCR-based method	Hummel and Feuerpfeil (2003)
Faecal streptococci	Real distribution system	PCR-based method	Schwartz et al. (1998a)
Escherichia coli	Real distribution system, experimental flow-through system	PNA-FISH	Juhna et al. (2007), Williams and Braun-Howland (2003)
Helicobacter pylori	Real distribution system, plumbing system	PNA-FISH PCR-based method	Bragança et al. (2005) Park et al. (2001), Watson et al. (2004)
	Experimental flow-through system	PNA-FISH PCR-based method	Azevedo et al. (2003) Mackay et al. (1999), Exner and Rechenburg (2003)
Legionella pneumophila	Experimental flow-through system	FISH	Storey et al. (2004a), Långmark et al. (2005b), Moritz et al. (2010)
		PNA-FISH	Lehtola et al. (2007)
	Real distribution system	PCR-based method	Wullings et al. (2011)
Mycobacterium avium	Experimental flow-through system	PNA-FISH	Lehtola et al. (2007)
Pseudomonas aeruginosa	Experimental flow-through system	FISH	Moritz et al. (2010)

species are listed that were shown to be present in real or experimental drinking water biofilms, using culture-independent methods (Table 1). Significantly higher cell numbers of pathogens over longer times were frequently detected by culture-independent methods compared to colony counts determined on culture media as has, for example, been shown for *Camplyobacter jejuni* (Buswell et al. 1998), *Legionella pneumophila* (Lehtola et al. 2007) and *Pseudomonas aeruginosa* (Moritz et al. 2010). These observations indicate that at least part of the populations of bacterial pathogens may persist in a VBNC state in drinking water biofilms, although this biofilm-associated non-culturable state and the potential of resuscitation of these bacteria have not yet been characterized in detail. From a health perspective, the detection of pathogens in the VBNC state is important, since they can retain their virulence and are able to initiate infection when they revert to the culturable state under favourable environmental conditions (Oliver 2010), thus representing an infectious potential when present in biofilms of man-made water systems.

Different approaches have been used to study pathogens in biofilms of man-made water systems. A limited number of field studies have investigated the presence of pathogens directly in biofilms of pipes and other parts of real distribution systems in regular operation. A more commonly used strategy for an efficient monitoring of biofilm formation and pathogen occurrence in real systems is the installation of biofilm-forming devices as a by-pass or directly connected to a water distribution system. This allows more easy sampling of coupons for microbiological procedures compared to the investigation of pipe section whose removal from water systems is rather complex. Another common approach for the study of water-related pathogens in biofilms is the use of laboratory biofilm reactors operated as continuous flow-through systems or occasionally under batch conditions. These experimental systems permit the investigation of pure culture or mixed-population biofilms under controlled conditions, and have often been employed to study the integration and the fate of pathogens in established biofilms, the influence of physical and chemical factors on these processes, the interaction of pathogens with other biofilm organisms and the protection of pathogens in biofilms from disinfectants. An overview of literature data relating to the incorporation and persistence of hygienically relevant microorganisms in biofilms of man-made water systems obtained on the basis of seeding experiments is given in Table 2.

3.1 Bacterial Pathogens of Faecal Origin

Important waterborne bacterial pathogens, which can infect the gastrointestinal tract of humans or warm-blooded animals and are excreted with the faeces into the environment include *Salmonella enterica* (e.g. serovar Typhi and Typhimurium), *Shigella* spp., *Vibrio cholerae*, pathogenic *E. coli* variants (e.g. enterotoxigenic *E. coli*, enterohaemorrhagic *E. coli* O157:H7), *Yersinia enterocolitica*, *Campylobacter* spp. and *Helicobacter pylori*. These pathogens have in common that they are transmitted by ingestion of faecally contaminated water, are involved in enteric infections and can cause gastrointestinal (diarrhoeal) diseases.

3.1.1 *Salmonella* and *Shigella*

S. enterica serovars, *Shigella* spp. and *V. cholerae* are obligate pathogens, which have been known since the nineteenth century to be involved in waterborne epidemics worldwide. Field observations suggest that these pathogens are not normally present in biofilms of properly maintained municipal drinking water networks. Thus, neither *S. enterica* nor *Shigella* spp. were detected culturally in biofilms of galvanized iron pipes in a pilot system at the periphery of the Seoul (Korea) (Lee and Kim 2003) or in biofilms from various South African DWDS (September et al. 2007). In biofilms of the drinking water network within the Stockholm area (Sweden), salmonellae and other enterobacteria were not detected by FISH (Långmark et al. 2005a). However, seeding experiments suggest that members of these pathogens have the potential to survive in biofilms of distribution

Table 2 Persistence of hygienically relevant microorganisms in biofilms after inoculation of pre-established biofilms in seeding experiments under environmental conditions found in real man-made water systems

Organism	Type of established biofilms and substratum	Persistence in biofilms	References
Bacterial pathogens of faecal origin			
Salmonella enterica serovar Typhimurium	Drinking water biofilm on glass coupons	43 days (24°C), decline; 31 days (36°C), growth	Armon et al. (1997)
	Drinking water biofilm on polycarbonate slide, established after primary attachment	20 days (20°C); decline	Camper et al. (1998)
Yersinia enterocolitica	Drinking water biofilm on paraffin-coated glass coupons	14 days (ambient temperature), decline	Hummel and Feuerpfeil (2003)
Campylobacter jejuni,	Drinking water biofilm on glass	28 days (30°C), 42 days (4°C)	Buswell et al. (1998)
Campylobacter coli	Drinking water biofilm on paraffin-coated glass coupons	28 days (ambient temperature)	Hummel and Feuerpfeil (2003)
	Drinking water biofilm on PVC coupons	1 week, probably up to 3 weeks (15°C)	Lehtola et al. (2006a)
Helicobacter pylori	Drinking water biofilms on paraffin-coated glass and PE coupons (14 days old)	14 days (ambient temperature), no growth	Exner and Rechenburg (2003)
Pathogenic Escherichia coli	Biofilms of autochthonous river water bacteria on GAC	14 days (20°C–22°C), decline	Camper et al. (1985)
	Drinking-water biofilms	Several weeks (10°C and 20°C)	Keevil (2002)
Index/indicator bacteria			
Escherichia coli	Biofilms of autochthonous river water bacteria on GAC	14 days (20°C–22°C), decline	Camper et al. (1985)
	Drinking water biofilms on bitumen-painted steel and glass in a chemostat	13 days (20°C–25°C)	Robinson et al. (1995)
	Drinking water biofilms on PVC plates in pilot system	18 days (20°C), limited growth	Fass et al. (1996)
	Biofilms of drinking water-derived microorganisms on glass and ductile iron	10 days (room temperature)	Williams and Braun-Howland (2003)
	Biofilms of groundwater-derived microorganisms on glass coverslips	15 days (23°C), decline over time	Banning et al. (2003)
	Drinking water biofilms (1 month old) on PVC coupons	4 days (15°C)	Lehtola et al. (2007)

(continued)

Table 2 (continued)

Organism	Type of established biofilms and substratum	Persistence in biofilms	References
Coliform bacteria: *Klebsiella* spp. and *Enterobacter* spp.	Drinking water biofilm on polycarbonate, established after primary attachment	8 weeks (20°C), decline depending on growth rate of inoculum	Camper et al. (1996)
Klebsiella pneumoniae	Drinking water biofilm on polycarbonate, established after primary attachment	2 months (20°C), slow decline	Camper et al. (1998)
	Drinking water biofilms (30 days old) on corroded iron coupons	16 days (21°C–23°C), decline depending on inoculum size, growth conditions of preculture, chlorine concentration	Szabo et al. (2006)
Opportunistic bacterial pathogens of environmental origin			
Legionella pneumophila	Drinking water biofilm on glass and PVC coupons (flow-through system)	43 days at 24°C and 36°C, almost no decline	Armon et al. (1997)
	Biofilm of *P. aeruginosa*, *K. pneumoniae* and *Flavobacterium* sp. on stainless steel grown in drinking water (flow-through system)	15 days at 30°C, persistence in the absence of amoebae, replication in the presence of amoebae	Murga et al. (2001)
	Drinking water biofilm on glass slides (flow-through system)	38 days at 5.0°C–8.5°C, decline: rapid for culturable cells, slow for FISH-positive cells	Långmark et al. (2005b)
	Tap water biofilms on polypropylene (static system)	50 days at 28°C	Vervaeren et al. (2006)
	Drinking water biofilm (1 month old) on PVC coupons in Propella reactor	4 weeks (15°C), slow decline of culturable and FISH-positive cells	Lehtola et al. (2007)
	Drinking water biofilm on EPDM, PE-X and copper (14 days old)	28 days (19°C), slow decline of culturable cells and no or slow decline of FISH-positive cells	Moritz et al. (2010)
	Biofilm in recirculating model cooling tower water system on steel, copper, PVC, PP and PE coupons	180 days (29°C), no decline	Türetgen and Cotuk (2007)
Pseudomonas aeruginosa	Seeding of sterile PVC pipes	Colonization of pipe surfaces, survival of exposure to 10–15 mg/L free chlorine for 7 days	Vess et al. (1993)
	Biofilms of groundwater microorganisms on glass coverslips	40 days (23°C), growth	Banning et al. (2003)
	Drinking water biofilm on EPDM coupons	28 days (ambient temperature), decline of culturable cells	Bressler et al. (2009)
	Drinking water biofilms on EPDM and PE-X (14 days old)	28 days (19°C), decline of culturable cells, no or slow decline of FISH-positive cells	Moritz et al. (2010)

Mycobacterium xenopi	Drinking water biofilm (5 weeks old) on high-density PE in Propella reactor	9 weeks at 20°C with no decline (10^2–10^3 cfu/cm^2)	Dailloux et al. (2003)
	Drinking water biofilm (2 weeks old) on paraffin-coated glass slides	3 months (>10^3 cfu/cm^2)	Schulze-Röbbecke and Ilg (2003)
Mycobacterium avium	Drinking water biofilm (1 month old) on PVC coupons in Propella reactor	4 weeks (15°C), slow decline	Lehtola et al. (2006b, 2007)
Aeromonas hydrophila	Drinking water biofilm (bitumen-painted mild steel)	21 days, no decline	Keevil (2002)
	Recycled wastewater biofilm (stainless steel, glass)	30 days, decline	Bomo et al. (2004)
	Drinking water biofilm on polycarbonate, established after primary attachment	20 days (20°C), decline	Camper et al. (1998)
Enteric viruses			
Poliovirus	Drinking-water biofilm on PVC coupon (1 week old)	28 h	Quignon et al. (1997)
	Drinking-water biofilm on paraffin-coated glass slides (2–9 weeks old)	7 days (cell culture) or 14 days (RT-PCR) at 20°C–25°C	Botzenhart and Hock (2003)
Canine calicivirus (surrogate for human norovirus)	Drinking water biofilm (1 month old) on PVC coupons in Propella reactor	4 weeks (15°C), slow decline	Lehtola et al. (2007)
Intestinal protozoan parasites			
Cryptosporidium parvum oocysts	Drinking water biofilm	Several months (20°C)	Keevil (2002, 2003)
Giardia lamblia cysts	Drinking water biofilm (7 months old)	34 days	Helmi et al. (2008)

The data referring to the persistence of the target organisms in the biofilms are based on analyses by cultural and/or molecular detection methods system (Table 2). Thus, the integration and survival of *S. enterica* serovar Typhimurium in drinking water biofilms for at least 43 days have been reported; at 36°C, even multiplication of these bacteria was observed (Armon et al. 1997). In annular reactors initially seeded with pure cultures of *S. enterica* serovar Typhimurium, and then continuously challenged with heterotrophic drinking water microorganisms at 20°C, the target bacteria were detectable in the biofilms until the end of the 60-day experiment, in the absence and presence of low levels of chlorine (0.05 and 0.2 mg/L) (Camper et al. 1998). *S. enterica* is also able to form biofilms in pure culture (Bridier et al. 2010). In a laboratory reactor, the bacteria formed substantial biofilms on polyvinyl chloride (PVC) pipe surfaces in a low-nutrient medium at 22°C within 24 h (Jones and Bradshaw 1996).

Filter media used for water treatment are usually covered by biofilms composed of heterotrophic water bacteria. In laboratory experiments, *S. enterica* serovar Typhimurium and *Shigella sonnei* have been shown to attach to granular activated carbon (GAC) particles from an operating water treatment plant, which were colonized by HPC bacteria (LeChevallier et al. 1984). After attachment, these bacteria survived in the biofilms in the presence of 2 mg/L of chlorine for 60 min with only small decreases in culturable numbers, while planktonic bacteria were undetectable after 5 min of contact. In another laboratory study, *S. enterica* serovar Typhimurium readily colonized sterile GAC in pure culture and maintained populations of ca. 10^5–10^7 cfu/g for 14 days (Camper et al. 1985). When added to a GAC column that had been pre-colonized with HPC bacteria by circulating river water through the system at 20°C–22°C for at least 2 weeks, the bacteria also attached to the biofilms, but at a lower level, and subsequently decreased over 8 days to undetectable levels. These observations indicate that biofilms in filters of water treatment plants may be a temporary reservoir for enteric bacterial pathogens, which can be released from filter beds and then enter water distribution systems.

Taken together, at least *S. enterica* serovar Typhimurium can survive for prolonged periods, and perhaps even multiply under favourable conditions in mixed-population drinking water biofilms either as a primary colonizer or after integration in pre-existing biofilms. It remains to be investigated whether other salmonellae, *Shigella* spp. and *V. cholerae* are also capable to survive in biofilms of water distribution systems.

3.1.2 Yersinia enterocolitica

Y. enterocolitica is considered a relevant waterborne pathogen transmitted through ingestion of faecally contaminated water (WHO 2008). In a study of German DWDS, this organism was not detected by culture or PCR-based methods in established biofilms in 18 pipe sections (2–99 years old) and from biofilm reactors with stainless steel, copper, PVC and polyethylene (PE) coupons exposed to unchlorinated drinking water for 24 months (Hummel and Feuerpfeil 2003). However, seeding experiments showed that *Y. enterocolitica* persisted in a culturable form in drinking water biofilms for 14 days after inoculation with declining numbers (Hummel and Feuerpfeil 2003). Camper et al. (1985) found that *Y. enterocolitica* was able to attach to biofilms of indigenous water bacteria on GAC with a subsequent slow decline of the concentrations of culturable bacteria over 2 weeks after inoculation. These observations indicate that *Y. enterocolitica* is not normally present in DWDS biofilms, but in case of contamination events has the potential to transiently colonize biofilms in filters of water treatment and in water distribution systems and to persist there without multiplication before the bacteria are eliminated within days.

3.1.3 Campylobacters

Campylobacter spp. (mainly *Campylobacter jejuni* and *Campylobacter coli*) are a major cause of acute human gastroenteritis worldwide and have been involved in waterborne epidemics.

Campylobacter spp. were not detected, neither by culture nor by PCR-based methods, in established biofilms in 18 pipe sections (2–99 years old) from German drinking water systems and in biofilm reactors with stainless steel, copper, PVC and PE coupons exposed to unchlorinated drinking water for 24 months (Hummel and Feuerpfeil 2003). However, laboratory experiments indicate that campylobacters can form monospecies biofilms (Joshua et al. 2006) and survive after introduction into pre-established mixed-population biofilms (Table 2). In a two-stage continuous culture biofilm model system containing an established biofilm of autochthonous water microflora, different isolates of *C. jejuni* and *C. coli* seeded into the system were found to survive in these biofilms (Buswell et al. 1998). Using FISH with *Campylobacter*-specific 16S rRNA probes, it was demonstrated that *Campylobacter* spp. persisted up to the termination of the experiments after 28 days and 42 days of incubation at 30 and 4°C, respectively. The survival times determined by culture were significantly shorter, indicating that the bacteria possibly entered a VBNC state. In addition, the background autochthonous water microflora of the biofilms seemed to enhance survival of the pathogen. *C. jejuni* was associated with the bacterial microcolonies of the biofilms, suggesting that it was incorporated within the biofilm matrix (Buswell et al. 1998; Keevil 2002). It was speculated that this location was probably preferred to obtain metabolites from the biofilm consortium and also to seek lower oxygen environments favourable for the microaerophilic physiology of *Campylobacter* (Keevil 2002). Lehtola et al. (2006a) studied the survival of *C. jejuni* after injection in a flow-through reactor (Propella™ reactor) with 4-week-old drinking water biofilms on PVC coupons. By culture methods, *C. jejuni* was detectable for only 1 day, while the bacteria were found for at least 1 week in the biofilms and for 3 weeks in the reactor outlet when using the FISH technique. In another seeding experiment, *C. coli* was added to a flow-through reactor containing 14-day-old drinking water biofilms on paraffin-coated glass coupons (Hummel and Feuerpfeil 2003). In the biofilms, *C. coli* was detected culturally for only 1 h after inoculation, but at the end of the experiment (28 days) the bacteria could still be detected by a PCR-based method. These investigations indicate that drinking water biofilms can be a transient reservoir for *Campylobacter* spp. in case of contamination. Culture methods underestimate the real number of *Campylobacter* in these biofilms; culture-independent methods may prove to be useful as additional and more sensitive tools for detection of these pathogenic organisms, which may survive in a VBNC state in drinking water biofilms, preferentially at lower water temperatures.

3.1.4 *Helicobacter pylori*

H. pylori is an etiological agent of chronic gastritis, peptic and duodenal ulcer disease. The exact mode of transmission remains unclear, but the water-associated faecal–oral route has been suggested as one potential transmission pathway (Percival and Thomas 2009). So far, no reports have been published on the successful cultivation of *H. pylori* from drinking water biofilms. There is evidence that *H. pylori* occurs in a VBNC state in drinking water environments. Thus, culture-independent molecular methods (immunofluorescence techniques, FISH and PCR-based detection methods) allowed the detection of *H. pylori* in water distribution system biofilms. *Helicobacter* DNA was found in a biofilm of a cast-iron pipe from a Scottish drinking water distribution system (Park et al. 2001) and in biofilms of plumbing systems of schools and domestic environments in England (Watson et al. 2004). In the latter study, a higher detection rate for *Helicobacter* DNA was obtained in biofilms (42% of samples) compared to water (26% of samples). *H. pylori* DNA was detected by real-time PCR in drinking water biofilms generated on silicone tube surfaces (Linke et al. 2010). Using FISH, *H. pylori* was found on PVC coupons in flow cells placed in a by-pass to a Portuguese drinking water distribution system (Bragança et al. 2005).

In laboratory experiments, *H. pylori* formed monospecies biofilms at air-liquid interfaces (Stark et al. 1999; Cole et al. 2004), and on the inner surfaces of silicone tubes, where the bacteria developed cell shapes typical of the VBNC state of *H. pylori* within 2–3 weeks (Linke et al. 2010). *H. pylori* rapidly colonized plumbing materials including stainless steel, PVC, polypropylene (PP) and copper, reaching steady state after 96 h (Azevedo et al. 2006a). High shear stress negatively influenced the adhesion of *H. pylori* to stainless steel and PP, while temperature (4, 23 and 37°C) and inoculation concentration appeared to have no influence on adhesion (Azevedo et al. 2006b). Thus, sites in water systems with low shear stress and high residence time such as well surfaces and water storage reservoirs may promote the attachment of *H. pylori* to solid surfaces and an association with biofilms (Azevedo et al. 2006b). The integration of *H. pylori* in pre-existing biofilms was described in some studies (Table 2). The bacteria bound to established drinking-water biofilms grown on stainless steel where they persisted for up to 8 days after inoculation (Mackay et al. 1999; Azevedo et al. 2003). Seeding experiments in a flow-through reactor demonstrated the persistence of *H. pylori* in low numbers in established drinking water biofilms on paraffin-coated glass slides and on PE coupons for 14 days (end of the experiment) after inoculation (Exner and Rechenburg 2003), indicating the capability of these bacteria to attach to existing biofilms and persist in this environment for days without multiplication. Both field investigations and laboratory seeding experiments indicate that *H. pylori* can integrate and survive, possibly in a VBNC state, in drinking water biofilms for extended periods of time, supporting a potential waterborne route of transmission for this pathogen.

3.2 Faecal Index and Indicator Organisms

Important index/indicator organisms include coliform bacteria (total coliforms, *E. coli*) and faecal streptococci/enterococci (Payment et al. 2003). According to the WHO (2008), *E. coli* is the parameter of choice for monitoring drinking water quality. Coliforms other than *E. coli* may also indicate the presence of faecal pollution, but they could also originate from a non-faecal source. However, their presence indicates an undesirable contamination of water systems due to treatment deficiencies or lack of water system integrity. Enterococci are used as an additional parameter of faecal pollution. Long-term survival and regrowth of the index/indicator bacteria in biofilms may contribute to the contamination of water distribution systems by these organisms in the absence of known contamination events (LeChevallier et al. 1987). From a public health perspective, this phenomenon is of importance since contamination of drinking water with coliforms from biofilms in distribution systems can interfere with their function to indicate faecal or other undesirable exogenous contaminations and mask true failures in water treatment and maintenance of the network. In addition, some index/indicator organisms can also be relevant as pathogens in water-related diseases.

3.2.1 Escherichia coli

E. coli is a harmless and common inhabitant of the human intestinal tract, but some variants of *E. coli* are obligate pathogens such as enterotoxigenic and enterohaemorrhagic *E. coli*, which have been involved in waterborne outbreaks. *E. coli* was not found in biofilms of DWDS in Germany (Wingender and Flemming 2004) and Sweden (Långmark et al. 2005a) using cultural detection methods. In contrast, Juhna et al. (2007) detected *E. coli* culturally in the biofilm of one of five cast iron pipes that had been removed from DWDS in France, England, Latvia and Portugal; all pipes were *E. coli* positive when analyzed by PNA-FISH. In the same study, *E coli* was detected by PNA-FISH on 56% of cast iron, PVC and stainless steel coupons exposed to drinking water distribution networks in France and Latvia for 1–6 months. The approximate bacterial numbers at several sites were in the range of 200–500 cells/cm^2. The *E. coli* cells were demonstrated to be viable based on their capability of cell elongation after resuscitation in low-nutrient medium supplemented with the DNA gyrase inhibitor pipemidic acid, suggesting that the bacteria were present in the biofilms in a metabolically active, but unculturable state. The *E. coli* cells were observed as single cells and not in microcolonies supporting the general assumption that these bacteria are most likely not multiplying in drinking water environments (Juhna et al. 2007). On the occasion of a coliform occurrence in a drinking water distribution system in New Jersey, USA, biofilms on different surfaces were examined; among the coliforms isolated, *E. coli* was identified on iron tubercles (LeChevallier et al. 1987). In these field investigations, the sources of contamination were not known, but the observations indicate the possibility of a sporadic appearance of viable *E. coli* in DWDS biofilms, which is expected to occur following an ingress of faecally polluted water into the distribution system.

E. coli is capable to form monospecies biofilms independent of the strain (Bridier et al. 2010), and under low-nutrient and low-temperature conditions relevant to drinking water systems (Jones and Bradshaw 1996), but can also become incorporated into established drinking water biofilms of autochthonous water bacteria. In a flow-through biofilm reactor which was used to simulate biofilm formation in pipes, *E. coli* was shown to attach to chlorinated PVC pipe surfaces under laminar flow conditions in a low-nutrient medium at an average temperature of 22°C (Jones and Bradshaw 1996). The bacteria formed a substantial biofilm population of more than 10^6 cells/cm^2 within 24 h with concomitant exopolysaccharide production which was maximal after 4 days.

A number of studies demonstrated that *E. coli* can colonize mixed-population biofilms both on filters used for water treatment and on materials used in DWDS and domestic plumbing (Table 2). When added to a GAC column that had been pre-colonized with HPC bacteria by circulating river water through the system for at least 2 weeks, *E. coli* attached to the biofilm, and subsequently decreased slowly in numbers over 14 days to culturally undetectable levels (Camper et al. 1985). The fate of *E. coli* labelled with green fluorescent protein (GFP) was investigated after introduction of the cells into a biofilter with glass beads as a filter medium on which a steady-state biofilm of a microbial community from a coal/sand filter from a full-scale water treatment plant had been established (Silva et al. 2006; Li et al. 2006). *E. coli* cells were able to survive inside the filter biofilm matrix for a prolonged period after the contamination event; cells were mainly present as microcolonies intertwined with other biofilm components. A practical implication of the location of *E. coli* microcolonies in deeper parts of biofilms in water filtration systems may be protection from disinfectants such as chlorine applied during the backwash process (Li et al. 2006).

Banning et al. (2003) reported the incorporation of GFP-labelled *E. coli* into biofilms formed by indigenous groundwater microorganisms in a flow-through laboratory-scale reactor. The numbers of *E. coli* increased over the first 3 days and subsequently declined over 15 days, probably due to death and/or detachment processes. The interaction with the biofilms seemed to slow the washout effect. Removal of attached *E. coli* was slower with low-nutrient groundwater flow-through compared with high-nutrient treated effluent, suggesting that an increase in available nutrients may reduce *E. coli* survival potential due to enhanced competition for nutrients and/or enhanced antagonism by the indigenous microbial biofilm populations (Banning et al. 2003). Using a chemostat model of a drinking water distribution system fed with tap water, Robinson et al. (1995) found that an environmental strain of *E. coli* became established in biofilms of naturally occurring drinking-water bacteria and was present in these biofilms as microcolonies for the duration of the experiments (13 days). Using *lacZ* and *luxAB* as reporter genes, these microcolonies were found to reside in deeper parts of the biofilms where an anaerobic environment existed. The fate of experimentally injected *E. coli* strains was investigated in a drinking water pilot system whose surfaces were largely colonized by autochthonous bacteria in several weeks old biofilms at quasi-steady state (Fass et al. 1996). The bacteria rapidly attached to the surfaces; during the first days the number of *E. coli* decreased more quickly than the theoretical washout, but subsequently the number of the bacteria slowly increased. During 7 days at 20°C, only a small fraction of the injected bacteria (0.001%) was able to adapt and start to grow. The data indicated that the laboratory and environmental *E. coli* strains used were able to grow at 20°C in the biofilms of the drinking water pilot system, but growth was insufficient to sustain the presence of *E. coli* in this system. A similar observation was reported by Sibille et al. (1998) when *E. coli* was added to an experimental distribution system, which was free of protozoa. In the presence of protozoa, there was an enhanced elimination of *E. coli*, indicating that removal of *E. coli* from biofilms can be accelerated by the grazing activitiy of protozoa. Williams and Braun-Howland (2003) observed that environmental *E. coli* inoculated into 2-week-old drinking water biofilms survived for at least 10 days. Using FISH, the cells were shown to maintain detectable amounts of rRNA indicating their metabolically active state within the biofilms where they were protected from exposure to high levels of chlorine and monochloramine and remained still viable.

There are also indications that pathogenic *E. coli* strains can persist in biofilms of water systems. Enterotoxigenic *E. coli* was shown to colonize both sterile GAC particles and GAC which had been precolonized by HPC bacteria (Camper et al. 1985). Szewzyk et al. (1994) demonstrated that when added to a fixed bed continuous flow reactor colonized by a heterotrophic bacterium, a pathogenic *E. coli* strain was able to grow in the biofilm and subsequently re-enter the free water phase. Individual cells as well as microcolonies of *E. coli* were observed in the biofilm. After injection of enterohaemorrhagic *E. coli* O157:H7 into a reactor with glass beads pre-colonized with drinking water biofilms for 2 or 3 weeks, significantly more bacteria were retained compared to uncolonized reactors and washout of cells was prolonged (Baumann et al. 2009). In another study *E. coli* O157 was shown to colonize drinking water biofilms and persist there for several weeks at low temperatures (10°C and 20°C) (Keevil 2002). Higher numbers were detected in biofilms on polybutylene compared to copper and stainless steel, indicating that the type of plumbing material may influence *E. coli* persistence in biofilms. In accord with this observation, Yu et al. (2010) also reported that colonization of drinking water biofilms by *E. coli* was dependent on pipe material. In a static test system, biofilm formation in drinking water inoculated with *E. coli* was followed at 25°C over 90 days. After this period, the concentrations of culturable *E. coli* in the biofilms were different on the materials in the order of copper < chlorinated PVC < PE < polybutylene < stainless steel < steel coated with zinc.

As a conclusion, it can be expected that faecal contamination of drinking water systems may lead to colonization of biofilms by *E. coli* in a transient manner. The potential of limited persistence and growth of *E. coli* in water network biofilms usually does not call into question the correlation between the presence of this species and its indication of faecal contamination. However, the growth of *E. coli* in biofilms and its later release and detection in the water phase may interfere with the diagnosis of when the contamination started. Recently, growth experiments showed that *E. coli* released mechanically from monospecies biofilms by ultrasonication revealed a relatively short lag time before initiating multiplication that was the same as the lag time of suspended exponential cells and significantly less than suspended stationary cells (Caubet et al. 2006). Practical implications are that *E. coli* cells mechanically mobilized from biofilms, for example by hydrodynamic forces, which may be applied in cleaning procedures can be an important source of contamination and display a high potential to rapidly recolonize new environments.

3.2.2 Coliform Bacteria Other than *E. coli*

Coliforms such as *Citrobacter*, *Enterobacter* and *Klebsiella* species are not only common inhabitants of the intestinal tract of warm-blooded animals, including humans, but can also be found in natural environments such as soil, vegetation or surface waters (Leclerc et al. 2001). A number of field observations indicate that coliform bacteria other than *E. coli* are occasionally present in low numbers in biofilms of DWDS. Coliform bacteria were detected only on one (1 coliform per cm^2) of 20 pipe coupons samples exposed in different DWDS in the USA (LeChevallier et al. 1987). Different coliform species (e.g., *Citrobacter freundii*, *Enterobacter agglomerans*, *Enterobacter sakazakii*) were detected in biofilms of exhumed water distribution pipes in the UK (Sartory and Holmes 1997). Wingender and Flemming (2004) found coliform bacteria in biofilms from 2 of 18 pipes removed from German DWDS. Batté et al. (2006) also detected coliforms only in 2 of 57 biofilm samples from cast-iron coupons exposed in French DWDS; *E. coli* was not found. Several coliform species were detected in biofilms on galvanized iron pipes at the periphery of the Seoul water distribution system (Lee and Kim 2003). *Enterobacter*, *Klebsiella* and *Pantoea* were identified in biofilms from various South African DWDS (September et al. 2007). On the basis of the microbiological analysis of cast-iron coupons in a French drinking water network, Gatel et al. (2000) estimated the average number of total coliforms at 0.2 coliforms per cm^2 of pipeline surface. Coliforms were often found in soft deposits recovered from pipelines in Finnish drinking water distribution networks, but not from water samples (Zacheus et al. 2001). Disturbances such as changes in pressure may release these deposits containing coliforms into the water; thus, deposits can be regarded as a source and reservoir for coliforms and possibly other hygienically relevant microorganisms. LeChevallier et al. (1987) reported high densities (>160 bacteria per g) of coliform organisms (*C. freundii*, *E. agglomerans* and even *E. coli*) associated with iron tubercles in a distribution system with long-term coliform problems within the drinking water network, while treatment plant effluents were negative for coliform bacteria.

In the context of coliform episodes, biofilms on rubber-coated valves in different German DWDS were identified as the source and reservoir of coliform bacteria (Kilb et al. 2003). In 15 of 21 biofilms, coliforms (mainly *Citrobacter* species) were detected with levels up to 5×10^3 organisms per cm^2. A possible explanation for the extensive biofilm formation on rubber-coated valves may be the availability of utilizable carbon compounds, for example low-molecular-weight additives such as paraffins used as softeners that leached from the coating material, providing nutrients at the valve surfaces (Kilb et al. 2003).

Coliform bacteria are able to survive and may multiply in biofilms even under low-nutrient and low-temperature conditions of distribution systems (LeChevallier et al. 1987; Camper et al. 1996). The incorporation of coliforms, including *C. freundii*, *Klebsiella pneumoniae* and *Klebsiella oxytoca*, into pre-existing mixed-population biofilms grown under drinking-water conditions has been described by several authors (Robinson et al. 1995; Camper et al. 1996; Packer et al. 1997; Szabo et al. 2006; Table 2). In laboratory experiments, addition of coliforms, including *Klebsiella* spp. and *Enterobacter* spp., in the presence of heterotrophic drinking water bacteria to an annular reactor showed that the growth rate of planktonic coliforms had a long-term impact on their ability to effectively compete in a mixed-population biofilms (Camper et al. 1996). Coliforms successfully colonized the surfaces (mild steel or polycarbonate) and persisted in the biofilms at densities of 10^3–10^4 cfu/cm^2 for 8 weeks only when they were previously adapted to low-nutrient conditions. Szabo et al. (2006) showed that *K. pneumoniae* inoculated into a drinking water biofilm on corroded iron surfaces in an annular reactor systems persisted longer (16 days) in the biofilms when the cells were pre-cultured in a low-nutrient environment compared to bacteria that were grown in a nutrient-rich medium (persistence for 7 days). These observations indicate that slowly growing coliforms from environmental sources are more likely to colonize surfaces and form biofilms in water systems under oligotrophic conditions compared to bacteria which are adapted to elevated nutrient levels.

Coliforms are also able to form biofilms as primary colonizers without the presence of an autochthonous sessile microflora. For example, using a laboratory reactor for simulation of biofilm formation in water pipes, Jones and Bradshaw (1996) demonstrated the ability of *K. pneumoniae* to form a dense and metabolically active single-species biofilm on chlorinated PVC pipe surfaces. In reactor experiments, clean coupon surfaces introduced into a reactor with already contained pre-established biofilms with coliforms were also colonized by coliform bacteria, suggesting that in practice, shedding of coliforms from a biofilm may result in the colonization of surfaces further downstream in a water system (Camper et al. 1996).

In laboratory experiments, it was found that *K. pneumoniae* biofilms were 150 times more resistant to chlorine than were unattached cells (LeChevallier et al. 1988). Keevil et al. (1990) reported that coliform bacteria in a biofilm survived prolonged exposure to 12 mg/L of free chlorine. *K. pneumoniae* persisted in drinking water biofilms on corroded iron coupons in the presence of free chlorine (0.6–1 mg/L) due to ineffective transport of the disinfectant to the surface, and increase in pH value at the iron surface decreasing the potency of chlorine (Szabo et al. 2006).

A recently published manual of the American Water Works Association states that under some relatively specific conditions, a water utility using surface water sources for drinking water production may experience coliform biofilm problems, which are characterized by the inability to maintain a disinfectant residual greater than 0.2 mg/L, relatively high levels of assimilable organic carbon (>100 μg/L) and TOC (>2 mg/L), warm water temperatures ($\geq 15°C$), a significant percentage of iron pipe in service for more than 75 years and lack of a regular flushing program for dead-end areas (Fox and Reasoner 2006).

In biofilms of man-made water systems with elevated water temperatures and sufficient nutrient levels, coliform bacteria can be found in high numbers as, for example, in water circulation systems of paper machines, where coliforms have been reported to occur at a concentration of 2×10^5 cfu/g dry weight of biofilm mass (Gauthier and Archibald 2001). Coliform bacteria seem to be common in submerged and unsubmerged slimes on surfaces in paper mills (Rättö et al. 2006). A characteristic property seems to be their ability to produce significant amounts of exopolysaccharides (e.g. colanic acid), which may be involved in biofilm formation. In these environments, the index and indicator function of coliform bacteria is invalid, since they represent a natural habitat for coliform growth and are not necessarily related to faecal or other pollution.

In conclusion, field and laboratory observations demonstrate that coliforms have the potential to form biofilms, become incorporated into biofilms and can multiply within biofilms, although a permanent colonization of biofilms does not occur. Increased tolerance to disinfectants of coliforms in biofilms in combination with chlorine-neutralizing environments can be contributing factors in the persistence of these bacteria in water systems. Biofilms at specific, discrete locations within a distribution system can then act as point sources of water contamination.

3.2.3 Faecal Streptococci/Enterococci

Biofilms in artificial water systems have only rarely been analyzed for faecal streptococci/enterococci. Using a PCR-based method, faecal streptococci were occasionally detected on coupons (PE, PVC) exposed in a German waterworks treating bank-filtered raw water and in the corresponding DWDS (Schwartz et al. 1998a). Similarly, *Enterococcus* spp. was only sporadically cultured from biofilms on galvanized iron pipes at the periphery of the Seoul water distribution system (Lee and Kim 2003). In these studies, the enterococci were not identified to the species level, so it was not clear whether the bacteria belonged to intestinal enterococci of faecal origin or were members of environmental species. In other studies of French, German and Swedish water systems, enterococci were not detected in biofilms of DWDS (Emtiazi et al. 2004; Långmark et al. 2005a; Batté et al. 2006). Based on these data, it can be concluded that, similar to *E. coli*, intestinal enterococci are usually not present in biofilms of DWDS in regular operation. Under laboratory conditions, *E. faecalis* is capable of forming monospecies biofilms (Kristich et al. 2004; Bridier et al. 2010). This may be an indication that enterococci can be primary colonizers of surfaces. However, it is still unknown whether enterococci have the potential of colonizing surfaces under conditions typical of artificial water systems and whether they can become integrated into established mixed-population biofilms of water systems in case of a contamination event.

3.3 Environmental Biofilm Bacteria with Pathogenic Properties

A number of opportunistic bacterial pathogens naturally occur in aquatic and soil environments and are able to persist and grow in biofilms of man-made water systems. These bacteria include *Acinetobacter calcoaceticus*, *Aeromonas* spp., some coliforms (e.g. *Citrobacter* spp., *Enterobacter* spp., *K. pneumoniae*), *Legionella* spp., *Mycobacterium* spp. and *Pseudomonas aeruginosa*. Clinically relevant strains of these organisms display relatively high infective doses (10^6–10^8) for healthy individuals and are mostly harmless for them (Rusin et al. 1997). However, these organisms are especially critical for the increasing proportion of sensitive human populations such as infants, the very elderly, hospitalized individuals, immunocompromised persons and those with other underlying diseases and under medical treatment, which can be highly susceptible to infection by these bacteria. Depending on the organism, the route of transmission leading to a water-related disease is ingestion, inhalation of aerosols or exposure to skin (e.g. through wounds), ears and eyes. At present, *Legionella pneumophila* and some other *Legionella* species, *Pseudomonas aeruginosa* and non-tuberculous mycobacteria are regarded as the most relevant opportunistic bacterial pathogens linked to water-related diseases.

3.3.1 *Legionella pneumophila* and Other Legionellae

Bacteria of the genus *Legionella* are ubiquitous in natural and technical aquatic environments. Important sources of infection are sites where aerosols are generated such as cooling towers, evaporative condensers, whirlpools and water outlets such as showerheads or taps. The main transmission pathway is inhalation of contaminated aerosols, which can cause severe pneumonia (Legionnaires' disease) or the mild flu-like disease of Pontiac fever. More than 50 *Legionella* species are known to date, and about half of them are associated with disease in humans, with *L. pneumophila* as the clinically most important organism. Accordingly, the vast majority of studies relating to the interaction of legionellae with water biofilms is focused on this species (for review, see Taylor et al. 2009).

Legionellae are found in biofilms on all types of surfaces of water abstraction, treatment, distribution and storage within DWDS. They were identified in biofilms in production wells (Riffard et al. 2001; Pryor et al. 2004), on filter materials (sand filters, activated carbon filters, swimming pool filters) (Loret and Greub 2010), on pipe walls of municipal DWDS (Pryor et al. 2004; Wullings et al. 2011), in plumbing systems of public and private buildings (Bonadonna et al. 2009) and in sediments and corrosion products of water storage tanks. Especially critical is the occurrence of legionellae in biofilms of hospital water systems (Walker et al. 1995b; Doleans et al. 2004) and other health-care facilities, where legionellae present a risk of nosocomial infections. In dental-unit water lines *L. pneumophila* were often recovered from tubing biofilms (Barbeau et al. 1998; Mavridou et al. 2006). Legionellae can also occur in biofilms of industrial settings such as cooling towers of power plants, where they have been found in surface-attached biofilms and in floating biofilms at the air-liquid interface of water basins (Declerck et al. 2007a).

The inclusion and persistence of naturally occurring *L. pneumophila* at cold-water temperatures and multiplication of these bacteria at warm-water temperatures in biofilms of autochthonous water bacteria have been observed on various plumbing materials including ethylene-propylene, PE-X, unplasticized PVC and copper (Keevil 2002; Rogers et al. 1994a, b; van der Kooij et al. 2002, 2005; Gião et al. 2009). Increased concentrations of *L. pneumophila* are preferentially found in warm water systems where the temperature is favourable for multiplication of these bacteria ($25°C–45°C$; Wadowsky et al. 1985). In model drinking water biofilms containing mixed populations of heterotrophic bacteria, *L. pneumophila* was shown to be most abundant at $40°C$, but was absent from biofilms at $60°C$ (Rogers et al. 1994a). Under cold water conditions, *L. pneumophila* and other legionellae can also survive in biofilms. Thus, the persistence of *L. pneumophila* in drinking water biofilms has been observed in the temperature range of $5.0°C–8.5°C$ (Långmark et al. 2005b) and $15°C–20°C$ for several weeks (Gião et al. 2009; Lehtola et al. 2007; Moritz et al. 2010; Rogers et al. 1994a). *Legionella* spp. have been described to proliferate in biofilms of PVC pipes from unchlorinated water supplies at temperatures below $18°C$ (Wullings et al. 2011).

Biofilms in man-made water systems often harbour multiple *Legionella* species either in the absence or in the presence of *L. pneumophila*. Thus, only non-pneumophila *Legionella* were detected in 3-week-old biofilms on different materials (PE, PVC, steel) installed in a waterworks and at house branch connections within a distribution network (Schwartz et al. 1998a). Pryor et al. (2004) also detected *Legionella* species other than *L. pneumophila* in biofilms of production wells ($23°C–25°C$) and in a chlorinated municipal drinking water system ($21°C–31°C$ during the year). *L. pneumophila* and other *Legionella* species were identified by FISH and fluorescent-antibody staining in 2-week-old biofilms composed of drinking water-derived populations in a flow-through reactor system, where they were present before and after treatment with recommended levels of disinfectants (hypochlorous acid, monochloramine) (Williams and Braun-Howland 2003). Analysis with quantitative PCR of biofilms on PVC pipes from the DWDS of two different Dutch groundwater supplies resulted in the detection of a large variety of *Legionella* spp. in 14 out of 15 biofilm samples (Wullings et al. 2011). *L. pneumophila* was observed in part of these biofilms. Culturable legionellae were not detected in any of the biofilm samples. Using real-time PCR, both *L. pneumophila* and *Legionella* spp. have been detected in biofilms of cooling tower water basins (Declerck et al. 2007a).

Seeding experiments indicate that *L. pneumophila* becomes incorporated into biofilms and can survive or multiply within biofilms, depending on the environmental conditions (Table 2). Armon et al. (1997) found that *L. pneumophila* survived in biofilms of heterotrophic drinking water bacteria for more than 40 days at 24°C and 36°C. At lower water temperatures of 5.0°C–8.5°C, addition of *L. pneumophila* to 8-week-old drinking water biofilms on glass coupons in a pilot-scale water distribution system provided with chlorinated or UV-treated and chloraminated water also resulted in the persistence of these bacteria over the experimental period of 38 days (Långmark et al. 2005b). The concentration of *L. pneumophila* determined by a standard culture method declined over the observation period and often represented a small fraction of the cell numbers that were enumerated by FISH; despite the absence of culturable legionellae at the end of the experimental period, FISH-positive cells were still detected. In another study, *L. pneumophila* inoculated into reactors containing 1-month-old drinking water biofilms on PVC coupons was shown to persist in the biofilms under high-shear turbulent flow at 15°C for at least 4 weeks and could also be found in the water during this period (Lehtola et al. 2007). There was only a slow decrease in the numbers of the legionellae determined by culture methods and FISH. Similarly, in a flow-through reactor system with cold tap water (mean temperature 18.8°C), Moritz et al. (2010) observed the persistence of *L. pneumophila* for 28 days after inoculation of drinking water biofilms on various plumbing materials (ethylene propylene diene monomer (EPDM) rubber, PE-X and copper) with concentrations of FISH-positive cells sometimes several orders of magnitude higher than those of culturable cells. Thus, culturable legionellae seemed to represent only a small fraction of FISH-positive cells indicating loss of culturability in the biofilm mode of existence and the possibility of an underestimation of the incidence of legionellae determined by culture methods. These observations suggest that *L. pneumophila* possibly enters a VBNC state in drinking water biofilms, at least under cold water conditions, which do not allow multiplication of the bacteria.

In a model cooling water system, *L. pneumophila* persisted in biofilms of heterotrophic bacteria on different metal and plastic materials at 29°C for the experimental period of 180 days without any decline in the concentrations of culturable bacteria after inoculation of the recirculating water system (Türetgen and Cotuk 2007). The highest concentrations of approximately 10^4 cfu/cm^2 were observed on galvanized steel, one log unit lower amounts were detected on copper, stainless steel, PVC, PP and PE. These results demonstrate the potential of *L. pneumophila* to persist in biofilms of industrial cooling towers associated with air conditioning and industrial processes independent of the materials used in these systems.

Legionella colonization of biofilms in man-made water systems seems to be dependent on or at least supported by the presence of other heterotrophic organisms, which establish the biofilms and thus, provide the habitat for *Legionella* colonization and allow interactions between legionellae and other biofilm inhabitants to occur (Taylor et al. 2009). An important mechanism of *L. pneumophila* growth in aquatic biofilms seems to be their interaction with protozoa and their replication within protozoan hosts (Lau and Ashbolt 2009). For example, intracellular occurrence of *Legionella* in amoebae within biofilms has been demonstrated in situ in domestic cold water plumbing systems (Kalmbach et al. 1997). In a number of studies, replication of *L. pneumophila* within surface-attached and floating biofilms of heterotrophic bacteria grown under static conditions or in flow-through systems was only observed in the presence of amoebae such as *Hartmannella vermiformis* (Murga et al. 2001; Kuiper et al. 2004) or *Acanthamoeba castellanii* at 35°C (Declerck et al. 2007b). In a quantitative risk assessment for distributed drinking water, experiments showed that the interaction of *L. pneumophila* with thermophilic acanthamoebae and their accumulation within biofilms resulted in an increased resistance to thermal and chlorine disinfection (Storey et al. 2004b). In combination with detachment from biofilms, these interactions were important ecological factors that significantly increased the risk of legionellosis presented to the general human population.

However, a number of studies suggest that *L. pneumophila* can exist extracellularly within biofilms in the absence of protozoa. Legionellae have been reported to occur as distinct microcolonies within the stacks of a heterogeneous drinking water biofilm grown at 40°C without any evidence of amoebae (Rogers and Keevil 1992; Keevil 2002). Murga et al. (2001) found that *L. pneumophila* persisted in the biofilm matrix as single viable cells, which were unable to replicate in the absence of amoebae and without any evidence of microcolony formation.

The integration of *L. pneumophila* in biofilms can also be influenced by the types of bacterial species present in the biofilm community. In continuous flow chamber experiments, attachment of *L. pneumophila* to 2-day-old monospecies biofilms and persistence within these biofilms for up to 14 day was found in biofilms of *Empedobacter breve*, *Microbacterium* sp., and to a lesser extent *Acinetobacter baumanii*, while no attachment to *P. aeruginosa*, *Pseudomonas putida*, *Corynebacterium glutamicum*, or *K. pneumoniae* biofilms was observed (Mampel et al. 2006).

Nutrients for *L. pneumophila* growth may be supplied by other biofilm bacteria, which excrete compounds serving as a carbon and energy source for legionellae. Sediments in water systems can stimulate the growth of aquatic microflora, which in turn can stimulate growth of *L. pneumophila* (Stout et al. 1985). Recently, it was reported that *L. pneumophila* can grow on heat-killed bacterial and amoebal cells, and heat-treated biofilms (nectotrophic growth) indicating that dead biomass still attached to surfaces may be a nutrient source for *L. pneumophila* (Temmerman et al. 2006). Growth yields were lower compared to replication within amoeba suggesting that the major route of replication is by means of protozoa, and nectrotrophy is an alternative or supplemental mode of *Legionella* growth. Practical implications are that remedial action against *Legionella* must be targeted not only to kill the bacteria and protozoa in the biofilms, but also to clean the surfaces of water systems with the aim to remove dead biomass. Thus, the colonization of biofilms by *L. pneumophila* can be favoured by interactions with certain environmental bacteria pre-existing in the biofilms. However, antagonistic relationships may also occur; for example, certain heterotrophic bacteria such as *Pseudomonas fluorescens*, *P. aeruginosa*, *A. hydrophila*, *Burkholderia cepacia* and *Stenotrophomonas maltophilia* have been shown to interfere with the survival and persistence of *L. pneumophila* in biofilms through the production of bacteriocin-like substances (Guerrieri et al. 2008).

In natural environments and man-made water systems, *L. pneumophila* is supposed to be mainly a secondary colonizer of pre-existing biofilms. However, recent laboratory experiments demonstrated that under static conditions, *L. pneumophila* and 37 other non-*L. pneumophila* species were able to form monospecies biofilms on glass, polystyrene and polypropylene and at air-liquid interfaces in nutrient-rich media (Piao et al. 2006; Mampel et al. 2006; Konishi et al. 2006). *L. pneumophila* showed the strongest biofilm formation. At 25°C and 35°C, the biofilm cells were rod shaped, while at 37°C and 42°C, biofilms of *L. pneumophila* were mycelial mat-like and were composed of filamentous multinucelate cells. The filamentous form seems to allow the bacteria to multiply more rapidly in the initial stage of growth under appropriate nutrient conditions compared to the normal rod-shaped form (Piao et al. 2006). This may represent a strategy to compete with other environmental bacteria for available nutrients. Mampel et al. (2006) found that surface colonization by *L. pneumophila* under static conditions in nutrient-rich medium seemed to occur only by adhesion; replication did not occur on the surface, but biofilm formation was due to growth of planktonic bacteria rather than to growth of sessile bacteria. The relevance of the formation of monospecies *L. pneumophila* biofilms in real water distribution systems is still unclear.

In summary, *L. pneumophila* primarily colonizes pre-existing biofilms in man-made water systems. *L. pneumophila* undergoes a life cycle within the biofilms consisting of the uptake by certain protozoa, transient persistence within the amoebal trophozoite or cyst, subsequent intracellular proliferation within the protozoa, lysis of the host cells, release of bacterial cells, survival and possibly also replication as planktonic cells, recolonization of biofilms and re-infection of new protozoan host cells (Lau and Ashbolt 2009). The function of the VBNC state of *L. pneumophila* in this life cycle is still unclear. On the basis of laboratory experiments, it is known that the VBNC state can be induced when *L. pneumophila* is exposed to chlorine (Dusserre et al. 2008) or when the bacteria are heat-treated (Allegra et al. 2008), and they can be resuscitated when they undergo a passage through amoebal host cells (Steinert et al. 1997). Possibly, extracellular *L. pneumophila* cells, either in the water or in the biofilms, may undergo the transition to the VBNC state triggered by adverse environmental factors, which have yet to be defined for man-made water systems, and the interactions with protozoa represent an option for the VBNC cells to resuscitate and regain their virulence.

3.3.2 Pseudomonas aeruginosa

P. aeruginosa is an opportunistic pathogen causing a broad spectrum of local and systemic illnesses, including infections of the skin, ears and eyes, urinary and respiratory tract, and sepsis. The main transmission route is by exposure of damaged skin and mucous membranes to contaminated water, and in more rare cases, the inhalation of *P. aeruginosa*-containing aerosols. *P. aeruginosa* is a typical biofilm organism, and has become one of the best studied model organisms in biofilm research (McDougald et al. 2008).

In municipal DWDS, *P. aeruginosa* is not (Wingender and Flemming 2004; September et al. 2007) or only sporadically found in biofilms on pipe surfaces by culture methods. For example, it was found in 1 of 12 samplings in a biofilm on galvanized iron surfaces at a pilot plant connected to the periphery of the drinking water system in Seoul (Republic of Korea) (Lee and Kim 2003). Similarly, *P. aeruginosa* was detected sporadically on steel coupons exposed in a German drinking water system (Emtiazi et al. 2004), and in 1 of 13 biofilms on rubber-coated valves exhumed from different German DWDS (Kilb et al. 2003). However, under certain conditions such as during repair works, installation of new pipelines, storage of water, and in certain areas such as plumbing systems, tubes in dental unit water systems *P. aeruginosa* has been shown to occur with higher frequency and can persist and multiply in biofilms (Botzenhart and Döring 1993; Barbeau et al. 1998). Especially important is the occurrence of *P. aeruginosa* in water systems of hospitals and other health-care facilities where *P. aeruginosa* can be the cause of nosocomial disease and outbreaks with biofilms being the source of infection (Anaissie et al. 2002; Ortolano et al. 2005; Exner et al. 2005). In plumbing systems of hospitals, contamination with *P. aeruginosa* has often been observed to be limited the colonization of water outlets (e.g. taps, showerheads), which were associated with transmission to patients. A special case was reported where *P. aeruginosa* was isolated from microbial slimes, which developed on drinking water taps exposed to atmospheres contaminated with vapours of volatile organic liquids such as ethanol (Poynter and Mead 1964); evidence was presented that the ethanol-utilizing slime bacteria originated from the water supply system. This observation indicates that volatile substances can act as nutrients for colonization of water outlets by hygienically relevant bacteria such as *P. aeruginosa*.

P. aeruginosa has been detected in biofilms on various materials. When four different materials were exposed to drinking water for 3 months, *P. aeruginosa* was identified in biofilms on epoxide, but not on paraffin, bitumen and chlorinated rubber materials (Dott and Schoenen 1985). *P. aeruginosa* constituted only 1% of the bacteria identified, whereas other *Pseudomonas* species were more abundant. In a chemostat model of a plumbing system inoculated with sludge from a calorifier, biofilm formation was studied on the surfaces of different plumbing materials

(Rogers et al. 1994b). Among the microorganisms identified, *P. aeruginosa* was detected after 24 h on PP and PE at concentrations of 1.9 and 260 cfu/cm^2, respectively; after 21 days, *P. aeruginosa* was only identified on the surface of mild steel (30 cfu/cm^2). Biofilms of *P. aeruginosa* on the interior surface of PVC pipes were visualized by scanning electron microscopy; the bacteria were shown to be embedded in extracellular material (Carr et al. 1996). In addition, *P. aeruginosa* has been identified among the organisms of biofilms in corroded copper pipes in water systems of a hospital (Wagner et al. 1992) and a nuclear power plant (Wallace et al. 1994). These observations indicate that *P. aeruginosa* is among the pioneering organisms developing aquatic biofilms and can persist in mixed-population biofilms on different materials under conditions relevant to conditions found in artificial water systems. However, in a culturable form, they usually seem to constitute only a minor fraction of the biofilm communities.

Seeding experiments show that *P. aeruginosa* can both colonize new materials and became incorporated into pre-existing biofilms (Table 2). Inoculation of PVC pipes with *P. aeruginosa* suspended in sterile water and incubation for 8 weeks under water stagnation resulted in the colonization of the interior surface of the pipes in the form of cell aggregates embedded in extracellular material (Vess et al. 1993). After subsequent exposure of the pipes with water containing 10–15 mg/L free chlorine for 7 days, *P. aeruginosa* could be recovered from the water, indicating bacterial survival in the presence of the relatively high levels of disinfectant in the biofilms, which then served as a reservoir for contamination of the water phase. These results suggest that under conditions of stagnation or slow flow rate colonization and biofilm formation with EPS production on the inner surfaces of distribution pipes is promoted and can be a source of continuous water contamination.

The incorporation of *P. aeruginosa* into established biofilms has been investigated in some cases. Banning et al. (2003) reported the incorporation of GFP-labelled *P. aeruginosa* into biofilms formed by indigenous groundwater microorganisms in a flow-through laboratory-scale reactor. With treated effluent flow-through, the attached bacteria could be detected for at least 40 days and persisted for longer than the theoretical time for cells to fall below the detection limits, suggesting that *P. aeruginosa* was growing under these high-nutrient conditions. Seeding experiments in a flow-through reactor demonstrated the integration and persistence of *P. aeruginosa* at densities of approximately 10^2 cfu/cm^2 in established drinking water biofilms on EPDM coupons for 28 days (Bressler et al. 2009). In a similar laboratory-reactor system, spiking of 14-day-old drinking water biofilms grown on EPDM and PE-X materials with *P. aeruginosa* resulted in the incorporation and persistence of the organisms in the biofilms for up to 29 days (end of experiments) (Moritz et al. 2010). The numbers of *P. aeruginosa* determined by FISH were often several orders of magnitude higher than the concentrations determined culturally. A decrease of culturable *P. aeruginosa* occurred over 4 weeks, while the concentration of FISH-positive remained constant. These observations suggest that *P. aeruginosa* can enter a VBNC state in drinking water biofilms.

In summary, *P. aeruginosa* represents an opportunistic organism which can act as a primary colonizer of diverse materials in artificial water systems, and can also colonize established biofilms. *P. aeruginosa* may occur in a culturable state in biofilms on materials, which are commonly employed in DWDS and domestic plumbing, mostly at relatively low concentrations compared to the general biofilm microflora. However, *P. aeruginosa* may also be present in a VBNC state at significantly higher concentrations suggesting that culture-based methods may significantly underestimate the actual occurrence of *P. aeruginosa* in drinking water biofilms. *P. aeruginosa* can persist at least for weeks in these biofilms, which then represent a reservoir for the pathogen. In mature biofilms, culturable *P. aeruginosa* usually seems to be only a minor fraction of the microbial biofilm communities, but proliferation to hygienically relevant levels may occur under certain conditions such as water stagnation, elevated water temperatures or nutrient leaching from materials.

3.3.3 Non-tuberculous (Environmental) Mycobacteria

Non-tuberculous mycobacteria (NTM) such as *M. avium*, *M. gordonae*, *M. kansasii*, *M. lentiflavum*, *M. intracellulare* and *M. xenopi* are gaining importance as opportunistic human pathogens. They can cause various diseases involving the respiratory and gastrointestinal tract, the skeleton, the skin and soft tissues, and are often involved in nosocomial infections. NTM are natural inhabitants of aquatic and terrestrial environments as well as engineered water systems worldwide and as a consequence of their ubiquitous distribution, humans are ubiquitously surrounded by these pathogens (Falkinham 2009). Transmission from environmental sources occurs by ingestion, inhalation and injuries of skin.

Mycobacterial cell surface hydrophobicity due to a lipid-rich outer membrane is regarded as a major determinant of surface adherence, biofilm formation, acrosolization as well as disinfectant and antibiotic resistance (Falkinham 2009). Hydrophobicity-driven surface attachment prevents flushing the cells from water systems. Water disinfection kills off competing microorganisms, selecting for NTM. In addition, biofilm formation increases tolerance to disinfectants. These are factors that likely contributed to the increase in *M. avium* concentrations in DWDS with the distance from the treatment plant (Falkinham et al. 2001). NTM are ubiquitous in biofilms and deposits of water treatment plants, DWDS and domestic plumbing systems (Schwartz et al. 1998b; Vaerewijck et al. 2005; Tsitko et al. 2006; Thomas et al. 2008; Corsaro et al. 2010). Reported densities are frequently in the range of 10^2–10^5 cfu/cm^2 in biofilms and approximately 2–4 × 10^5 cfu/g dry weight in loose deposits (for example, Schulze-Röbbecke et al. 1992; Falkinham et al. 2001; Torvinen et al. 2004). Mycobacteria have been recovered from 90% of 50 biofilms samples from water treatment plants and domestic water systems with densities ranging from 10^3 to 10^4 cfu per cm^2 (Schulze-Röbbecke et al. 1992). *M. kansasii* and *M. flavescens* were detected in concentrations of 2 × 10^5 cfu/cm^2 and 7 × 10^4 cfu/cm^2, respectively, in a 10-month-old biofilm on the inner surface of a silicone tube perfused with warm water of 25°C (Schulze-Röbbecke and Fischeder 1989). *M. avium*, *M. intracellulare* and other slowly growing mycobacteria were detected in biofilms recovered from eight DWDS throughout the USA; *M. intracellulare* predominated and was found in densities between 1.3 and 2.9×10^3 cfu per cm^2 (average 600 cfu per cm^2; Falkinham et al. 2001). There were no statistically significant associations between mycobacterial colony counts in the biofilms and systems characteristics, including residual disinfectant concentration, assimilable organic carbon (AOC), biodegradable dissolved organic carbon (BDOC), nitrate, phosphate, alkalinity and hardness. September et al. (2004) reported mycobacterial concentrations up to 4.6×10^5 cfu per cm^2 in biofilms of South African DWDS. In an investigation of 18 drinking water pipes from German DWDS, Schulze-Röbbecke and Ilg (2003) found only low densities of mycobacteria (maximal number 11 cfu/cm^2). However, in the same study higher mycobacterial levels of 10^2 to over 10^3 cfu/cm^2 were detected on different materials exposed for 24 months at a waterworks and within a distribution network. The colonization was independent of the type of material used (PVC, PE, stainless steel, copper). Using a PCR-based method, Schwartz et al. (1998b) detected NTM in 3-week-old biofilms on coupons of PE, PVC and steel installed in a waterworks and distribution system supplied with bank-filtered drinking water. Torvinen et al. (2004) reported the appearance of mycobacteria in drinking water biofilms on PVC tubes installed at various network sites during the first 6 weeks with maximal numbers of 10^2 cfu/cm^2 after 10 weeks of exposure. *Mycobacterium* spp. were identified from a DNA-based clone library of biofilms grown for 2 months in a drinking water-fed biofilm reactor (Keinänen-Toivola et al. 2006). In parallel, a 16S rRNA-based clone library generated to characterize the active bacterial fraction also contained a clone classified as *Mycobacterium* spp., indicating that these bacteria belonged to the metabolically active fraction within the biofilm community. Analysis of 16S rDNA sequences from 45 showerheads from 9 cities in the USA showed that non-tuberculous mycobacteria were ubiquitous and enriched in the showerhead biofilms compared to the water; the most common sequences observed corresponded to *M. gordonae* and *M. avium* (Feazel et al. 2009). *M. avium* detected in water and a biofilm of a bathroom showerhead was linked to pulmonary disease, suggesting that biofilms in showers may serve as a reservoir and source of infection by this bacterium (Falkinham et al. 2008). Mycobacteria can also to be present in biofilms of other types of water systems. For example, mycobacterial gene sequences nearly identical to that of *M. avium* were detected in biofilms on the side walls of a hospital hydrotherapy pool, also with an enrichment of mycobacteria in the biofilms compared to the pool water (Angenent et al. 2005).

In experimental laboratory systems, mycobacteria readily form monospecies biofilms as has been demonstrated for *M. avium*, *M. intracellulare*, *M. kansasii*, *M. flavescens*, *M. chelonae*, *M. fortuitum* and *M. phlei* (Carter et al. 2003; Steed and Falkinham 2006). *M. avium* establishes more stable biofilms in the presence of Ca^{2+}, Mg^{2+} and Zn^{2+} (Carter et al. 2003). Mycobacteria reveal enhanced resistance to antimicrobial agents including chlorine compared to planktonic cells (Carter et al. 2003; Steed and Falkinham 2006). *M. chelonae* was recovered from stagnant water after exposure to 10–15 mg/L free chlorine for 7 days in a PVC pipe whose surfaces had been precolonized by the mycobacteria, indicating bacterial survival in the biofilm in the presence of relatively high concentrations of chlorine and contamination of the water phase by the mycobacteria (Vess et al. 1993).

Seeding experiments indicate that challenge of established biofilms with mycobacteria result in their incorporation and persistence for a prolonged time (Table 2). This ability to colonize drinking water biofilms has been shown for *M. xenopi* (Dailloux et al. 2003; Schulze-Röbbecke and Ilg 2003) and *M. avium* (Lehtola et al. 2006b, 2007). After introduction of *M. xenopi* into a 14-day-old drinking water biofilm on paraffin-coated coupons in a flow-through reactor, the mycobacteria were found to become incorporated and persisted at densities > 10^3 cfu/cm^2 for 3 months in the biofilms (Schulze-Röbbecke and Ilg 2003). Lehtola et al. (2006b, 2007) spiked a 1-month-old drinking-water biofilm on PVC coupons with *M. avium*; these bacteria could be detected over 4 weeks by culture and by PNA-FISH, with only a weak decline in cell concentrations. The bacterial numbers determined by PNA-FISH (about $4–8 \times 10^4$ cells/cm^2) were 1–4 log units higher than those determined by culture. Using the same experimental system, Torvinen et al. (2007) reported that water temperature and phosphorus (PO_4–P) influenced the culturability of NTM. Plate counts of *M. avium* were 2–20 times higher in biofilms at 20°C than those at 7°C, while the counts of FISH-positive cells were similar over the experimental period of 4 weeks, i.e. culturability was elevated at the higher temperature. When the phosphorus concentration was 4.2 µg/L, culturability was 2.7 times higher than with a higher phosphorus concentration of 13.8 µg/L. Since the higher phosphorus levels increased the number of heterotrophic biofilm bacteria, it was speculated that competition of the slow-growing *M. avium* with other biofilm organisms was the reason for the reduced culturability of the mycobacteria.

In man-made water systems, mycobacteria occur in multiple-species biofilms, where they may be associated with free-living amoebae and other protozoa as their hosts, but in contrast to legionellae, they are not dependent on amoebae for their multiplication within biofilms. Some mycobacterial species (*M. avium*, *M. fortuitum*, *M. matinum*) have been shown to multiply intracellularly in *Acanthamoeba castellanii*; growth within amoebae was accompanied with an increase in virulence of *M. avium* (Cirillo et al. 1997).

In summary, the combination of high cell surface hydrophobicity promoting attachment and the resistance to chlorine and heavy metals of mycobacteria such as *M. avium* and *M. intracellulare* suggest that these organisms may be biofilm pioneers in man-made water systems, but they also have the capability of colonizing established biofilms. Field observations indicate that biofilms can be significant sources of opportunistic mycobacteria in DWDS and other man-made water systems. The occurrence of mycobacteria in biofilms, their association with amoebae within biofilms, and their inherent and biofilm-related resistance to disinfectants are the major factors that contribute to the survival and long-term persistence of environmental mycobacteria, including species with pathogenic properties in man-made water systems. High ratios of FISH-positive cells to plate counts of NTM indicate that culture methods can be expected to underestimate the true numbers of NTM in biofilm of man-made water systems, because they may exist in a VBNC state in these environments.

3.3.4 Aeromonads

Among the aeromonads, the mesophilic species including *Aeromonas hydrophila*, *Aeromonas caviae* and *Aeromonas sobria* are human opportunistic pathogens, which may be important to public health (Percival et al. 2004). They can be involved in water-related gastroenteritis and associated diarrhoeal infections, and sometimes also wound infections. Accordingly, routes of exposure may be ingestion and dermal contact with contaminated water. Aeromonads occur widely in soil and natural waters, and were detected in drinking water networks (Holmes and Nicolls 1995), usually in low numbers. In particular, aeromonads were isolated towards the ends of surface-water supplied networks, and have been associated with aftergrowth in water systems (Holmes and Nicolls 1995). Because of their ability of utilizing a high number of different organic compounds, including amino acids, carbohydrates, carboxylic acids, peptides and long-chain fatty acids, mesophilic aeromonads were supposed to be ideally suited for growth in biofilms in water distribution systems (van der Kooij 1991; Percival et al. 2004).

A number of studies focused on the direct examination of aeromonads in water supply biofilms (Holmes and Nicolls 1995; Walker et al. 1995a; Chauret et al. 2001; September et al. 2007). Holmes and Nicolls (1995) investigated 20 pipe sections from a drinking water distribution system and determined an average *Aeromonas* concentration of 118 cfu/g (wet weight) in the biofilms from the pipe walls. After treatment of biofilms still attached to the pipes with 1 mg/L chlorine for 30 min, 10% of the pipe biofilms still contained aeromonads (average count of 51 cfu/g), indicating a certain degree of protection within the biofilm. Aeromonads were found to be able to multiply even at water temperatures as low as 4°C in laboratory experiments and have also been detected in distribution systems during colder months (Holmes and Nicolls 1995). These factors may contribute to the persistence and possibly regrowth in biofilms and appearance in drinking water. In another study, *A. hydrophila* was identified in 7.7% of biofilm samples from a drinking water network, but never in the disinfected bulk water of the distribution system, confirming disinfection efficiency on planktonic bacteria (Chauret et al. 2001). *Aeromonas* spp. were detected by culture in biofilms of various South African DWDS (September et al. 2007), and in a Swedish drinking water network, aeromonads were detected by FISH as 1–5% of total bacterial numbers (Långmark et al. 2005a).

Bomo et al. (2004) isolated aeromonads from biofilms on stainless steel and unplasticized PVC, which formed in potable and recycled water systems for 8–10 weeks, indicating their potential to colonize biofilms in these systems. The mean numbers of aeromonads in the system fed with potable water and recycled water ranged from 3.9 to 650 cfu/cm^2 and from 304 to 873 cfu/cm^2, respectively.

Seeding experiments revealed integration and persistence of *A. hydrophila* in established biofilms under oligotrophic and more nutrient-rich conditions. Challenging recycled wastewater biofilms with *A. hydrophila* resulted in the accumulation and persistence of the bacteria in the biofilms over the experimental period of 30 days (Bomo et al. 2004). The initial numbers (in the order of 10^3–10^4 *A. hydrophila* cells or cfu/cm^2) declined over time. At the end of the experiment, higher numbers were observed with the FISH technique than with a cultural method, indicating that high number of these bacteria persisted in the biofilms after this sub-population had ceased to be culturable. In another study, *A. hydrophila* was allowed to attach to surfaces in an annular biofilm reactor, before the system was continuously challenged with indigenous drinking water populations at elevated nutrient levels and 20°C, simulating conditions typically associated with increased risk for regrowth (Camper et al. 1998). *A. hydrophila* persisted for some time, but declined in numbers in the biofilm and bulk water, and was not detected after 2 weeks with culture methods. As to the influence of substratum, higher numbers of aeromonads were found on plastic materials (unplasticized PVC, polybutylene) compared with stainless steel (Assanta et al. 1998; Bomo et al. 2004).

Percival et al. (2004) gave the evaluation that *Aeromonas* spp. can colonize wells and water distribution systems for months and years, and can persist in biofilms; substantial growth can occur after disinfection. However, the actual role of drinking water and biofilms in the transmission of *Aeromonas* bacteria to humans and in water-related infections is still under discussion, and the exact human health significance of aeromonads is still unknown.

3.4 Enteric Viruses

Enteric viruses involved in water-related diseases cause acute gastrointestinal illness (e.g. noroviruses, rotaviruses) and can also affect other organs such as the liver (hepatitis A and E viruses) or the central nervous system (poliovirus). These viruses are excreted in the faeces of infected humans and are transmitted predominantly by ingestion.

In contrast to bacterial pathogens, relatively little information exists on the occurrence and survival of enteric viruses in biofilms of water distribution systems. However, there are some indications from field studies and laboratory experiments that pathogenic viruses can become incorporated into biofilms of DWDS, persist there and can be released again to represent a risk of infection (for review, see Skraber et al. 2005). Various strains of type B coxsackie viruses were detected in a biofilm colonizing the outlet mains of a drinking-water treatment plant (vanden Bossche and Krietemeyer 1995). Laboratory-scale seeding experiments demonstrated the attachment of polioviruses to drinking water biofilms grown on

PVC (Quignon et al. 1997), on polycarbonate where infectious viruses were detected up to 6 days and the viral genome up to 34 days (Helmi et al. 2008) or on paraffin-coated glass slides, where the viruses could still be detected after 7 days by cell culture and 14 days by RT-PCR (Botzenhart and Hock 2003). After inoculation of canine calicivirus (used as a surrogate for human norovirus) into a Propella reactor with 1-month-old drinking water biofilms on PVC coupons, the viruses could be detected over the experimental period of 4 weeks both in the biofilms and in the water (Lehtola et al. 2007). The decrease in virus numbers was only slow in biofilms and occurred more quickly in water; the decline was slower than the theoretical washout rate, indicating that the viruses were trapped in the biofilm and were shed with a delay. Bacteriophages used as model enteric viruses persisted for >30 days after incorporation into drinking water distribution pipe biofilms formed on PVC and stainless steel surfaces (Storey and Ashbolt 2001).

Recently, enteroviruses and noroviruses were found by RT-PCR to be present in wastewater biofilms, which had grown for 1 month to more than 2 years on polyethylene carriers in a moving-bed biofilm reactor of a wastewater treatment plant (Skraber et al. 2009). The viruses could be detected in the biofilms also at a time when their concentrations were under the detection limit in wastewater, suggesting the ability of these viruses to persist in the biofilms. In laboratory experiments, the persistence of noroviruses in naturally contaminated biofilms on PVC coupons taken from the wastewater treatment plant was observed at 4°C and 20°C for at least 2 months without a significant decline of the genome concentrations.

Thus, virion particles obviously have the potential to accumulate within biofilms of man-made water systems, which then act as reservoirs of virus particles. In addition, viruses in the biofilm seem to be protected against disinfectants such as chlorine compared to viruses in the water phase (Quignon et al. 1997). These processes represent a public health risk when virus-containing biofilm clusters are detached and released into the water phase, especially considering the low infective dose of one to only a few infectious virus particles.

3.5 Intestinal Protozoan Parasites

Important protozoan parasites, which have been involved in waterborne outbreaks of gastrointestinal disease due to the contamination of DWDS and swimming pool water systems include *Cryptosporidium* spp. (mainly *Cryptosporidium parvum* and *Cryptosporidium hominis*) and *Giardia lamblia*. They are obligate parasites, multiply within human or animal hosts and are excreted in the faeces in a fully infective form as oocysts (*Cryptosporidium* spp.) or cysts (*G. lamblia*). These transmissible stages can remain viable outside their hosts in aqueous environments for weeks to months and are highly resistant to chlorine and chloramine. A need for a more thorough understanding of the fate of oocysts, especially their interaction with

DWDS biofilms, was mentioned as essential for risk assessment of waterborne disease (Angles et al. 2007). There is no published information about the occurrence of these stages in biofilms of real DWDS, but laboratory investigations on the interaction of *C. parvum* and *G. lamblia* with biofilms of natural mixed populations and pure cultures indicate that oocysts or cysts can become incorporated and persist in biofilms, and be released back into the water.

Angles et al. (2007) described a study where the majority of *Cryptosporidium* oocysts introduced to a pipe-rig constructed from exhumed drinking water pipes (about 70 years old) became integrated in the existing biofilms and were released sporadically again. In seeding experiments, *C. parvum* oocysts were shown to attach to 16-day-old potable water mixed-population biofilms and persist in the biofilms over several months at 20°C (Keevil 2002, 2003). After this time, oocysts recovered from the biofilms were still viable and infective in an animal model. Epifluorescence microscopy of fluorochrome-labelled *C. parvum* oocysts showed that biofilm-associated oocysts occurred in clusters, so that biofilm sloughing might release a small but effective dose of oocysts into the water phase; it was speculated that this transient sloughing from the biofilm might explain sporadic disease cases of unknown origin (Keevil 2002). Following the spiking of 7-month-old drinking water biofilms, the persistence of *C. parvum* oocysts, but also of *G. lamblia* cysts in drinking water biofilms and their continuous release over 34 days was observed, demonstrating that not only *C. parvum* oocysts, but also *G. lamblia* cysts are capable of interacting with pre-existing drinking water biofilms (Helmi et al. 2008). *C. parvum* oocysts or *G. lamblia* cysts were also shown to attach to and being released from wastewater biofilms (Helmi et al. 2008) and experimental biofilms composed of natural stream microorganisms (Wolyniak et al. 2009) suggesting that a wider range of environmental biofilms can represent a reservoir for the infectious stages of these protozoan parasites.

A few studies examined the interaction of *C. parvum* oocysts with pure culture biofilms. White et al. (1999) reported the attachment of *C. parvum* oocysts to a three-species biofilm of *P. aeruginosa*, *Bacillus* sp. and *Acidovorax* sp., which developed on stainless steel in a flow cell supplied with diluted tryptic soy broth. After injection of oocysts into the system, they were retained by the biofilms at twice the amount compared to the bare surface without biofilm over the 3-day experimental period. Searcy et al. (2006) examined the capture and retention of *C. parvum* oocysts by monospecies biofilms of *P. aeruginosa* on glass surfaces in laboratory flow cells. The oocysts were present in the biofilms for more than 24 h and were not released after a 40-fold increase in flow rate. More oocysts were retained on the biofilm-coated surfaces compared with biofilm-free glass surfaces. To assess the role of bacterial EPS in the capture of oocysts by biofilms, a wild-type strain of *P. aeruginosa* whose EPS consist primarily of glucose- and rhamnose-containing polysaccharide as well as nucleic acids was compared with a mucoid strain overproducing the extracellular polysaccharide alginate. There was slightly less oocyst capture by the alginate-containing mucoid biofilms, although this trend was not statistically significant. It was concluded that biofilm architecture and surface–chemical interactions were more important that EPS composition in controlling oocyst deposition from suspension into *P. aeruginosa* biofilms (Searcy et al. 2006).

Overall, biofilms seem to represent a potentially significant, long-term reservoir of *Cryptosporidium* oocysts and *Giardia* cysts that can be released back into the surrounding water, thus posing a public health threat. Practical implications may be the appearance of oocysts in water distribution systems long after a contamination event. Thus, biofilm involvement was supposed to be the reason for ongoing recoveries of oocysts from a drinking water distribution system, following a waterborne cryptosporidiosis outbreak in England (Howe et al. 2002).

3.6 Free-Living Amoebae

Free-living amoebae are prevalent in man-made water systems and are common members of biofilm communities in sand and activated-carbon filters of water treatment plants, DWDS, plumbing systems and cooling towers (Block et al. 1993; Barbeau et al. 1998; Pedersen 1990; Hoffmann and Michel 2003; Thomas et al. 2008; Loret and Greub 2010). In different German DWDS, free-living amoebae, including hygienically relevant thermophilic *Acanthamoeba* species, were detected in biofilms recovered from pipe surfaces at densities between 2 and over 300 amoebae per cm^2 (Hoffmann and Michel 2003).

Free-living amoebae can present a risk for water quality (Thomas et al. 2010). Some free-living amoebae such as *Naegleria* and *Acanthamoeba* species are intrinsically pathogenic for humans and have been associated with a number of water-related diseases (e.g. encephalitis, meningoencephalitis or keratitis). Another important aspect of health significance is the phenomenon that so-called amoebae-resisting bacteria (ARB) can be taken up by free-living amoebae, survive and grow inside the amoebal cell, and then exit the protozoan host (Brown and Barker 1999). In a recent review, a total of 102 recognized and 27 suspected pathogenic bacterial species have been described as ARB, including the majority of hygienically relevant water-related bacteria such as *E. coli*, coliform bacteria, *V. cholerae*, *Shigella* spp., *Campylobacter jejuni*, *Legionella* spp. and *Mycobacterium* spp. (Thomas et al. 2010). In a number of studies, free-living amoebae infected with water-related pathogens such as *Legionella* spp. or *Mycobacterium* spp. have been identified in biofilms, sediments and sludges of drinking water treatment plants and DWDS (Corsaro et al. 2010; Loret and Greub 2010), and of cooling towers (Berk et al. 2006), indicating the ubiquity of the association of hygienically relevant bacteria with amoebae in biofilms of man-made water systems.

Intracellular survival within protozoa protects bacteria from deleterious environmental conditions as, for example, the presence of disinfectants (King et al. 1988; Thomas et al. 2010). Thus, after uptake of coliform bacteria (*C. freundii*, *E. cloacae*) and a number of pathogens (*S. enterica* serovar Typhimurium, *Shigella sonnei*, *Y. enterocolitica*, *C. jejuni* and *Legionella gormanii*) by amoebae and ciliates the bacteria were shown to display 30–120-fold increases in resistance to chlorine compared with free bacteria. The bacteria survived chlorine concentrations of 2–10 mg/L inside the protozoa, while under the same conditions, freely suspended bacteria were inactivated by 99% in the presence of 1 mg/L chlorine (King et al. 1988). Thus, the interaction between pathogenic bacteria and protozoa can contribute to the persistence of these bacteria in chlorinated water systems, especially when this association occurs within the protected environment of biofilms. Protozoa may also be involved in the elimination of hygienically relevant bacteria. As an example, it was shown that *E. coli* added to experimental water distribution systems disappeared more rapidly in a system containing protozoa compared with a system without protozoa, presumably due to the grazing activity of protozoa (Sibille et al. 1998). These observations show that protozoa in artificial water systems not only control general biofilm populations (Pedersen 1990), but can also have the function to regulate the presence of bacterial pathogens in biofilms.

An additional public health concern is based on the observation that the passage of some opportunistic pathogens such as *L. pneumophila* and *M. avium*, after ingestion by *Acanthamoeba* and other aquatic protozoa, results in an increase in the infectivity for human cells and also in further resistance to biocides (Cirillo et al. 1994, 1997; Donlan et al. 2005). When this process takes place within biofilms, it can be expected that following multiplication of these opportunistic pathogens bacteria with enhanced infectious potential are produced and released into the water.

In summary, protozoa in biofilms can be of hygienic relevance, since (1) some species are pathogenic for man, (2) they provide increased protection for pathogenic bacteria from disinfectants, and (3) they allow some opportunistic bacteria to multiply intracellularly with the result of enhanced infectivity of these bacteria.

3.7 Fungi

Hygienic problems associated with fungi in drinking water systems are the generation of undesirable taste and odour, and the production of mycotoxins. Opportunistic fungi are an increasingly important cause of nosocomial infections and can be life threatening in immunocompromised individuals. In addition, they can be the cause of human allergies. In comparison with other water-related microbial pathogens, limited information exists on the relationship between fungal occurrence in water systems and risk to human health. Some fungi have been implicated in nosocomial infections related to the hospital water supply. *Aspergillus fumigatus* and other *Aspergillus* species are especially important in causing nosocomial invasive aspergillosis with life-threatening infections among immunosuppressed patients. In addition to air, water is now supposed to be a source of these fungi in hospitals (Anaissie and Costa 2001). The transmission routes may be ingestion and, more importantly, inhalation of aersosols containing fungal conidia and hyphal fragments during aerosolization of tap or shower water, and during reaerosolization of fungi from shower walls.

Fungi are supposed to be common constituents of water distribution systems (Doggett 2000); they were isolated from water of municipal water distribution networks and from hospital plumbing systems (Rosenzweig et al. 1986; Arvanitidou et al. 1999; Anaissie et al. 2003; Warris et al. 2003; Hageskal et al. 2006). However, only a few studies focused on biofilms as a potential reservoir of fungi. In an investigation of biofilms from a municipal water distribution system in the USA (Doggett 2000), fungi were cultured from iron and PVC pipe surfaces. Densities of yeasts and filamentous fungi ranged from 0 to 8.9 cfu cm^2 and 4.0 to 25.2 cfu/cm^2, respectively. Among the most common filamentous fungi were species of *Aspergillus* and *Penicillium*. The fungi existed predominantly as spores, but not as hyphae or vegetative cells on the pipe walls. Fungi (*Penicillium*, *Phialophora*, *Cladosporium*, *Acremonium*) were also generally present in soft pipeline deposits investigated in different DWDS in Finland; their numbers varied between 1.5×10^3 and 5.0×10^6 cfu/L (Zacheus et al. 2001).

Interior surfaces in hospital water systems have been investigated using surface swabs (Anaissie et al. 2003). Fungi were recovered on the interior surfaces of waterlines, sink plumbing lines, showerhead surfaces and shower drains. The molds included *Aspergillus*, *Paecilomyces* and *Fusarium* species with pathogenic properties. Since the same genera and species were recovered from municipal water and hospital water systems, these molds were supposed to become part of the hospital water system biofilms (Anaissie et al. 2003).

Taken together, these observations suggest that biofilms of drinking water distribution networks and hospital plumbing systems can be a reservoir of fungi with pathogenic properties. Contamination of tap water by release of these fungi from biofilms can then be a potential cause of community- and hospital-acquired infections, especially in immunocompromised persons.

4 Outlook

In recent years, it has become obvious that biofilms in man-made water systems can act as a reservoir for pathogenic microorganisms, including various faecal and environmental bacteria, enteric viruses, faecally derived parasitic protozoa, free-living amoebae and fungi. Under epidemiological and ecological aspects, biofilms can be regarded as temporary or long-term reservoirs and habitats for pathogens, whose biofilm mode of existence may even represent part of their natural life cycle (Fig. 1). Most research into the distribution of hygienically relevant microorganisms in real man-made water systems and the integration of pathogens in biofilms under laboratory conditions in flow-through experiments has focused on the description of the fate of water-related pathogens. Thus, based on the knowledge of the biology and ecology of the single pathogen species and their behaviour

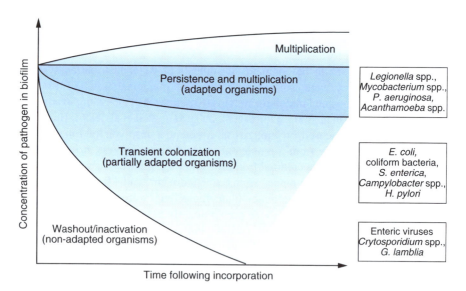

Fig. 2 Fate of hygienically relevant microorganisms after introduction into established biofilms of man-made water systems (modified according to Batté et al. 2003)

in biofilms summarized in this review, specific modes of persistence can be attributed to the different types of pathogens after their attachment to or incorporation into biofilms of man-made water systems (Fig. 2).

Environmental microorganisms such as legionellae, mycobacteria or *P. aeruginosa* adapted to oligotrophic conditions can persist over long times in aquatic biofilms and possibly even multiply in these environments. A transitory persistence for a few days to a few weeks seems to be possible for bacteria of faecal origin, while enteric viruses and the oocysts or cysts of parasitic protozoa are largely eliminated by washout from the biofilms. All these organisms can persist in biofilms or are released also at a time when they are not normally circulating in the water phase, so that their detection can lead to false assumptions as to the origin of the pathogens and impair risk assessment.

Biofilms in man-made water systems are usually multispecies biofilms. These types of biofilms are characterized by complex relationships involving intraspecies and interspecies interactions between the autochthonous aquatic microorganisms (Simões et al. 2007). Antagonism (e.g. production of inhibitory substances such as bacteriocins, antibiotics or oxidants), competetion for substrates or cooperation in the utilization of carbon and energy sources are interspecies interactions that determine the survival and proliferation of biofilm inhabitants. Intercellular communication via small diffusible signal molecules can be involved in the control of these interactions. Physical and chemical environmental factors such as hydrodynamic conditions, temperature, nutrient availability and the presence of disinfectant residuals additionally influence the biological processes of biofilm development. As outlined in this review, many studies have been performed to gather information on the fate of water-related pathogens in biofilms, mainly of DWDS under different physical and chemical conditions. However, the mechanisms of the biological interactions between pathogenic microorganisms and complex biofilm communities and the response of biofilm organisms to the invasion of pathogens into the biofilms under conditions of real man-made water systems are still largely unknown. There is still incomplete knowledge about the molecular processes, which are involved in the fate of pathogens in biofilms, including the attachment to biofilms, the processes governing the incorporation into the pre-existing biofilm community and the interactions between the pathogens and the autochthonous biofilm members as well as the factors influencing the release of pathogens from biofilms, especially under the environmental conditions of biofilms in man-made water systems.

Generally, the heterogeneous architecture of biofilms containing pores, channels and voids is supposed to contribute to the capture and retention of microbial cells. Cell surface molecules such as proteins, lipopolysaccharides, cell wall polysaccharides, EPS and cell appendages such as flagella and pili which are usually involved as adhesins in the attachment of abiotic and biotic surfaces can also be expected to be involved in the adherence and integration of pathogens in biofilms. In the case of *L. pneumophila*, pili or other unidentified adhesins mediate the colonization of biofilms, but only in the absence of protozoa (Lucas et al. 2006). Coaggregation, the specific recognition and adherence of genetically distinct bacteria to one another, was first discovered to mediate biofilm formation in dental plaque, but has also been found to be relevant in the sequential integration of autochthonous bacterial species into freshwater biofilms (Simões et al. 2007). However, it is unknown whether coaggregation interactions are also involved in the colonization of biofilms by water-related pathogens.

Once integrated in a multispecies biofilm of man-made water systems, the invading organisms are expected to be involved in multiple interactions with other biofilm members. Well-studied are the interactions between *L. pneumophila* and free-living protozoa, but also beneficial and antagonistic interactions of legionellae with various bacterial species seem to be possible in biofilms (Taylor et al. 2009; Guerrieri et al. 2008). However, little is known about interactions of other hygienically relevant microorganisms with the constituent biofilm organisms. Another limitation is that studies of these interactions are usually performed in co-culture experiments with selected organisms under laboratory conditions, so that a gap exists in the knowledge of the actual in situ interactions between pathogens and other organisms within biofilms in man-made water systems. Interactions within biofilm communities are significantly regulated by cell-to-cell communication (quorum sensing) processes, which are triggered at the typically high cell densities of biofilms. It is almost completely unknown how the fate of hygienically relevant bacteria is influenced by signalling molecules in biofilms of man-made water systems, and whether and how they could became integrated in or adapt to the existing communication web within these biofilms. Another aspect of interactions of pathogens and biofilm organisms is the observation that the life cycle within the biofilm can lead to a change in the properties of some bacterial pathogens such as the increase in biocide resistance or enhanced infectivity of *L. pneumophila* and mycobacteria triggered by their passage in amoebae within biofilms. For some pathogens, it has been shown that biofilm cells are more resistant to antimicrobial agents including disinfectants used in practice for water treatment. Cells released from biofilms still retain at least part of the enhanced resistance acquired during their passage in amoebae within biofilms compared to planktonically grown cells as has been shown for legionellae and mycobacteria (Steed and Falkinham 2006). Thus, survival of cells released from biofilms into the water is enhanced and adherence to other surface locations downstream in the water system and initiation of biofilm formation is probable.

The introduction of culture-independent methods in the analysis of water-related bacterial pathogens revealed that in many cases the organisms in biofilms lose culturability, entering a VBNC state, and thus, represent only a fraction of those which are detected by culture-independent methods. The human health significance of non-culturable pathogens is unclear. More research is needed to evaluate the pathogenic potential of those VBNC organisms and to define the factors relevant in man-made water systems, which trigger the VBNC state and induce resuscitation to the culturable and infectious state.

In general, knowledge of the ecology specific for each type of water-related pathogen in combination with the epidemiological situation relevant for each pathogen is necessary for risk assessment applied to DWDS (Szewzyk et al. 2000) and other types of man-made water systems. It has become clear that the biofilm mode of existence of pathogens is an important factor that has to be included in risk assessment applied to water-related pathogens as a basis for the proper operation and maintenance of water systems to ensure the provision of microbiologically safe drinking water and other types of water with the aim to minimize the disease burden of the human population potentially emanating from man-made water systems.

References

Allegra S, Berger F, Berthelot P, Grattard F, Pzzetto B, Riffard S (2008) Use of cytometry to monitor *Legionella* viability. Appl Environ Microbiol 74:7813–7816
Anaissie EJ, Costa SF (2001) Nosocomial aspergillosis is waterborne. Clin Infect Dis 33:1546–1548
Anaissie EJ, Penzak SR, Dignani MC (2002) The hospital water supply as a source of nosocomial infections: a plea for action. Arch Intern Med 162:1483–1492
Anaissie EJ, Stratton SL, Dignani MC, Lee C, Summerbell RC, Rex JH, Monson TP, Walsh TJ (2003) Pathogenic molds (including *Aspergillus* species) in hospital water distribution systems: a 3-year prospective study and clinical implications for patients with hematologic malignancies. Blood 101:2542–2546
Angenent LT, Kelley ST, Amand A St, Pace NR, Hernandez MT (2005) Molecular identification of potential pathogens in water and air of a hospital therapy pool. Proc Nat Acad Sci USA 102:4860–4865

Angles ML, Chandy JP, Cox PT, Fisher IH, Warnecke MR (2007) Implications of biofilm-associated waterborne *Cryptosporidium* oocysts for the water industry. Trends Parasitol 23:352–356

Armon R, Starosvetzky J, Arbel T, Green M (1997) Survival of *Legionella pneumophila* and *Salmonella typhimurium* in biofilm systems. Water Sci Technol 35(11–12):293–300

Arvanitidou M, Kanellou K, Constantinides TC, Katsouyannopoulos V (1999) The occurrence of fungi in hospital and community potable waters. Lett Appl Microbiol 29:81–84

Assanta MA, Roy D, Montpetit D (1998) Adhesion of *Aeromonas hydrophila* to water distribution system pipes after different contact times. J Food Prot 61:1321–1329

Azevedo NF, Vieira MJ, Keevil CW (2003) Establishment of a continuous model system to study *Helicobacter pylori* survival in potable water systems. Water Sci Technol 47(5):155–160

Azevedo NF, Pacheco AP, Keevil CW, Vieira MJ (2006a) Adhesion of water stressed *Helicobacter pylori* to abiotic surfaces. J Appl Microbiol 101:718–724

Azevedo NF, Pinto AR, Reis NM, Vieira MJ, Keevil CW (2006b) Shear stress, temperature, and inoculation concentration influence the adhesion of water-stressed *Helicobacter pylori* to stainless steel 304 and polypropylene. Appl Environ Microbiol 72:2936–2941

Bachmann RT, Edyvean RGJ (2005) Biofouling: an historic and contemporary review of its causes, consequences and control in drinking water distribution systems. Biofouling 2:197–227

Bagh LK, Albrechtsen H-J, Arvin E, Ovesen K (2004) Distribution of bacteria in a domestic hot water system in a Danish apartment building. Water Res 38:225–235

Banning N, Toze S, Mee BJ (2003) Persistence of biofilm-associated *Escherichia coli* and *Pseudomonas aeruginosa* in groundwater and treated effluent in a laboratory model system. Microbiology 149:47–55

Barbeau J, Gauthier C, Payment P (1998) Biofilms, infectious agents, and dental unit waterlines: a review. Can J Microbiol 44:1019–1028

Batté M, Appenzeller BMR, Grandjean D, Fass S, Gauthier V, Jorand F, Mathieu L, Boualam M, Saby S, Block JC (2003) Biofilms in drinking water distribution systems. Rev Environ Sci Bio Tech 2:147–168

Batté M, Féliers C, Servais P, Gauthier V, Joret J-C, Block J-C (2006) Coliforms and other microbial indicators occurrence in water and biofilm in full-scale distribution systems. Water Sci Technol 54(3):41–48

Baumann WJ, Nocker A, Jones WL, Camper AK (2009) Retention of a model pathogen in a porous media biofilm. Biofouling 25:229–240

Berk SG, Gunderson JH, Newsome AL, Farone AL, Hayes BJ, Redding KS, Uddin N, Williams EL, Johnson RA, Farsian M, Reid A, Skimmyhorn J, Farone MB (2006) Occurrence of infected amoebae in cooling towers compared with natural aquatic environments: implications for emerging pathogens. Environ Sci Technol 40:7440–7444

Block JC, Haudidier K, Paquin JL, Miazga J, Levi Y (1993) Biofilm accumulation in drinking water distribution systems. Biofouling 6:333–343

Bomo A-M, Storey MV, Ashbolt NJ (2004) Detection, integration and persistence of aeromonads in water distribution pipe biofilms. J Water Health 2:83–96

Bonadonna L, Briancesco R, Libera SD, Lacchetti I, Paradiso R, Semproni M (2009) Microbial characterization of water and biofilms in drinking water distribution systems at sport facilities. Cent Eur J Public Health 17:99–102

Botzenhart K, Döring G (1993) Ecology and epidemiology of *Pseudomonas aeruginosa*. In: Campa M, Bendinelli M, Friedman H (eds) *Pseudomonas aeruginosa* as an opportunistic pathogen. Plenum, New York, pp 1–18

Botzenhart K, Hock C (2003) Auftreten von obligat und fakultativ pathogenen Organismen in Trinkwasser-Biofilmen: Viren. In: Flemming H-C (ed) Erfassung des Wachstums und des Kontaminationspotentials von Biofilmen in der Verteilung von Trinkwasser. GmbH, Mülheim an der Ruhr, pp 160–184

Bragança SM, Azevedo NF, Simões LC, Vieira MJ, Keevil CW (2005) Detection of *H. pylori* formed in a real drinking water distribution system using peptide nucleic acid fluorescence *in situ* hybridization. In: McBain A, Allison D, Pratten J, Spratt D, Upton M, Verran J (eds) Biofilms: persistence and ubiquitity. Biofilm Club, Manchester, pp 231–239

Bressler D, Balzer M, Dannehl A, Flemming H-C, Wingender J (2009) Persistence of *Pseudomonas aeruginosa* in drinking-water biofilms on elastomeric material. Water Sci Technol Water Supp 9:81–87

Bridier A, Dubois-Brissonet F, Boubetra A, Thomas V, Briandet R (2010) The biofilm architecture of sixty opportunistic pathogens deciphered using a high throughput CLSM method. J Microbiol Meth 82:64–70

Brown MRW, Barker J (1999) Unexplored reservoirs of pathogenic bacteria: protozoa and biofilms. Trends Microbiol 7:46–59

Buswell CM, Herlihy YM, Lawrence LM, McGuiggan JTM, Marsh PD, Keevil CW, Leach SA (1998) Extended survival and persistence of *Campylobacter* spp. in water and aquatic biofilms and their detection by immunofluorescent-antibody and -rRNA staining. Appl Environ Microbiol 64:733–741

Camper AK, LeChevallier MW, Broadway SC, McFeters GA (1985) Growth and persistence of pathogens in granular activated carbon filters. Appl Environ Microbiol 50:1378–1382

Camper AK, Jones WL, Hayes JT (1996) Effect of growth conditions and substratum composition on the persistence of coliforms in mixed-population biofilms. Appl Environ Microbiol 62:4014–4018

Camper AK, Warnecke M, Jones WL, McFeters GA (1998) Pathogens in model distribution system biofilms. AWWA Research Foundation and American Water Works Association, Denver

Carr JH, Anderson RL, Favero MS (1996) Comparison of chemical dehydration and critical point drying for the stabilization and visualization of aging biofilm present on interior surfaces of PVC distribution pipe. J Appl Bacteriol 80:225–232

Carter G, Wu M, Drummond DC, Bermudez LE (2003) Characterization of biofilm formation by clinical isolates of *Mycobacterium avium*. J Med Microbiol 52:747–752

Caubet R, Pedarros-Caubet F, Quataert Y, Lescure A, Moreau JM, Ellison WJ (2006) Assessing the contamination potential of freshly extracted *Escherichia coli* biofilm cells by impedancemetry. Microb Ecol 52:239–243

Chauret C, Volk C, Creason R, Jarosh J, Robinson J, Warnes C (2001) Detection of *Aeromonas hydrophila* in a drinking-water distribution system: a field and pilot study. Can J Microbiol 47:782–786

Cirillo JD, Falkow S, Tompkins LS (1994) Growth of *Legionella pneumophila* in *Acanthamoeba castellanii* enhances invasion. Infect Immun 62:3254–3261

Cirillo JD, Falkow S, Tompkins LS, Bermudez LE (1997) Interaction of *Mycobacterium avium* with environmental amoebae enhances virulence. Infect Immun 65:3759–3767

Cole SP, Harwood J, Lee R, She R, Guiney DG (2004) Characterization of monospecies biofilm formation by *Helicobacter pylori*. J Bacteriol 186:3124–3132

Corsaro D, Pages GS, Catalan V, Loret J-F, Greub G (2010) Biodiversity of amoebae and amoeba-associated bacteria in water. Int J Hyg Environ Health 213:158–166

Costerton JW, Cheng K-J, Geesey GG, Ladd TI, Nickel JC, Dasgupta M, Marrie TJ (1987) Bacterial biofilms in nature and disease. Annu Rev Microbiol 41:435–464

Dailloux M, Albert M, Laurain C, Andolfatto S, Lozniewski A, Hartemann P, Mathieu L (2003) *Mycobacterium xenopi* and drinking water biofilms. Appl Environ Microbiol 69: 6946–6948

Declerck P, Behets J, van Hoef V, Ollevier F (2007a) Detection of *Legionella* spp. and some of their amoeba host in floating biofilms from anthropogenic and natural aquatic environments. Water Res 41:3159–3167

Declerck P, Behets J, van Hoef V, Ollevier F (2007b) Replication of *Legionella pneumophila* in floating biofilms. Curr Microbiol 55:435–440

Doggett MS (2000) Characterization of fungal biofilms within a municipal water distribution system. Appl Environ Microbiol 66:1249–1251

Doleans A, Aurell H, Reyrolle M, Lina G, Freney J, Vandenesch F, Etienne J, Jarraud S (2004) Clinical and environmental distributions of *Legionella* strains in France are different. J Clin Microbiol 42:458–460

Donlan RM, Forster T, Murga R, Brown E, Lucas C, Carpenter J, Fields B (2005) *Legionella pneumophila* associated with the protozoan *Hartmannella vermiformis* in a model multispecies biofilm has reduced susceptibility to disinfectants. Biofouling 21:1–7

Dott W, Schoenen D (1985) Qualitative und quantitative Bestimmung von Bakterienpopulationen aus aquatischen Biotopen. 7. Mitteilung: Entwicklung der Aufwuchsflora auf Werkstoffen im Trinkwasser. Zbl Bakt Hyg I Abt Orig B 180:436–447

Dusserre E, Ginevra C, Hallier-Soulier S, Vandenesch F, Festoc G, Etienne J, Jarraud S, Molmeret M (2008) A PCR-based method for monitoring *Legionella pneumophila* in water samples detects viable but noncultivable legionellae that can recover their cultivability. Appl Environ Microbiol 74:4817–4824

Eboigbodin KE, Seth A, Biggs CA (2008) A review of biofilms in domestic plumbing. J Am Water Works Assoc 100(10):131–138

Emtiazi F, Schwartz T, Marten SM, Krolla-Sidenstein P, Obst U (2004) Investigation of natural biofilms formed during the production of drinking water from surface water embankment filtration. Water Res 38:1197–1206

Exner M, Rechenburg A (2003) Auftreten von obligat und fakultativ pathogenen Organismen in Trinkwasser-Biofilmen: *Helicobacter pylori*. In: Flemming H-C (ed) Erfassung des Wachstums und des Kontaminationspotentials von Biofilmen in der Verteilung von Trinkwasser. IWW Rheinisch-Westfälisches Institut für Wasserforschung gemeinnützige GmbH, Mülheim an der Ruhr, pp 144–159

Exner M, Kramer A, Lajoie L, Gebel J, Engelhart S, Hartemann P (2005) Prevention and control of health care-associated waterborne infections in health care facilities. Am J Infect Control 33: S26–S40

Falkinham JO III (2009) Surrounded by mycobacteria: nontuberculous mycobacteria in the human environment. J Appl Microbiol 107:356–367

Falkinham JO III, Norton CD, LeChevallier MW (2001) Factors influencing numbers of *Mycobacterium avium*, *Mycobacterium intracellulare*, and other mycobacteria in drinking water distribution systems. Appl Environ Microbiol 67:1225–1231

Falkinham JO III, Iseman MD, de Haas P, van Solingen D (2008) *Mycobacterium avium* in a shower linked to pulmonary disease. J Water Health 6:209–213

Fass S, Dincher ML, Reasoner DJ, Gatel D, Block J-C (1996) Fate of *Escherichia coli* experimentally injected in a drinking water distribution pilot system. Water Res 30:2215–2221

Feazel LM, Baumgartner LK, Peterson KL, Frank DN, Harris JK, Pace NR (2009) Opportunistic pathogens enriched in showerhead biofilms. Proc Natl Acad Sci USA 106:16393–16399

Flemming H-C (2002) Biofouling in water systems: cases, causes and countermeasures. Appl Microbiol Biotechnol 59:629–640

Flemming H-C, Wingender J (2010) The biofilm matrix. Nat Rev Microbiol 8:623–633

Flemming H-C, Percival SI, Walker JT (2002) Contamination potential of biofilms in water distribution systems. Water Sci Technol Water Supp 2(1):271–280

Fox KR, Reasoner DJ (2006) Water quality in source water, treatment, and distribution systems. In: Christensen M (ed) Waterborne pathogens. AWWA Manual M48, 2nd edn. American Water Works Association, Denver, pp 21–34

Gatel D, Servais P, Block JC, Bonne P, Cavard J (2000) Microbiological water quality management in the Paris suburbs distribution system. J Water Suppl Res Tech Aqua 49:231–241

Gauthier F, Archibald F (2001) The ecology of "fecal indicator" bacteria commonly found in pulp and paper mill water systems. Water Res 35:2207–2218

Gião MS, Wilks SA, Azevedo NF, Vieira MJ, Keevil CW (2009) Comparison between standard cultures and peptide nucleic acid 16S rRNA hybridization quantification to study the influence of physico-chemical parameters on *Legionella pneumophila* survival in drinking water biofilms. Biofouling 25:335–343

Guerrieri E, Bondi M, Sabia C, de Niederhäusern S, Borella P, Messi P (2008) Effect of bacterial interference on biofilm development by *Legionella pneumophila*. Curr Microbiol 57:532–536

Hageskal G, Knutsen AK, Gaustad P, de Hoog GS, Skaar I (2006) Diversity and significance of mold species in Norwegian drinking water. Appl Environ Microbiol 72:7586–7593

Hall-Stoodley L, Costerton JW, Stoodley P (2004) Bacterial biofilms: from the natural environment to infectious diseases. Nat Rev Micobiol 2:95–108

Harris A (1999) Problems associated with biofilms in cooling tower systems. In: Keevil CW, Godfree A, Holt D, Dow C (eds) Biofilms in the aquatic environment. The Royal Society of Chemistry, Cambridge, UK, pp 139–144

Helmi K, Skraber S, Gantzer C, Willame R, Hoffmann L, Cauchie H-M (2008) Interactions of *Cryptosporidium parvum*, *Giardia lamblia*, vaccinal poliovirus type 1, and bacteriophages ΦX174 and MS2 with a drinking water biofilm and a wastewater biofilm. Appl Environ Microbiol 74:2079–2088

Hoffmann R, Michel R (2003) Auftreten von obligat und fakultativ pathogenen Organismen in Trinkwasser-Biofilmen: Freilebende Amöben (FLA). In: Flemming H-C (ed) Erfassung des Wachstums und des Kontaminationspotentials von Biofilmen in der Verteilung von Trinkwasser. IWW Rheinisch-Westfälisches Institut für Wasserforschung gemeinnützige GmbH, Mülheim an der Ruhr, pp 216–232

Holmes P, Nicolls LM (1995) Aeromonads in drinking-water supplies: their occurrence and significance. J CIWEM 9:464–469

Howe AD, Forster S, Morton S, Marshall R, Osborn KS, Wright P, Hunter PR (2002) *Cryptosporidium* oocysts in a water supply associated with a cryptosporidiosis outbreak. Emerg Infect Dis 8:619–624

Hummel A, Feuerpfeil I (2003) Auftreten von obligat und fakultativ pathogenen Organismen in Trinkwasser-Biofilmen: *Campylobacter* und *Yersinia*. In: Flemming H-C (ed) Erfassung des Wachstums und des Kontaminationspotentials von Biofilmen in der Verteilung von Trinkwasser. IWW Rheinisch-Westfälisches Institut für Wasserforschung gemeinnützige GmbH, Mülheim an der Ruhr, pp 103–126

Jones K, Bradshaw SB (1996) Biofilm formation by the Enterobacteriaceae: a comparison between *Salmonella enteritidis*, *Escherichia coli* and a nitrogen-fixing strain of *Klebsiella pneumoniae*. J Appl Bacteriol 80:458–464

Joshua GW, Guthrie-Irons C, Karlyshev AV, Wren BW (2006) Biofilm formation in *Campylobacter jejuni*. Microbiology 152:387–396

Juhna T, Birzniece D, Larsson S, Zulenkovs D, Sharipo A, Azevedo NF, Ménard-Szczebara F, Castagnet S, Féliers C, Keevil CW (2007) Detection of *Escherichia coli* in biofilms from pipe samples and coupons in drinking water distribution networks. Appl Environ Microbiol 73:7456–7464

Kalmbach S, Manz W, Szewzyk U (1997) Dynamics of biofilm formation in drinking water: phylogenetic affiliation and metabolic potential of single cells assessed by formazan reduction and in situ hybridization. FEMS Microbiol Ecol 22:265–279

Keevil CW (2002) Pathogens in environmental biofilms. In: Bitton G (ed) Encyclopedia of environmental microbiology, vol 4. Wiley, New York, pp 2339–2356

Keevil CW (2003) Rapid detection of biofilms and adherent pathogens using scanning confocal laser microscopy and episcopic differential interference contrast microscopy. Water Sci Technol 47(5):105–116

Keevil CW, Mackerness CW, Colbourne JS (1990) Biocide treatment of biofilms. Int Biodet 26:169–179

Keinänen-Toivola M, Revetta RP, Santo Domingo JW (2006) Identification of active bacterial communities in a model drinking water biofilm system using 16S rRNA-based clone libraries. FEMS Microbiol Lett 257:182–188

Kilb B, Lange B, Schaule G, Flemming H-C, Wingender J (2003) Contamination of drinking water by coliforms from biofilms grown on rubber-coated valves. Int J Hyg Environ Health 206:563–573

King CH, Shotts EB Jr, Wooley RE, Porter KG (1988) Survival of coliforms and bacterial pathogens within protozoa during chlorination. Appl Environ Microbiol 54:3023–3033

Konishi T, Yamashiro T, Koide M, Nishizono A (2006) Influence of temperature on growth of *Legionella pneumophila* biofilm determined by precise temperature gradient temperature gradient incubator. J Biosci Bioeng 101:478–484

Kristich CJ, Li Y-H, Cvitkovitch DG, Dunny GM (2004) Esp-independent biofilm formation by *Enterococcus faecalis*. J Bacteriol 186:154–163

Kuiper MW, Wullings BA, Akkermans ADL, Beumer RR, van der Kooij D (2004) Intracellular proliferation of *Legionella pneumophila* in *Hartmannella vermiformis* in aquatic biofilms grown on plasticized polyvinyl chloride. Appl Environ Microbiol 70:6826–6833

Långmark J, Storey MV, Ashbolt NJ, Stenström TA (2005a) Biofilms in urban water distribution system: measurement of biofilm biomass, pathogens and pathogen persistence within the Greater Stockholm area. Sweden Water Sci Technol 52(8):181–189

Långmark J, Storey MV, Ashbolt NJ, Stenström TA (2005b) Accumulation and fate of microorganisms and microspheres in biofilms formed in a pilot-scale water distribution systems. Appl Environ Microbiol 71:706–712

Lau HY, Ashbolt NJ (2009) The role of biofilms and protozoa in *Legionella* pathogenesis: implications for drinking water. J Appl Microbiol 107:368–378

LeChevallier MW, Hassenauer TS, Camper AK, McFeters GA (1984) Disinfection of bacteria attached to granular activated carbon. Appl Environ Microbiol 48:918–923

LeChevallier MW, Babcock TM, Lee RG (1987) Examination and characterization of distribution system biofilms. Appl Environ Microbiol 53:2714–2724

LeChevallier MW, Cawthon CP, Lee RG (1988) Inactivation of biofilm bacteria. Appl Environ Microbiol 54:2492–2499

Leclerc H, Mossel DAA, Edberg SC, Struijk CB (2001) Advances in the bacteriology of the coliform group: their suitability as markers of microbial water safety. Annu Rev Microbiol 55:201–234

Lee D-G, Kim S-J (2003) Bacterial species in biofilm cultivated from the end of the Seoul water distribution system. J Appl Microbiol 95:317–324

Lehtola MJ, Pitkänen T, Miebach L, Miettinen IT (2006a) Survival of *Campylobacter jejuni* in potable water biofilms: a comparative study with different detection methods. Water Sci Technol 54(3):57–61

Lehtola MJ, Torvinen E, Miettinen IT, Keevil CW (2006b) Fluorescence in situ hybridization using peptide nucleic acid probes for rapid detection of *Mycobacterium avium* subsp. *avium* and *Mycobacterium avium* subsp. *paratuberculosis* in potable-water biofilms. Appl Environ Microbiol 72:848–853

Lehtola M, Torvinen E, Kusnetsov J, Pitkänen T, Maunula L, von Bonsdorff C-H, Martikainen PJ, Wilks SA, Keevil CW, Meittinen IT (2007) Survival of *Mycobacterium avium*, *Legionella pneumophila*, *Escherichia coli*, and caliciviruses in drinking water-associated biofilms grown under high-shear turbulent flow. Appl Environ Microbiol 73:2854–2859

Li J, McLellan S, Ogawa S (2006) Accumulation and fate of green fluorescent labeled *Escherichia coli* in laboratory-scale drinking water biofilters. Water Res 40:3023–3028

Linke S, Lenz J, Gemein S, Exner M, Gebel J (2010) Detection of *Helicobacter pylori* in biofilms by real-time PCR. Int J Hyg Environ Health 213:176–182

Loret J-F, Greub G (2010) Free-living amoebae: biological by-passes in water treatment. Int J Hyg Environ Health 213:167–175

Lucas CE, Brown E, Fields BS (2006) Type IV pili and type III secretion play a limited role in *Legionella pneumophila* biofilm colonization and retention. Microbiology 152:3569–3573

Mackay WG, Gribbon LT, Barer MR, Reid DC (1999) Biofilms in drinking water systems: a possible reservoir for *Helicobacter pylori*. J Appl Microbiol Symp Suppl 85:52S–59S

Mampel J, Spirig T, Weber SS, Haagensen JAJ, Molin S, Hilbi H (2006) Planktonic replication is essential for biofilm formation by *Legionella pneumophila* in a complex medium under static and dynamic flow conditions. Appl Environ Microbiol 72:2885–2895

Martiny AC, Jørgensen TM, Albrechtsen H-J, Arvin E, Molin S (2003) Long-term succession of structure and diversity of a biofilm formed in a model drinking water distribution system. Appl Environ Microbiol 69:6899–6907

Mavridou A, Kamma J, Mandilara G, Delaportas P, Komioti F (2006) Micobial risk assessment of dental unit water systems in general practice in Greece. Water Sci Technol 54(3): 269–273

McDougald D, Klebensberger J, Tolker-Nielsen T, Webb JS, Conibear T, Rice SA, Kirov SM, Matz C, Kjelleberg S (2008) *Pseudomonas aeruginosa*: a model for biofilm formation. In: Rehm BHA (ed) Pseudomonas: model organism, pathogen, cell factory. Wiley-VCH, Weinheim, pp 215–253

Mittelman MW (1995) Biofilm development in purified water systems. In: Lappin-Scott HM, Costerton JW (eds) Microbial biofilms. Cambridge University Press, Cambridge, UK, pp 133–147

Moritz MM, Flemming H-C, Wingender J (2010) Integration of *Pseudomonas aeruginosa* and *Legionella pneumophila* in drinking water biofilms grown on domestic plumbing materials. Int J Hyg Environ Health 213:190–197

Murga R, Forster TS, Brown E, Pruckler JM, Fields BS, Donlan RM (2001) Role of biofilms in the survival of *Legionella pneumophila* in a model potable-water system. Microbiology 147:3121–3126

Norton CD, LeChevallier MW (2000) A pilot study of bacteriological population changes through potable water treatment and distribution. Appl Environ Microbiol 66:268–276

Oliver JD (2010) Recent findings on the viable but nonculturable state in pathogenic bacteria. FEMS Microbiol Rev 34:415–425

Ortolano GA, McAlister MB, Angelbeck JA, Schaffer J, Russell RL, Maynard E, Wenz B (2005) Hospital water point-of-use filtration: a complementary strategy to reduce the risk of nosocomial infection. Am J Infect Control 33:S1–S19

Packer PJ, Holt DM, Colbourne JS, Keevil CW (1997) Does *Klebsiella oxytoca* grow in the biofilm of water distribution systems? In: Kay D, Fricker C (eds) Coliforms and E. coli. Problem or solution? The Royal Society of Chemistry, Cambridge, UK, pp 189–194

Park SR, Mackay WG, Reid DC (2001) *Helicobacter* sp. recovered from drinking water biofilm sampled from a water distribution system. Water Res 35:1624–1626

Payment P, Waite M, Dufour A (2003) Introducing parameters for the assessment of drinking water quality. In: Dufour A, Snozzi M, Koster W, Bartram J, Ronchi E, Fewtrell L (eds) Assessing microbial safety of drinking water: improving approaches and methods. WHO, OECD, London, pp 47–77

Pedersen K (1990) Biofilm development on stainless steel and PVC surfaces in drinking water. Water Res 24:239–243

Percival SL, Thomas JG (2009) Transmission of *Helicobacter pylori* and the role of water and biofilms. J Water Health 7:469–477

Percival S, Chalmers R, Embrey M, Hunter P, Sellwood J, Wyn-Jones P (2004) Microbiology of waterborne diseases. Elsevier/Academic Press, Amsterdam

Piao Z, Sze CC, Barysheva O, Iida K-I, Yoshida S-I (2006) Temperature-regulated formation of mycelial mat-like biofilms by *Legionella pneumophila*. Appl Environ Microbiol 72:1613–1622

Poynter SFB, Mead GC (1964) Volatile organic liquids and slime production. J Appl Bacteriol 27:182–195

Pryor M, Springthorpe S, Riffard S, Brooks T, Huo Y, Davis G, Sattar SA (2004) Investigation of opportunistic pathogens in municipal drinking water under different supply and treatment regimes. Water Sci Technol 50(1):83–90

Quignon F, Sardin M, Kiene L, Schwartzbrod L (1997) Poliovirus-1 inactivation and interaction with biofilm: a pilot-scale study. Appl Environ Microbiol 63:978–982

Rättö M, Verhoef R, Suihko M-L, Blanco A, Schols HA, Voragen AGJ, Wilting R, Siika-aho M, Buchert J (2006) Colanic acid is an exopolysaccharide common to many enterobacteria isolated from paper-machine slimes. J Ind Microbiol Biotechnol 33:359–367

Riffard S, Douglass S, Brooks T, Springthorpe S, Filion LG, Sattar SA (2001) Occurrence of *Legionella* in groundwater: an ecological study. Water Sci Technol 43(12):99–102

Robinson PJ, Walker JT, Keevil CW, Cole J (1995) Reporter genes and fluorescent probes for studying the colonisation of biofilms in a drinking water supply line by enteric bacteria. FEMS Microbiol Lett 129:183–188

Rogers J, Keevil CW (1992) Immunogold and fluorescein immunolabelling of *Legionella pneumophila* within an aquatic biofilms visualized by using episcopic differential interference contrast microscopy. Appl Environ Microbiol 58:2326–2330

Rogers J, Dowsett AB, Dennis PJ, Lee JV, Keevil CW (1994a) Influence of temperature and plumbing material selection on biofilm formation and growth of *Legionella pneumophila* in a model potable water system containing complex microbial flora. Appl Environ Microbiol 60:1585–1592

Rogers J, Dowsett AB, Dennis PJ, Lee JV, Keevil CW (1994b) Influence of plumbing materials on biofilms formation and growth of *Legionella pneumophila* in potable water systems. Appl Environ Microbiol 60:1842–1851

Rosenzweig WD, Minnigh H, Pipes WO (1986) Fungi in potable water distribution systems. J Am Water Works Assoc 78:53–55

Rusin PA, Rose JB, Haas CN, Gerba CP (1997) Risk assessment of opportunistic bacterial pathogens in drinking water. Rev Environ Contam Toxicol 152:57–83

Sartory DP, Holmes P (1997) Chlorine sensitivity of environmental, distribution system and biofilm coliforms. Water Sci Technol 35(11–12):289–292

Schulze-Röbbecke R, Fischeder R (1989) Mycobacteria in biofilms. Zbl Hyg 188:385–390

Schulze-Röbbecke R, Ilg B (2003) Auftreten von obligat und fakultativ pathogenen Organismen in Trinkwasser-Biofilmen: Mykobakterien. In: Flemming H-C (ed) Erfassung des Wachstums und des Kontaminationspotentials von Biofilmen in der Verteilung von Trinkwasser. IWW Rheinisch-Westfälisches Institut für Wasserforschung gemeinnützige GmbH, Mülheim an der Ruhr, Germany, pp 185–203

Schulze-Röbbecke R, Janning B, Fischeder R (1992) Occurrence of mycobacteria in biofilm samples. Tub Lung Dis 73:141–144

Schwartz T, Hoffmann S, Obst U (1998a) Formation and bacterial composition of young, natural biofilms obtained from public bank-filtered water systems. Water Res 32:2787–2797

Schwartz T, Kalmbach S, Hoffmann S, Szewzyk U, Obst U (1998b) PCR-based detection of mycobacteria in biofilms from a drinking water distribution system. J Microbiol Meth 34:113–123

Searcy KE, Packman AI, Atwill ER, Harter T (2006) Capture and retention of *Cryptosporidium parvum* oocysts by *Pseudomonas aeruginosa* biofilms. Appl Environ Microbiol 72:6242–6247

September SM, Brözel VS, Venter SN (2004) Diversity of nontuberculoid *Mycobacterium* species in biofilms of urban and semiurban drinking water distribution systems. Appl Environ Microbiol 70:7571–7573

September SM, Els FA, Venter SN, Brözel VS (2007) Prevalence of bacterial pathogens in biofilms of drinking water distribution systems. J Water Health 5:219–227

Servais P, Laurent P, Randon G (1995) Comparison of the bacterial dynamics in various French distribution systems. J Water SRT Aqua 44:10–17

Sibille I, Sime-Ngando T, Mathieu L, Block JC (1998) Protozoan bacterivory and *Escherichia coli* survival in drinking water distribution systems. Appl Environ Microbiol 64:197–202

Silva M, McLellan S, Li J (2006) The removal of green fluorescent labelled *Escherichia coli* by pilot scale drinking water biofilters. In: Gimbel R, Graham NJD, Collins MR (eds) Recent progress in slow sand and alternative biofiltration processes. IWA, London, pp 337–344

Simões LC, Simões M, Vieira MJ (2007) Microbial interactions in drinking water biofilms. In: Gilbert P, Allison D, Brading M, Pratten J, Spratt D, Upton M (eds) Biofilms: coming of age. The Biofilm Club, Manchester, UK, pp 43–52

Skraber S, Schijven J, Gantzer C, de Roda Husman AM (2005) Pathogenic viruses in drinking-water biofilms: a public health risk. Biofouling 2:105–117

Skraber S, Ogorzaly L, Helmi K, Maul A, Hoffmann L, Cauchie H-M, Gantzer C (2009) Occurrence and persistence of enteroviruses, noroviruses and F-specific RNA phages in natural wastewater biofilms. Water Res 43:4780–4789

Stark RM, Gerwig GJ, Pitman RS, Potts LF, Williams NA, Greenman J, Weinzweig IP, Hirst TR, Millar MR (1999) Biofilm formation by *Helicobacter pylori*. Lett Appl Microbiol 28:121–126

Steed KA, Falkinham JO III (2006) Effect of growth in biofilms on chlorine susceptibility of *Mycobacterium avium* and *Mycobacterium intracellulare*. Appl Environ Microbiol 72:4007–4011

Steinert M, Emödy L, Amann R, Hacker J (1997) Resuscitation of viable but nonculturable *Legionella pneumophila* Philadelphia JR32 by *Acanthamoeba castellanii*. Appl Environ Microbiol 63:2047–2053

Stoodley P, Sauer K, Davies DG, Costerton JW (2002) Biofilms as complex differentiated communities. Annu Rev Microbiol 56:187–209

Storey MV, Ashbolt NJ (2001) Persistence of two model enteric viruses (B40-8 and MS-2 bacteriophages) in water distribution pipe biofilms. Water Sci Technol 43(12):133–138

Storey MV, Långmark J, Ashbolt NJ, Stenström TA (2004a) The fate of legionellae within distribution pipe biofilms: measurement of their persistence, inactivation and detachment. Water Sci Technol 49(11–12):269–275

Storey MV, Ashbolt NJ, Stenström TA (2004b) Biofilms, thermophilic amoebae and *Legionella pneumophila*: a quantitative risk assessment for distributed water. Water Sci Technol 50(1):77–82

Stout JE, Yu VL, Best MG (1985) Ecology of *Legionella pneumophila* within water distribution systems. Appl Environ Microbiol 49:221–228

Szabo JG, Rice EW, Bishop PL (2006) Persistence of *Klebsiella pneumoniae* on simulated biofilm in a model drinking water system. Environ Sci Technol 40:4996–5002

Szewzyk U, Manz W, Amann R, Schleifer KH, Stenström T-A (1994) Growth and in situ detection of a pathogenic *Escherichia coli* in biofilms of a heterotrophic water-bacterium by use of 16S- and 23S-rRNA-directed fluorescent oligonucleotide probes. FEMS Microbiol Ecol 13:169–176

Szewzyk U, Szewzyk R, Manz W, Schleifer K-H (2000) Microbiological safety of drinking water. Annu Rev Microbiol 54:81–127

Taylor M, Ross K, Bentham R (2009) *Legionella*, protozoa, and biofilms: interactions within complex microbial systems. Microb Ecol 58:538–547

Temmerman R, Vervaeren H, Noseda B, Boon B, Verstrate W (2006) Nectotrophic growth of *Legionella pneumophila*. Appl Environ Microbiol 72:4323–4328

Thomas V, Loret J-F, Jousset M, Greub G (2008) Biodiversity of amoebae and amoebae-resisting bacteria in a drinking water treatment plant. Environ Microbiol 10:2728–2745

Thomas V, McDonnell G, Denyer SP, Maillard J-Y (2010) Free-living amoebae and their intracellular pathogenic microorganisms: risks for water quality. FEMS Microbiol Rev 34:231–259

Torvinen E, Suomalainen S, Lehtola MJ, Miettinen IT, Zacheus O, Paulin L, Katila M-L, Martikainen PJ (2004) Mycobacteria in water and loose deposits of drinking water distribution systems in Finland. Appl Environ Microbiol 70:1973–1981

Torvinen E, Lehtola MJ, Martikainen PJ, Miettinen IT (2007) Survival of *Mycobacterium avium* in drinking water biofilms as affected by water flow velocity, availability of phosphorus, and temperature. Appl Environ Microbiol 73:6201–6207

Tsitko I, Rahkila R, Priha O, Ali-Vehmas T, Terefework Z, Soini H, Salkinoja-Salonen MS (2006) Isolation and automated ribotyping of *Mycobacterium lentiflavum* from drinking water distribution system and clinical specimens. FEMS Microbiol Lett 256:236–243

Türetgen I, Cotuk A (2007) Monitoring of biofilm-associated *Legionella pneumophila* on different substrata in model cooling tower system. Environ Monit Assess 125:271–279

Unhoch MJ, Vore RD (2005) Recreational water treatment biocides. In: Paulus W (ed) Directory of microbicides for the protection of materials: a handbook. Springer, Dordrecht, pp 141–155

Vaerewijck MJM, Huys G, Palomino JC, Swings J, Portaels F (2005) Mycobacteria in drinking water distribution systems: ecology and significance for human health. FEMS Microbiol Rev 29:911–934

Van der Kooij D (1991) Nutritional requirements of aeromonads and their multiplication in drinking water. Experientia 47:444–446

Van der Kooij D, Veenendaal HR, Slaats NPG, Vonk D (2002) Biofilm formation and multiplication of *Legionella* on synthetic pipe materials in contact with treated water under static and dynamic conditions. In: Marre R, Abu Kwaik Y, Bartlett C, Cianciotto NP, Fields BS, Frosch M, Hacker J, Lück PC (eds) *Legionella*. ASM, Washington, DC, pp 86–89

Van der Kooij D, Veenendaal HR, Scheffer WJH (2005) Biofilm formation and multiplication of *Legionella* in a model warm water system with pipes of copper, stainless steel and cross-linked polyethylene. Water Res 39:2789–2798

Vanden Bossche G, Krietemeyer S (1995) Detergent conditioning of environmental samples: the most sensitive method for the detection of viral activity. Presented at the IAWQ 17th Biennial international conference health-related water microbiology symposium, Budapest, Hungary, July 1994

Vervaeren H, Temmerman R, Devos L, Boon N, Verstraete W (2006) Introduction of a boost of *Legionella pneumophila* into a stagnant-water model by heat treatment. FEMS Microbiol Ecol 58:583–592

Vess RW, Anderson RL, Carr JH, Bond WW, Favero MS (1993) The colonization of solid PVC surfaces and the acquisition of resistance to germicides by water micro-organisms. J Appl Bacteriol 74:215–221

Wadowsky RM, Wolford R, McNamara AM, Yee RB (1985) Effect of temperature, pH, and oxygen level on the multiplication of naturally occurring *Legionella pneumophila* in potable water. Appl Environ Microbiol 49:1197–1205

Wagner D, Fischer W, Paradies HH (1992) Copper deterioration in a water distribution system of a county hospital in Germany caused by microbially influenced corrosion: II. Simulation of the corrosion process in two test rigs installed in this hospital. Werkst Korros 43:496–502

Walker JT, Mackerness CW, Rogers J, Keevil CW (1995a) Heterogenous mosaic biofilm-a haven for waterborne pathogens. In: Lappin-Scott HM, Costerton JW (eds) Microbial biofilms. Cambridge University Press, Cambridge, pp 196–204

Walker JT, Mackerness CW, Mallon D, Makin T, Williets T, Keevil CW (1995b) Control of *Legionella pneumophila* in a hospital water system by chlorine dioxide. J Ind Microbiol 15:384–390

Wallace WH, Rice JF, White DC, Sayler GS (1994) Distribution of alginate genes in bacterial isolates from corroded metal surfaces. Microb Ecol 27:213–223

Warris A, Klaassen CHW, Meis JFG, de Ruiter MT, de Valk HA, Abrahamsen TG, Gaustad P, Verweij PE (2003) Molecular epidemiology of *Aspergillus fumigatus* isolates recovered from water, air, and patients shows two clusters of genetically distinct strains. J Clin Microbiol 41:4101–4106

Watson CL, Owen RJ, Said B, Lai S, Lee JV, Surman-Lee S, Nichols G (2004) Detection of *Helicobacter pylori* by PCR but not culture in water and biofilm samples from drinking water distribution systems in England. J Appl Microbiol 97:690–698

White DC, Kirkegaard RD, Palmer RJ Jr, Flemming CA, Chen G, Leung KT, Phiefer CB, Arrage AA (1999) The biofilm ecology of microbial biofouling, biocide resistance and corrosion. In: Keevil CW, Godfree A, Holt D, Dow C (eds) Biofilms in the aquatic environment. The Royal Society of Chemistry, Cambridge, pp 120–130

WHO (2008) Guidelines for drinking-water quality: incorporating the first and second addenda, vol 1, 3rd edn, Recommendations. World Health Organization, Geneva

Williams MW, Braun-Howland EB (2003) Growth of *Escherichia coli* in model distribution system biofilms exposed to hypochlorous acid or monochloramine. Appl Environ Microbiol 69:5463–5471

Wingender J, Flemming H-C (2004) Contamination potential of drinking water distribution network biofilms. Water Sci Technol 49(11–12):277–286

Wolyniak EA, Hargreaves BR, Jellison KL (2009) Retention and release of *Cryptosporidium parvum* oocysts by experimental biofilms composed of a natural stream microbial community. Appl Environ Microbiol 75:4624–4626

Wullings BA, Bakker G, van der Kooij D (2011) Concentration and diversity of uncultered *Legionella* spp. in two unchlorinated drinking water supplies with different concentrations of natural organic matter. Appl Environ Microbiol 77:634–641

Yu J, Kim D, Lee T (2010) Microbial diversity in biofilms on water distribution pipes of different materials. Water Sci Technol 61:163–171

Zacheus OM, Lehtola MJ, Korhonen LK, Martikainen PJ (2001) Soft deposits, the key site for microbial growth in drinking water distribution networks. Water Res 35:1757–1765

Index

A
Abrasion, 4
Acanthamoeba, 223
Actinobacillus actinomycetemcomitans, 14
N-Acyl homoserine lactones (AHLs), 18, 33, 164
Additives, 94
Adhesion, 50
Adhesive failure, 121
Aeromonads, 219
Aeromonas, 51
Aeromonas hydrophila, 11, 19
Aging of groundwater wells, 66
Alginate, 13
Alginate lyase, 13
Alkyl-chinolinols, 32
Amoebae, 223
Anoxygenic photosynthesis, 64
Antibodies, 47
Antimicrobial strategy, 177
Anti-Staphylococcal biofilm approach, 141
Apparent Young's modulus, 131
Astrobiology, 65
Aufwuchs, 30
Autofluorescence of biomolecules, 101
Autoinducing peptide (AIP-I), 19
Autoinduction, 164

B
Bacillus, 52
Bacillus subtilis, 16, 165
Bacterial gossip, 163
Banded iron formation (BIF), 64
Beta-sheets, 42, 44, 45, 47
Biofilm(s), 29, 41, 82, 189
    detachment, 3
    development, 5
    dispersion, 5
    failure, 135
    mechanics, 111
    resilience, 163
Biofilm-attached tubes, 129
Biofilm mode of growth, 164
Biofilm negative, 171
Biofilm reactor in the wrong place, 93
Biofouling, 82, 190
Biosurfactants, 14
Bog iron, 65
Bryostatin, 36
Burger model, 120

C
Campylobacter, 201
Carbon storage regulator (Csr), 15
Catabolite control protein (CcpA), 16
Catabolite repression control (Crc), 16
Cell envelope, 53
Centrifugation, 128
Chaplins, 42, 52
Chemical communication, 33
Chemical crosstalk, 29
Chromobacterium violaceum, 34
Chronic biofilm-based nonhealing wounds, 142
Citrobacter, 42, 56, 206
Cleaning, 96
Clonothrix, 68
Clostridium thermocellum, 3
Cohesion forces, 97
Cohesive failure, 121
Coliform bacteria, 206

Colistin, 172
Colony forming units (cfu), 87
Competition by chemical weapons, 31
Compression testing, 130
Conditioning film, 88
Cooperation by chemical communication, 33
Copper-plating, 92
Costs of biofouling, 83
Couette–Taylor reactor, 126
Coupons, 98
Cross-kingdom signaling, 175
Cryptosporidium, 221
CsgA, 42, 46, 47, 50, 51, 55, 56
Curli, 42, 50
  bis-(3′–5′)-cyclic dimeric guanosine monophosphate (c-di-GMP), 16, 164, 170
Cystic fibrosis, 174
Cystic fibrosis transmembrane conductance regulator (CFTR), 174
Cytotoxin, 54

D

Daughter cells, 4
cis–2-Decenoic acid, 19
Defence mechanism by chemical weapons, 31
Delisea pulchra, 35
Derjaguin, Landau, Vervey, and Overbeck (DLVO) theory, 89
Desorption, 3
Detachment, 2
Detachment process, 112
Diabetes mellitus, 146
Diabetic foot ulcer, 145, 147
Diffusible signal factor (DSF), 21, 98
Digital images correlation (DIC), 124
Disaggregatase, 13
Disinfection, 95
Dispersin-B (DspB), 14
Dispersion, 2
DNA array-based analysis, 172
Drinking water, 189
Drinking water pipe, 67
DSF. *See* Diffusible signal factor (DSF)

E

Early warning capacity, 102
Early warning systems, 99
Elasticity, 114
Emerging cost factor, 84
Entanglement, 97
Enteric viruses, 220

Enterobacter, 206
Enterococci, 208
Enterococcus faecalis, 165
Environmentally responsive polymers, 92
Enzyme applications, 97
Epibiosis, 35
Epibiotic biofilms, 36
EPS. *See* Extracellular polymeric substances (EPS)
Erosion, 4
Escape, 2
Escherichia coli, 10, 41, 203
Eukaryotic response, 34
Exobiology, 65
Extracellular inducers of biofilm dispersion, 17
Extracellular polymeric substances (EPS), 3, 82, 168, 189, 190, 222

F

Ferric iron, 63
Ferrous iron, 64
Fibrillation, 45
Fimbriae, 50
FISH, 71
Flow-cell, 122
Fluid dynamic gauging (FDG), 124
Fluorescence staining, 95
Flustra foliacea, 36
Formic acid, 51
Fouling tolerance, 85
FTIR-ATR-spectroscopy, 101
Full-field measurement, 134
Functional equivalent pathogroups (FEPs), 156
Fungi, 224
Furuncle, 148

G

Gallionella, 63
Geology, 64
Giardia lamblia, 221
Gluconacetobacter xylinus, 17
Good housekeeping, 102
Gordonia, 53
Grazing, 4
Groundwater, 73

H

Hamamelitannin, 149
Harpins, 54
Heat exchangers, 84
Helicobacter pylori, 202

Heterotrophic plate count (HPC), 192
Humic substances, 72
Hydrodynamic loading, 122
Hydrogen bonding, 97
Hydroides elegans, 35
Hydrophobins, 52
Hygienically relevant microorganisms, 189, 193, 197
Hyphomicrobium, 63

I
Indicator organisms, 203
Indole, 18
Industrial bioreactors, 112
Infected nonhealing surgical wound, 148
Interaction, 46, 47, 50, 51
Intracelluar regulation of biofilm dispersion, 15
Intracellular signaling, 170
Iron, 64
Iron-depositing bacteria, 63
Iron-depositing microorganisms, 63
Iron ore, 63

J
Juvenile onset diabetic with renal transplant, 146

K
Klebsiella, 206
Klebsiella pneumoniae, 17

L
Legionellae, 209
Legionella pneumophila, 209
Leptothrix, 63
Level 2 monitoring, 101
Level 3 monitoring, 101
Level 1 monitoring devices, 100
Level of interference, 85
Living colloids, 89
Loss modulus, 120
Lotus effect, 90
Low-fouling surfaces, 87

M
Manganese, 64
Man-made water systems, 189
Mechanical cleaning, 96

Medical devices, 112
Methicillin-resistant S. aureus (MRSA), 144
Methicillin-resistant *S. epidermidis* (MRSE), 148
Methanosarcina mazei, 13
Microbial biofouling, 81
Microcantilever technique, 129
Micro-indentation, 131
Microscopical observation, 101
Migration, 2
Mucoid, 13
Mushroom-shaped multicellular structures, 169
Mycobacteria, 216
Mycobacterium, 51
Mycobacterium tuberculosis pili (MTP), 51
Mycolata, 53, 55
Mycothrix, 68

N
Natural antifouling compounds, 89
Naumanniella, 68
Neisseria subflava, 8
Nitric oxide (NO), 18
NMR imaging, 102
Nucleation, 46, 47, 55

O
Ochrous depositions, 65
Opportunistic human pathogen, 166

P
Pain threshold, 85
Pathogens, 189, 193
Pedomicrobium, 63
Phage enzymes, 97
Phylogeny, 74
Pigs, 96
Pili, 50
Planctomyces, 68
Plantar diabetic foot ulcer, 148
Plasticity, 120
Plastic strain, 121
Podsolic soils, 64
Poisson's ratio, 117
Polymorphonuclear neutrophils (PMNs), 173, 174
Pre-dispersion behavior, 5
Protofibril, 46, 54
Protozoan parasites, 221

Pseudoalteromonas tunicata, 31
Pseudomonas, 56, 63
   P. aeruginosa, 3, 34, 165, 166, 168, 172, 214
   P. fluorescens, 3
   P. putida, 6
Pseudomonas quinolone signal (PQS), 167
Pulling testing shear, 133

Q
QscR receptor, 167
QS inhibitors (QSI), 173, 177
Quaternary ammonium compounds (QACs), 157
Quorum sensing (QS), 143, 166

R
Race to the surface, 88
Regulation, 50, 51
Relaxation function, 118
Relaxation time, 118
Release of cells, 2
Release of extracellular enzymes, 12
Replacement, 84
Rhamnolipid, 14, 175
RhlR receptor, 166
Rhodobacter, 64
Rhodobacter sphaeroides, 19
RNAIII activating protein (RAP), 144
RNAIII inhibiting peptide (RIP), 141, 143
   bacterial stress response, 155
   biofilm survival, 154
   functional equivalent pathogroups and synergistic treatment protocols, 156
   molecular mechanisms of, 150
   RIP downregulates toxin genes, 151
   stress response genes regulation by, 152
   toxin production reduction, 154
Robbins device, 98
Rotating disk biofilm reactor (RDBR), 126

S
Salmonella, 45, 50, 56, 196
Seeding dispersal, 5
Serratia marcescens, 18, 34
Shear forces, 94
Shear modulus, 117
Shear tests, 132
Shewanella oneidensis, 8, 17
Shigella, 196

Siderocapsa, 66
Siderococcus, 68
Signaling and EPS matrix, 171
Signaling molecules, 98
Silver-coated surfaces, 90
Sloughing, 4
Spinning disk rheometry, 132
Spores, 50, 52, 53
Staphylococcal biofilm, 141, 142
Staphylococcus aureus, 119, 142, 165
Staphylococcus epidermidis, 14
Starvation, 3
Storage modulus, 120
Streamers, 117
Streptococcus mutans, 13
Streptomyces, 42
Structural heterogeneity, 114
Surface-bound biocides, 90
Surface monitoring, 98
Surface protein-releasing enzyme (SPRE), 13

T
Tambjamine alkaloid, 31
Target of RNAIII activating protein (TRAP), 144
Teeth, 102
Tensile strength, 121
Tensile testing, 128
Thioflavin T, 47, 52
Total cell counts, 192
Transcutaneous oximetry measurements (TCOMs), 150
Transition from a flowing system to a batch system, 8
Tributyl tin, 89

U
Ultimate strength, 121
Ultrasonic treatment, 96
Ulva intestinalis, 35
Ulva lactuca, 36
Uniaxial compressive test, 130
Universal biofilm disperser, 98
UV irradiation, 91

V
van der Waals interactions, 97
Viable but nonculturable (VBNC), 98, 194
Vibrio cholerae, 34
Violacein, 32

Virtual field method (VFM), 135
Viscoelasticity, 117
Volvulus, 147
Von Mises criterion, 120

W
Waste water treatment, 112
Water samples, 87
Water wells, 63
Weak ionic interactions, 97
Well clogging, 66

Wetlands, 64
Wound healing, 141

X
Xanthomonas campestris, 21

Y
Yersinia enterocolitica, 200
Yersinia pestis, 14
Young's modulus, 116